Springer Tracts in Advanced Robotics 94

For further volumes:
http://www.springer.com/series/5208

STAR (Springer Tracts in Advanced Robotics) has been promoted under the auspices of EURON (European Robotics Research Network)

Florian Röhrbein · Germano Veiga
Ciro Natale
Editors

Gearing Up and Accelerating Cross-Fertilization between Academic and Industrial Robotics Research in Europe

Technology Transfer Experiments from the ECHORD Project

Second Edition

Springer

Editors
Florian Röhrbein
Department of Informatics VI
Technische Universität München
Boltzmannstr. 3
85748 Garching
Germany
florian.roehrbein@in.tum.de

Germano Veiga
Inesc Porto
Feup Campus
Rua Dr. Roberto Frias, 378
4200 - 465 Porto
Portugal
germano.veiga@inescporto.pt

Ciro Natale
Dipartimento di Ingegneria
Industriale e dell'Informazione
Seconda Università degli
Studi di Napoli
Via Roma 29
81031 Aversa
Italy
ciro.natale@unina2.it

ISSN 1610-7438 ISSN 1610-742X (electronic)
ISBN 978-3-319-34474-4 ISBN 978-3-319-03838-4 (eBook)
DOI 10.1007/978-3-319-03838-4
Springer Cham Heidelberg New York Dordrecht London

Printed on acid-free paper

Springer is part of Springer Science+Business Media (www.springer.com)

Foreword

Robotics is undergoing a major transformation in scope and dimension. From a largely dominant industrial focus, robotics is rapidly expanding into human environments and vigorously engaged in its new challenges. Interacting with, assisting, serving, and exploring with humans, the emerging robots will increasingly touch people and their lives.

Beyond its impact on physical robots, the body of knowledge robotics has produced is revealing a much wider range of applications reaching across diverse research areas and scientific disciplines, such as: biomechanics, haptics, neurosciences, virtual simulation, animation, surgery, and sensor networks among others. In return, the challenges of the new emerging areas are proving an abundant source of stimulation and insights for the field of robotics. It is indeed at the intersection of disciplines that the most striking advances happen.

The *Springer Tracts in Advanced Robotics (STAR)* is devoted to bringing to the research community the latest advances in the robotics field on the basis of their significance and quality. Through a wide and timely dissemination of critical research developments in robotics, our objective with this series is to promote more exchanges and collaborations among the researchers in the community and contribute to further advancements in this rapidly growing field.

The monograph by Florian Röhrbein, Germano Veiga and Ciro Natale is an edited collection of fifteen authoritative contributions in the area of robot technology transfer from industry to academia which stemmed from the results of ECHORD (European Clearing House for Open Robotics Development), a large-scale integrating project funded by the European Commission within the 7th Framework Programme from 2009 to 2013.

ECHORD developed the metaphor of the structured dialogue into a powerful interaction instrument between the stakeholders. This was particularly important for the broad goal-directed discussion of a future research and development agenda that reflects industry needs and interests of researchers. To that purpose, the ECHORD Coordination Committee was assisted by an Advisory Board, consisting of nine distinguished international experts from both research and industry, who gave recommendations for general research parameters and areas, goals, and new challenges during the course of the project. Excerpts from their testimonies are collected in the next section.

The results described in the volume are expected to shed a new light on innovation and technology transfer from academia to industry in the field of robotics. A fine addition to STAR!

September 2013
Naples, Italy

Bruno Siciliano
STAR Editor

Testimonies

"Economic health is a Europe wide priority. Europe is proud of its record in robotics research covering a broad portfolio of topics. Research is expensive and clearly it is important for present and future economic growth, jobs and the development of new markets that the fruits of research programmes, and associated skills and expertise, are productively harnessed. The ECHORD initiative has provided a foundation for a pragmatic mechanism to link academic and industrial interests to stimulate real opportunities on the road to developing new products."

Chris Melhuish

Director of Bristol Robotics Laboratory, University of Bristol and the University of the West of England

"Robotics stems from a tension between technology and science ... The tension echoes the challenge of promoting coherent research and development programs in robotics. How to gather theorists and practitioners? How to stimulate a synergy between academia and industry? These are the fundamental questions to be answered to give robotics a real impact on the society. ECHORD has introduced the innovative concept of small and easy to manage experiments gathering one academic lab and one company."

Jean-Paul Laumond

Directeur de Recherche, LAAS-CNRS Toulouse

"Academia, by its very nature is the natural basin where these ideas come to life, but then it is in the industry that these find their application. Following this reasoning ECHORD is one of the most innovative ideas that have occurred in the last period, allowing a huge dock of diverse experiments to establish new industry-academia relations, permitting to carry out research in various areas of robotics from industry to service."

Gian Paolo Gerio

Performance Engineering Manager, COMAU S.p.A. BU Robotics & Service

"Collaboration between academia and industry is recognized as an important contributor to technology innovation in broad range of domains. This fact is certainly most critical in robotics, a field that is both multidisciplinary and technology-driven ... From the initial joint exploration of a new robotic application, through the research development and validation, to the rich bilateral transfer of knowledge, ECHORD enabled meaningful partnerships with many successful outcomes. The project has been also very successful at dissemination both in the public media and in the research community."

Oussama Khatib

Professor of Robotics, Stanford University

"An important point for the suppliers of equipment to ECHORD was the opportunity to expose young researchers to their equipment. Often, these young scientists of today may be the decision-making engineers of tomorrow, giving us a better degree of access to the community of users of industrial robots through improved familiarity with our products. This aspect is not only a competitive one … but in this way we can be optimistic that new applications of industrial robots will be developed with a higher intensity."

Björn Matthias

Senior Principal Scientist, Robotic Automation ABB AG

"The structured dialogue has further helped to strengthen relations between industry and academia. Especially the visits to research labs around the world and the intense discussions at many venues helped to bring together researchers and practitioners. With these elements, the ECHORD project has defined a new format for European publicly funded projects."

Lothar Baum

Robert Bosch GmbH Corporate Research

"I believe that ECHORD generated a technology push in robotics that was never seen before. It generated a lot of follow-up collaborations and I am sure that its impact will last for many years. I would like to congratulate on the coordinators for their vision and dedication, the European Commission for taking the risk with a novel funding model, and all the engineers and researchers for their outstanding results."

Roland Siegwart

Professor for Robotics and Vice President for Research and Corporate Relations, ETH Zürich

"As discussed in this volume, the human robot co-worker is a key enabler for future robot applications. Particularly in an advanced matured society like Europe, one must remove fences for robots to work side by side with humans. Yet, at present, European safety regulation does not allow removing fences. Countries that could solve social acceptance of human robot co-workers will acquire a fruit of novel working style of humans as well as robots in the aging society of the future."

Hirochika Inoue

Professor Emeritus, The University of Tokyo

"The experiments reported in this book describe the improvements in moving robot demonstrators towards practical usability, in addition to promising technology breakthroughs in the areas of sensors, actuators, advanced control, reasoning and task based systems that make robots safe to use with people in complex and dynamic environments. I highly recommend this book to readers who are interested in knowing how innovation can work in the field of robotics."

Alex Zelinsky

Chief Defence Scientist of Australia Defence Science & Technology Organisation

Preface

This volume is an edited collection of 15 high-value scientific contributions stemming from the results of 15 small scale projects, the so-called 'experiments', selected among those funded within the ECHORD project 'European Clearing House for Open Robotics Development' supported by the European Commission under the Seventh Framework Programme.

The idea of such a research project originates from several discussions within the European robotics community about the possibility of advancing European robotics research through a unique, new approach – one which involves intense interaction between academia and industry. Even though Europe has a very strong robot industry, significant world-class research potential, as well as technological knowledge spread throughout Europe, finding common ground between robot manufacturers and research institutions has been difficult in the past.

The ECHORD project was started with the goal of bridging this gap. Looking at previous successful collaborations between robot manufacturers and research institutions, one can identify three factors which led to successful technology transfer: (i) a concrete problem was both relevant to a manufacturer and scientifically interesting to researchers; (ii) the specific competences of both sides were really challenged; (iii) the manufacturer provided state of the art equipment, so that the researchers' work was carried out on the manufacturers' equipment.

Using this as a basis, ECHORD developed its two main concepts – the 'experiment' and the 'structured dialogue'. In order to fuel the knowledge exchange between researchers and manufacturers the path that would reap the most benefits would be to emulate the important aspects that made the previous collaborations successful, in other words, do what they did right, even expand, and improve on it. *Experiments* were small-scale funded projects lasting 12 to 18 months, in which researchers and manufacturers collaborated on a specific, concrete challenge – using state-of-the-art equipment provided by the manufacturer. The *structured dialogue* can be seen as an iterative process of successive information gathering and consensus finding between all stakeholders. The guiding principle was "information aggregation and densification". Ideas gathered in polls, web consultations, expert meetings and lab tours, were synthesised and re-distributed to stakeholders along with specific questions for discussion and then summarised in a white paper of working hypotheses.

The chapters in this volume are a testimony to the fact that such a funding strategy can actually be effective and, excitingly enough, this is the case independently of the particular application at hand. In fact, the 15 chapters cover almost all the topics nowadays considered 'hot' within the robotics community, from reliable object recognition to dexterous grasping, from speech recognition to intuitive robot programming, from mobile robot navigation to aerial robotics, from safe physical human-robot interaction to body extenders.

The chapters are organised into the following three parts:

Part I, *Future Industrial Robotics*, summarises the results of five experiments focussed on the challenges of industrial robotics. Industrial robots represent the vast majority of the operational robots worldwide nowadays but their use is still dominated by some traditional applications, like welding, assembly or handling. Furthermore, the majority of these setups still present low levels of autonomy/flexibility, relying heavily on programming tasks, with reduced human-robot interaction during operation. The results presented in the five chapters pave the way towards solutions that overcome such limitations.

Part II, *Robotic Grasping*, summarises the results of five experiments focussed on one of the most challenging research areas in robotics, i.e. grasping and manipulation. Besides the methodological and technological advancements that the consortia—most of them involved a cooperation of academic partners and a large industrial or SME partner—achieved in the course of their projects, the peculiarity of the research work conducted within such experiments is their strict relation to real-world applications, from bin picking in industrial environments to deboning operations in the food industry.

Part III, *Human-centred Robots*, covers five experiments that address different ways of interconnecting human and machine. This linkage is of a varying degree of closeness from wearable robots to ambient intelligence; it comprises haptics, multimodal interfaces and even empathic cognition.

We would like to take this opportunity to thank all authors for their valuable contributions and Dr. Thomas Ditzinger from Springer for his kind support during the development of this book. We are grateful to many people, institutions and the funding body, the EC, without whom this book would not have been possible. Among all, we want to express our warmest gratitude to the three co-coordinators of the ECHORD project, Alois Knoll, Bruno Siciliano and Norberto Pires, for having entrusted us with the responsibility of editing this volume. This gave us the opportunity to work together and with some of the most prominent research groups in the European robotics community. We hope you will enjoy the reading and find it exciting and inspiring, regardless of whether you are interested in the exploitation of one of the technological solutions therein or if you are thinking of a new research line.

September 2013

München, Germany — Florian Röhrbein
Porto, Portugal — Germano Veiga
Aversa, Italy — Ciro Natale

Contents

Part II: Robotic Grasping

Part III: Human-Centered Robots

The ECHORD Project: A General Perspective

Sascha Griffiths[1], Ciro Natale[2], Ricardo Araújo[3], Germano Veiga[3,4], Pasquale Chiacchio[5], Florian Röhrbein[1], Stefano Chiaverini[6], and Reinhard Lafrenz[1]

[1] Department of Informatics VI, Technische Universität München, Germany
[2] Dipartimento di Ingegneria dell'Informazione, Seconda Università degli Studi di Napoli Aversa, Italy
[3] Faculdade de Ciencias e Tecnologia da Universidade de Coimbra, Portugal
[4] INESC Porto, FEUP Campus, Portugal
[5] Dipartimento di Ingegneria dell'Informazione, Ingegneria Elettrica e Matematica Applicata, Università degli Studi di Salerno, Fisciano, Italy
[6] Dipartimento di Ingegneria Elettrica e dell'Informazione, Università degli Studi di Cassino e del Lazio Meridionale, Cassino, Italy
`http://echord.info`

Abstract. The European funded ECHORD project[1] European Clearing House for Open Robotics Development began in January 2009 with the ambitious goal of bringing together European robotics manufacturers with the excellent European research institutions. Europe has a very strong robot industry and there is significant research potential as well as technological knowledge. There has been a long history of outstanding research and development in both robot manufacturers and research institutes. However, finding common ground between manufacturers and the research community, especially when it comes to defining the future direction of robotics research, has proven difficult in the past. This is one of the recurring themes on both sides, and a new level of cooperation is long overdue. Thus, ECHORD acted as a clearing house to streamline successful know-how transfers.

1 Introduction

The idea of the ECHORD project was born before the economic crisis had its maximum impact on the robotics industry in 2008/2009. Therefore, the concept of a project with the clear goal to strengthen the collaboration between academia and industry was a good opportunity to support the industry by offering funding opportunities and fostering already existing networks and creating new partnerships with the academic world. The project itself was proposed as an Integrating Project (IP) in the Call ICT-2007.2.2 under the Challenge 2: Cognitive Systems, Interaction, Robotics and started in 2009.

[1] The research leading to the results presented in this book has received funding from the European Union through the 7th Framework Programme (FP7) under grant agreement number FP7-ICT-231143.

F. Röhrbein et al. (eds.), *Gearing Up and Accelerating Cross-Fertilization between Academic and Industrial Robotics Research in Europe*, Springer Tracts in Advanced Robotics 94,
DOI: 10.1007/978-3-319-03838-4_1,

2 The Concept of ECHORD

The ECHORD concept comprises two instruments to reach the overall goal: the first one is the funding of so-called experiments, small research and development projects, carried out by typically 2-3 partners, both from industry and academia. The second instrument is the so-called structured dialogue, a means to strengthen the relationship between academic institutions and industrial companies, and to identify and support the knowledge transfer.

The project started with only three core partners from academia:

- Technische Universität München, Prof. A. Knoll, acting as the coordinator of the project
- Università di Napoli Federico II, Prof. B. Siciliano, acting as a co-coordinator[2]
- Universidade de Coimbra, Prof. N. Pires, acting as a co-coordinator

This core group was intentionally kept small, just fulfilling the legal requirements, in order to allow for a wide participation of institutions active in robotics research and development, as the coordinating partners were explicitely excluded from proposal submission and participation in the experiments.

These experiments were funded by the European Commission according to their rules, but the organisation of all processes which showed up during their full life-cycle are mainly handled by the core consortium.

3 Creating Awareness and Open Calls

As the word about the new project format spread in the community, there was a need to give clarifications about the project, the funding opportunities, requirements, and to start networking activities to link potential partners interested in the same fields of R&D. Therefore, two information days were held, one as a public opening event for the ECHORD project, held in Munich, Sept. 2009 and one in Lyon, Feb. 2010 with the aim to raise awareness especially in the French robotics community, which was under-represented in the first call. During these information days, potential proposers were informed about the funding scheme within ECHORD, the application procedure, and some special rules which apply in the project. Question and answer sessions with members of the core consortium clarified specific issues, mainly related to administrative aspects of the proposal and execution phase of the experiments. Another main goal of these events was the possibility to present the own research idea and to search for partners with complementary skills and expertise.

Three rounds of open calls were performed, leading to 243 submitted proposals, where 51 of them were finally selected for funding. The details of the three calls are given in Table 1.

---

[2] Università di Napoli Federico II is supported by third-parties belonging to PRISMA (Progetti di Robotica Industriale e di Servizio Meccatronica e Automazione) which includes researchers from the Università della Basilicata, Università degli Studi di Cassino, Seconda Università degli Studi di Napoli, Università degli Studi Roma Tre and Università degli Studi di Salerno.

Table 1. Proposals and experiments

Call	Num. proposals	Experiments funded	Indicative budget
1	108	16	4.5 M€
2	70	20	5.5 M€
3	65	15	5.1 M€

3.1 Evaluation and Selection

The evaluation and selection procedures were handled by the coordinating partners and the same strict set of rules as they were applied as in regular EU framework proposals. International experts were asked to evaluate the proposals, taking potential conflicts of interest into account.

Three independent experts acted as evaluators for each proposal in the first evaluation step, which was perfomed remotely using web forms and a blog functionality for the discussion between the experts to come to a consolidated assessment. In one single physical meeting per call, a subset of the experts involved in the remote evaluations formed a panel with the task of calibrating the individual evaluation reports and scores and to set up a fully ranked list of all proposals which passed the thresholds and were therefore eligible for funding. As the number of these proposals exceeded the indicative budget reserved for each call, only the highest-ranked proposals according to this budget were finally selected for funding.

3.2 Proposal Analysis

This section presents a summary of the results extracted in [12]. The data resulting from the set of 242 ECHORD proposals (one full copy resubmission was excluded from the complete set of 243) involves a total of 509 proponents, 264 institutions and 26 countries. The ECHORD core consortium lowered the entrance barrier by allowing single partner and single country proposals. The results presented in Fig. 1 show that, even in small projects (typically 18 months and 300k€), single partner projects have worse quality than multi-partner proposals: 48.1% versus 25,5% ($\chi^2(1) = 8.014; p = 0.005$).

In fact, the probability of a proposal with two or more partners to have a score above or equal to 10 is 2.717 times higher than in case of a single partner proposal (Odds-Ratio=2.717, IC95%[1.389;5.316]).

In terms of international cooperation, the ratio of proponents that participated in proposals whose evaluation score is above 10 with international cooperation is similar to the ratio for single country proposals. From these values one might conclude that the common mandatory requirement of multi-country proposals for EU financing may indeed promote the participation from peripheral countries but does not have a direct influence on the quality of the proposals.

From the analysis of the network of institution-level co-authorship, see Fig. 2, two conclusions were identified: (1) the clusters and the key players (with a

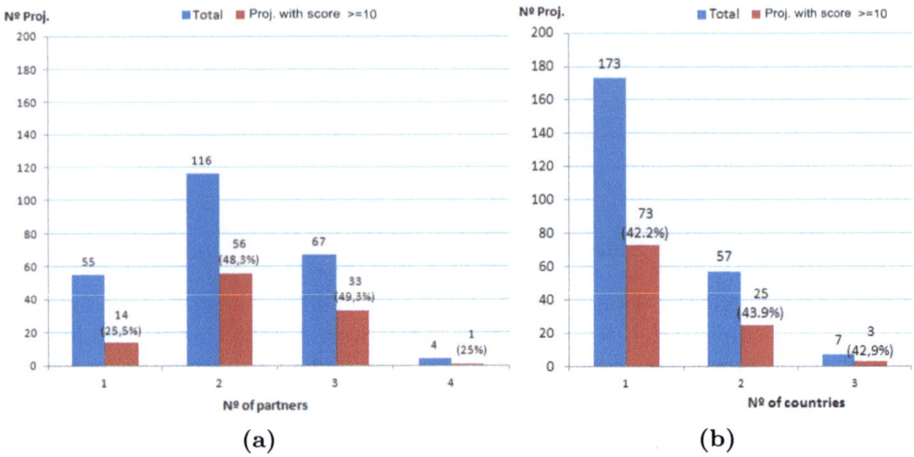

Fig. 1. Distribution of proposals per number of partners (a) and proposals per number of countries (b)

relevant number of connections or playing inter-cluster connections - labelled in Fig. 2) are organized geographically; (2) in addition to the major European robotics manufacturers (ABB, KUKA, REIS, COMAU, SCHUNK) there is a new group of key players that are small robot hardware suppliers with highly differentiating products coming from countries without significant tradition in this area, such as Spain, Austria and UK (Robotnik, FerRobotics, Shadow Robot Company).

In terms of the geographic distribution, the clustering of the proposals effecting the proposal production per inhabitant, the score and the international cooperation revealed that Belgium and Sweden constitute a cluster characterized by high proposal production, with very high scores and extensive international collaboration. Belgium also excels in another cluster analysis, being the only country for which 100% of proposals involved industry-academia cooperation and obtained scores above 10.

Industry-academia collaboration was one of the major foci of the ECHORD project. The collaboration patterns and the respective success are depicted in Fig 4 which reveals several facts: (1) the number of proponents from either industry or academia approximately doubles the number of proposals from research institutions. (2) research institutions collaborations with industry double their collaborations with academia. (3) partnerships for the proposals between research institutes have the highest quality judging by the score the proposals received.

Despite the high number of industry proponents in the ECHORD initiative, it is clear from the available data that usually industry plays a secondary role in the submission process: they usually do it together with universities or research institutes, and do not take the lead of the consortium. In fact, there were

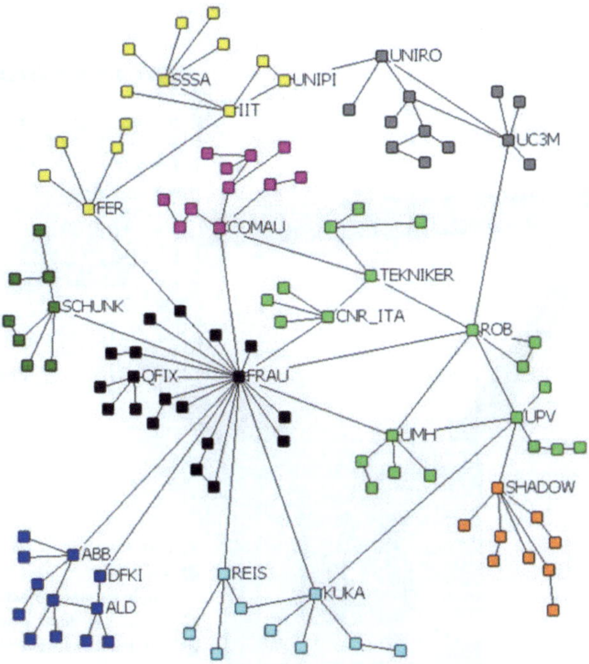

Fig. 2. Largest component of the institution-level co-authorship and clustering. Arrangement for visualization optimization. Adapted from [12].

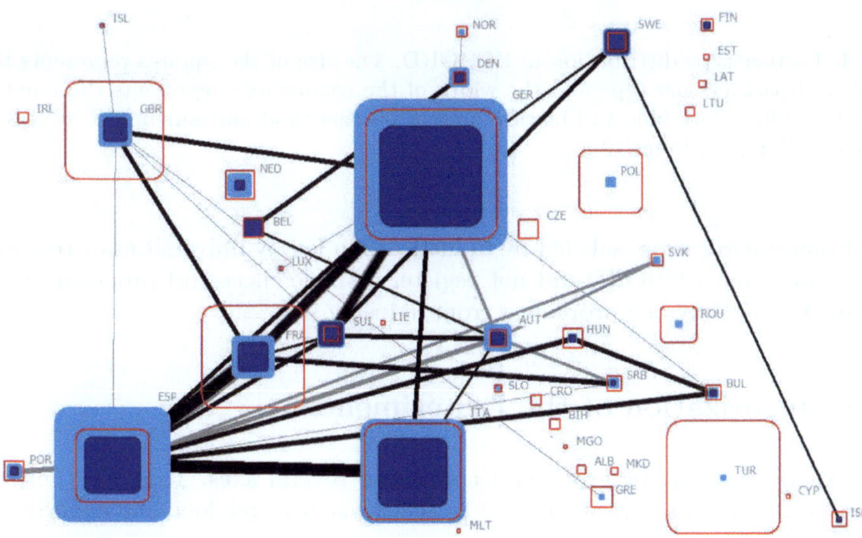

Fig. 3. Geographic distribution of ECHORD experiment proponents. The light blue in % is the number of proponents, the dark blue the number of proponents with reviews above 10 points and the red frame represents the population of the respective country in million inhabitants. Adapted from [12].

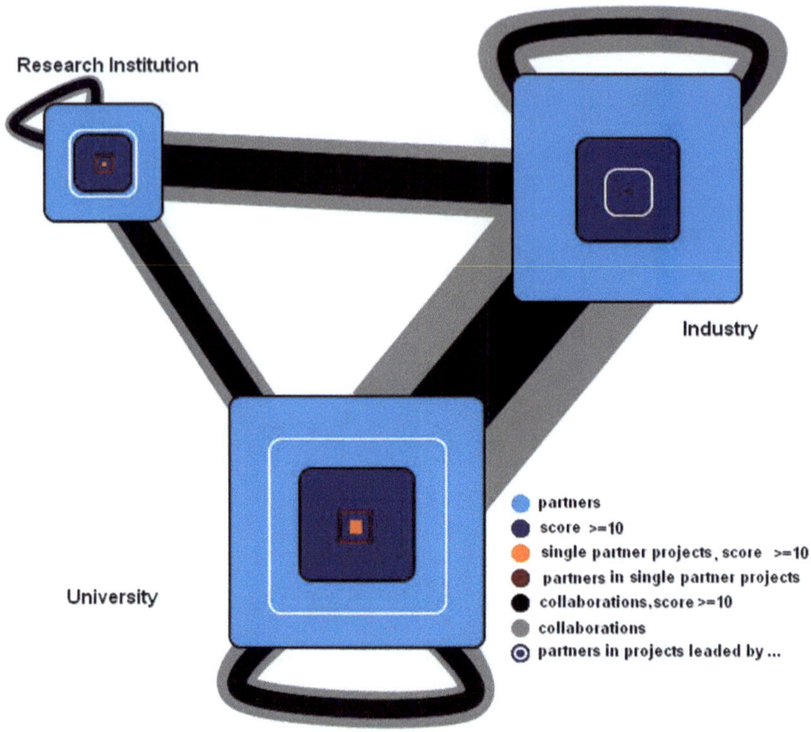

Fig. 4. Partner type distribution in ECHORD. The size of the squares represents the number of partners per type and the width of the connections represents the number of partnerships. Dark blue and black represent partners and partnerships of successful proposals. Adapted from [12].

significantly fewer proposals led by industry than led by universities or research institutions, and ECHORD did not register a single successful proposal (score above 10) with only one proponent from industry.

4 Categorization of the Experiments

The experiments can be grouped according to several axes. To guide the proposers, the proposals were structured by scenarios, research foci, and experiment types.

Three scenarios for likely future robot use were defined to outline the scope of research work to be performed in the experiments. These scenarios intended to make it possible for all stakeholders to get a clear picture if and how their proposed work and envisaged results can be embedded into a coherent vision of robotic applications. Thus, they describe the application context from an exterior view.

For breaking down the application-driven scenarios into concrete research and development, four research foci were identified. The research foci guide the research work. They were chosen to provide complete coverage of the relevant aspects of the scenarios.

4.1 Scenarios

Three scenarioswere identified which are both scientifically challenging and commercially relevant. They represent comprehensive sets of challenges in an illustrative way, so that robotics experts can easily relate their own research to them. The scenarios build on each other.

- The first scenario of ECHORD is the **human-robot co-worker**. In this scenario, the traditional idea of pre-programmed robots was dropped, and the robot interacts with a human towards achieving a common goal. This scenario is especially relevant for future industrial applications, where the (physical or sensor-based) fences between robots and humans disappear.
- The second scenario is the **hyper-flexible cells** scenario. This scenario envisages not only one or more highly dexterous and cooperative robots, but also the hardware and software integration of the robots with an automatic warehouse system and the other devices present in the cell.
- The third scenario is the **cognitive factory**. This scenario aimed at taking the classical concept of the flexible manufacturing systems to a new level. The final goal is to create environments which configure themselves and are fault-tolerant, and which contain autonomous robots jointly participating in the production process with their human counterparts.

4.2 Research Foci

Within the scenarios, different research foci have been identified. The research foci are reference points for the expected scientific progress of experiment proposals. They bring together mechanical design and controller technology from manufacturers with the knowledge and experience in sensing, cognition and behaviour control from the research community.

- The first research focus is on **human-robot interfacing and safety**. Here, the main goal of the experiments is to show that safe human-robot cooperation is possible, taking all kinds of sensor failures and inconsistencies into account.
- The second research focus is on **robot hands and complex manipulation**. Here, the experiments show the improvement of laboratory setups towards practical usability as well as promising breakthroughs in the areas of sensors and sensor-guided manipulation.

- The third research focus is on **mobile manipulators and cooperation**. Here, mobile manipulators solve concrete problems in dynamically changing environments with moving obstacles and interaction with humans.
- The fourth research focus is on **networked robots**. Here, two areas are possible: one is networked industrial robots, where we expect demonstrators that can only be built in collaboration between industry and academia, with industry providing controller architecture and academia contributing knowledge in advanced real-time networking technologies, as well as service-oriented architectures. The second area concerns more loosely coupled systems, where experiments with mobile robots are expected to establish new showcases, e.g. in the area of search and rescue with robots, new applications of robots in urban areas, and robot systems for monitoring tasks.

4.3 Types of Experiments

As a third axes, the so-called type of an experiment was defined in order to categorize the experiment's main contribution to future applications:

- experiments of type **joint enabling technology development** aim at co-operative development of components or systems with the main focus on finding solutions for specific technical problems.
- The experiment type **application development** covers experiments that create robust prototypes for standard tasks in new areas or new tasks in known areas.
- The **feasibility demonstration** experiment type explores the possibility of applying robot technologies in completely new domains.

These classifications took place during the proposal phase by a self-assessment of the proposers. For the funded experiments, the statistics of the assignment to scenarios, research foci, and experiment types are given in Table 2. The statistics clearly show that, for this specific funding scheme, smaller, individual problems as in the scenarios human-robot co-worker or the hyper-flexible cell are better suited than the complex scenario of a cognitive factory.

After the selection of the experiments, other classification criteria turned out to be useful in addition to the ones mentioned before, namely the application area, concrete research topics, and technologies. This classification was only possible after the analysis of the successful experiment proposals, as the calls did not aim at restricting the ideas of the proposers. These additional classifications are shown in Fig. 5 and give a good impression of the variety of ECHORD's research. Robotics technologies for medical application, for example, are of increasing interest, both scientifically and economically. Several ECHORD experiments address this application area, despite the fact that this could not be predicted at the time of the calls. Also, autonomous inspection and surveillance, often making use of unmanned aerial vehicles, is a developing field.

Table 2. Categorization of funded experiments

Scenario	Num. experiments
human-robot co-worker	30
hyper-flexible cells	18
cognitive factory	3
Research Focus	**Num. experiments**
human-robot interfacing and safety	16
robot hands and complex manipulation	20
mobile manipulators and cooperation	12
networked robots	3
Experiment type	**Num. experiments**
Joint enabling technology development	17
Application development	20
Feasibility demonstration	14

5 Overview of Individual Experiments

This section presents an overview of the scientific content of all the experiments funded by ECHORD but which did not produce an extended contribution to this volume. In the following, the abstracts of these experiments are reported in alphabetical order.

ALEXA (An advanced light-weight robot arm for flexible and mobile applications in hyper-flexible cells). The ALEXA experiment tested the capabilities of a new lightweight manipulator built using the innovative and low-cost robolink construction kit from igus. The robolink joints are cable-driven and connected with lightweight, carbon fiber-reinforced plastic links. The Fraunhofer IFF developed a 5-DOF robolink manipulator for use as an assistant in flexible cells and demonstrated its high portability in a common pick-and-place scenario.

AssRob T.I. (Semi-Autonomous Surgical Tool Instrumentor for Robot Co-Workers in Hip-Surgery). The main innovation of the experiment was to create a semi-autonomous two-arm assistant bone lever holding robot system and to demonstrate that it can take over the tiring lever holding and handing over task from a human assistant. The goal was to show the capability of the robot to autonomously recognize the bone lever in space and then approach it close enough, just like a human assistant does, to enable the surgeon to connect already in situ placed bone levers and retractors to it.

BABIR (A better audition for a better interaction with humanoid robot). In this experiment, a robust vocal interface between human and robot was developed and implemented on the small humanoid robot Nao. Speech processing algorithms have been developed not only to recognize the speech but also for the localization of the speaker which is necessary to get the attention of the robot. Experiments were performed in a domestic environment.

Fig. 5. Classification of experiments

BRACOG (Brain-Controlled Grasping). The research carried out in this experiment led to the development of a robotic arm that people with severe motor handicaps (due to traumatic injuries or strokes) can use to grasp and manipulate common objects, controlled by voluntarily changing their brain activity. The developed robot which is controlled by thoughts, was able to grasp unknown objects.

C-KOMPAI (Providing Cognitive capabilities to the KOMPA Robot with the addition of a Cognitive Brain). The experiment integrated into the KOMPAI robot, a modular system for interaction with people. The CBRAIN (Cognitive Brain for Service Robotics) system, includes the following functionalities and capabilities: high-level perception and representation of the environment, robust and reliable autonomous discovering of the environment and map building functionality, cognitive navigation, intelligent problem solving. Possible applications of the robotic system are assistance to elderly people and floor scrubbing in the household.

ContainerBot (Stochastic impact-triggered mobile manipulation for fast cycle time unloading of variable-sized boxes from unordered piles). The experiment proposed new ways to empty standard shipping containers with new robot technology. This experiment used one or two KUKA lightweight robot arms to pick up light (up to 5kg) or heavy (up to 10kg) boxes of various sizes and shapes in unordered stacks. Applications are automatic emptying of parcel containers, assisting persons with robots for lifting or positioning of heavy equipment, and flexible use of robots in factory work cells.

COWBOI (Cooperative Welding employing Robot Intelligence). This experiment targeted a system to increase the efficiency and usability of welding robots by focusing on the interactive task specification from the human to the robotic workmate and the fast acquisition of unknown work pieces (through integration of a laser scanner and a Kinect sensor), as well as the automation of the path planning (releasing the human from this responsibility).

EASYPRO (Accurate Manual Guided robot programming). The aim of the experiment was the integration of a breakthrough programming approach combining a universal Manual Guidance Device (MGD) for a fast, intuitive but rough tool path programming with a 3D visual data analysis system to adjust the obtained trajectories and to allow accurate end-effector positioning. The results were validated in a laser cladding task.

EduFill (Filling the Educational Gap in Service Robotics). The experiment intended to bridge the gap between frontier research and education, by introducing state of the art concepts and solutions in mobile manipulation into the classroom. The project prepared a practical robotics curriculum, focused on modern concepts and solutions for mobile manipulation, such that students can bring their robot knowledge to the industry. The approach of hands-on learning was adopted, based on the use of a novel software (and partly hardware) toolbox centred on state-of-the-art educational and research robots.

ERICA (Evaluating Human-Robot Interaction and Cooperation-based on Analysis of 3D Image Sequences). The experiment had the goal of achieving safe human-robot-cooperation and to advance a system based on real-time estimation of significant parameters of human body kinematics. The knowledge about parts of the human kinematics is a key issue for cognitive vision based systems, which deal with real cooperation between man and machine. This knowledge was used in the experiment to guarantee safety for the human co-worker by estimating the risk of a situation and adapting the robots behaviour accordingly.

FIDELIO (FIxtureless DEburring of wheeLs by human demonstration). The experiment investigated the feasibility (and the related advantages) of the programming-by-demonstration paradigm in an industrial application scenario, exemplified by a fixtureless wheel deburring task. To this aim, a robotized cell was set up, composed of an industrial manipulator equipped with both vision and force sensors and a workstation where aluminium wheels are placed for deburring.

Flexprass (Flexible Precision Assembly with Mobile Robots). The partners of the experiment combined human-friendly modular industrial desktop robotics with high-precision equipment in order to setup a flexible precision assembly system for highly demanding products such as laser-systems.

GISA (Gesture Based Instruction of Safe Mobile Robot Arm). The experiment aimed at providing information about the consequences and effects of using mobile robot arms in real manufacturing environments. A gesture-based human robot communication was developed that allows a shop floor worker to provide commands to the robot on-the-fly without the need of hardcoding them into a robot program.

GOP (Generating optimal paths for industrial and humanoid robots in complex environments). This experiment combined state-of-the-art developments of path planning and numerical optimal control research areas. The objective was to create the algorithmic foundations to tackle real-time optimal control problems of industrial and humanoid robots in cluttered environments, trying to overcome the limitations of both domains.

HERMES (Hyper-flexible bimanual robot manipulation and packing of deformable parts in footwear industry). The experiment addressed the automation of the packaging process in the footwear industry, which is a growing market. The HERMES experiment demonstrated the potential to apply robotic technologies to such processes through detection and bimanual manipulation of non-rigid objects.

HipRob (Robot-Assisted and Ultrasound-Guided Navigation for Hip Resurfacing Arthroplasty). The experiment set up a robotic system that has the ability to aid orthopaedic surgeons in performing Hip Resurfacing prosthesis surgery with consistent high accuracy and precision. The solution is based on variable impedance control for physical surgeon-robot interaction integrated with an ultrasound image sensor for non-invasive real-time bone tracking.

HUBRINA (HUman-roBot co-woRking IN Agricultural master-slave systems). A master-slave robot control for agricultural activities was developed in this experiment and its feasibility demonstrated. The solution sees the human taking over the non-robotized tasks of safety prevention and feedback on the quality of work performed by the robot. A robotic master-slave system applied in agriculture was demonstrated beyond just the level of simulation using two tractors working in the main filed area.

HUROBIN (Human-Robot Object Interaction). The experiment focused on the human-robot interface and safety when the human operator and the robot handle the same object: the robot is supposed to bear the load while the human leads the movement of the object in a co-operative pick and place task. Distributed position sensors, a torque/force sensor on the robot, as well as a tracker sensor that tracks human motions ensure safety of operation.

HYFLAM (A Hyper-Flexible Work Cell for Biochemical Laboratory Automation). The experiment focused on the automatic execution of several complex manipulating actions that occur frequently in a biochemical laboratory. Several actions, like opening different types of laboratory plastic-ware, tubes and glass vials, pipetting, that had to be performed manually, before, were performed with a sophisticated robot hand leading to a significant enhancement of safety, flexibility and efficiency.

HYROPA (Hyper-flexible robot cells using reconfigurable passive kinematics). The experiment demonstrated the possibility of reducing the technical effort for highly flexible robot cells using passive kinematics and state-of-the-art serial robotic arms. The core idea of the experiment was the application of state-of-the-art industrial robots together with fixable, passive kinematic arms, which have a large number of degrees of freedom, but without their own drives. The industrial robot takes care of the reconfiguration of the passive kinematics.

Insewing (Development of a robotic manipulator of human tubular tissues for suture and support in anastomosis surgery interventions). The aim of the experiment was to develop a surgical robotic manipulator device, focused on the improvement of surgical interventions with anastomosis. In particular, the developed device was successfully tested in the anastomosis of pig intestines, thus simulating the hypothetic behaviour with human tissue.

InterAID (Interactive Mobile Manipulators for Advanced Industrial Diagnostics). The experiment demonstrated the feasibility of applying mobile robots with manipulation capabilities to advanced diagnosis and quality control in industrial environments. A mobile manipulator equipped with a dexterous hand was applied in the reliability lab of a white goods factory producing washing machines. The system is able to carry out the repetitive tasks of product quality control, such as open washing machine doors, pushing buttons and collecting results of the quality control test.

JILAS (Jig-Less Airplane Assembly in low volume production by enhanced human robot interaction). The core of the experiment was to realize a scenario where a human worker and a force-controlled industrial robot assemble airplane components in a classic human-robot co-worker cooperation. The robot has the capability to be hand guided by grasping and moving the gripped work piece. With the help of this robot, the human worker can pick up a component and move it near the final assembly position. The final position is then reached by moving the robot based on the accurate measurement of a laser tracker.

KANMAN (KANban integrated, magnetic orientated modular mobile MANipulator). The experiment developed an application in which a robotic co-worker supports a Kanban production process (a control method for just-in-time production) by taking care of the flow of materials. While the Kanban process is controlled by crates containing the production materials, these crates are transported by a mobile service robot platform with a manipulator. The experiment initially tested a magnetic approach for safe mobile navigation and manipulation.

KOMPEYE (Enhancing the Visual Perception Capabilities of Kompa Robot Using Parallel Processing). The experiment research activities led to a substantial improvement in the level of perception of current robots, in terms of human presence detection, recognition of faces and facial expressions, recognition of the gestures of people who ask the robot for help. This was accomplished using synergistically enhanced computationally intensive vision capabilities, achieved in real time through GPUs.

MONROE (Hyper-Modular Open Networked RObot systems with Excellent performance). The goal of the experiment was to enhance and apply a new type of hyper-modular parallel robot that also enables a performance increase in terms of stiffness, precision, and bandwidth with respect to feedback from external sensors. Forecasted applications are: small scale milling of aluminium casts, low-cost finishing of small plastic parts, assembly of consumer goods, handling and pick-and-place of large volumes.

MUCE (Modular Underwater Cleaning Equipment). The research focus of the experiment was on robust underwater positioning. The overarching goal of this project was to make a new generation of pool cleaner that replaces heavy mechanics with sensors and intelligent control.

ODEUO (Inner Oscillation Detection and Evaluation of Unknown test Objects). The experiment was aimed at performing with a robot what is a relatively simple job for humans, i.e. the detection of spare or lose parts by shaking objects. ODEUO investigated the sensorless detection and evaluation of inner oscillations of unknown test objects mounted on a compliant test bench constituted by a pneumatic hexapod. Successful tests were conducted on real automotive motor suspensions.

OMNIWORKS (Omnidirectional Vision for HUMAN-UAV co-working). The aim of the experiment was to exploit the growing interest and convenience of use of small scale UAV by developing a series of self-enclosed specialized and complementary modules and applications suitable for a large variety of commercial UAV currently on the market. The developed systems are easy-to-manage by a non-skilled person who can exploit the modules to automate easily different processes, such as mosaic maps, visual navigation based on maps, video stabilization, image tracking and servoing.

PRADA (Parallel Robot with Adaptive Dynamic Accuracy). The experiment aimed at improving the performance of high-speed parallel robots in terms of dynamic accuracy along complex paths and adaptability to changes in operational conditions. This was achieved by combining three enabling technologies specifically adapted to industrial parallel robots: adaptive dynamic control, sensor based control and actuation redundancy.

REMAV (Remote Eye for Micro Aerial Vehicles). The main goal of the experiment was to demonstrate the possibility to operate Micro Aerial Vehicles (small autonomous helicopters) precisely and safely in a dynamically changing environment consisting of fixed obstacles, humans and other MAVs. In order to achieve this goal, extremely precise position and speed control was developed based on a new specifically adapted miniature optical-flow based speed sensor for measuring both position and speed of the vehicle.

RIVERWATCH (Cooperating robots for monitoring of riverine environments). The experiment developed an autonomous multi-robot system for ecological monitoring of riverine environments. The multi-robot system is composed of an autonomous surface vehicle (ASV) with an unmanned aerial vehicle (UAV) piggybacked on it. The UAV overcomes the limitations imposed from observing the environment from the low viewpoint provided by the ASV. Conversely, the ASV, being equipped with a solar panel, is able to perform energy harvesting for itself and for the UAV, which is a key factor in long-lasting operation.

RODIN (Robust Control of Human-Robot-Environment Dynamic Interaction for Natural Stone Carving). The aim of the experiment was to prove experimentally the feasibility of concepts that extend industrial robot usage to an intelligent power tools carrier and act as an active assistant supporting dynamic physical human-robot-environment interaction during the complex and demanding artistic work of hard natural stone carving. Due to the strict physical interaction of the human operator with many parts of the robot, a specific risk analysis was carried out and measures for risk reduction were taken in view of the EC directive on machinery[3].

SPEAKY (SPEAKY for Robots). The experiment aimed at fostering the definition and deployment of voice user interfaces (VUIs) in robotic applications where human-robot interaction is required. It intended to promote speech technologies transfer towards manufacturing processes, to provide semi-automatic speech-based interface development for robotic platforms. A novel Robotic Voice Development Kit was conceived as an interactive environment aiding designers to define the voice interface according to the desired application requirements.

SprayBot (a Robotic Spray Booth for the Automatic Painting of Bodyworks). The objective of the experiment was to demonstrate that a robotic system (in particular a mobile manipulator), of affordable cost for the small/medium workshop, is able to carry out the painting phase in small/medium body shops improving the quality of the service with a reasonable investment. An effective varnishing

[3] `http://ec.europa.eu/enterprise/sectors/mechanical/machinery/` [accessed: 08/20/2013]

control algorithm was devised based on vision feedback, which allows carrying out the varnishing task without a priori knowledge of the part shape, position and orientation.

TACTIP (Tactile fingertip for robots). The experiment developed a biologically inspired tactile robotic fingertip based on a silicone rubber deformable structure with edges and bumps and a miniature camera. The image data are then interpreted into contextual contact data, such as shape and direction of an edge or raised bump. The sensor was integrated on the Elu-2 robot hand to carry objects, providing the robot hand with new sensing abilities: identify when contact is made, monitor the status of a grip and detect when a slip occurs.

6 Results

The results of ECHORD can be described from different points of view: from a scientific perspective, the achievements of the research work are mainly visible in the form of publications, but also in the form of software which can be used by the community. From a commercial perspective, the creation of close-to-product prototypes, generation of patents and development of new business ideas, ideally resulting in setting up a new company or business unit, can be seen as a measure of success. In terms of public visibility and raising awareness for innovative solutions generated by the experiments, each individual experiment created a demonstrator or prototype to show the research and development results and produced a multimedia report in form of a video which can be seen on the project's YouTube channel "RoboticsEurope"[4].

Besides these aspects, one less measurable aspect is of eminent importance: ECHORD contributed to closing the gaps between academia and industry in the robotics area by fostering existing and creating new networks and partnership with a special focus on SMEs, where some of them participated for the first time in an international project. The entrance barriers for newcomers to European funding, mainly SMEs and smaller research institutions, were lowered by offering extensive support by the coordinating partners in handling the administrative issues. Seen as a new way of funding, the rules for participating in the experiments were relaxed in comparison to individual projects, STREPs or IPs. For example the requirement of a specific number of partners from different countries was already fulfilled by the core consortium and therefore, the experiments were free to choose their preferred cooperation partner(s), be it in the same city or in a completely different region of Europe. The fear of contact with unknown partners with a different cultural background, the day-to-day communication in a different language (English) were obstacles that were minimized in ECHORD which was appreciated in particular by the smaller companies.

For some experiments, already during their run-time or shortly thereafter, follow-up projects in the regular FP7 framework were proposed and granted. In these cases, ECHORD acted as an incubator for the further development of the European Robotics Community.

[4] `http://www.youtube.com/user/RoboticsEurope` [accessed: 08/20/2013]

7 The Second Pillar: The Structured Dialogue

In additon to the experiments, a second concept important to ECHORD is the structured dialogue. This is comprised of systematic exchanges between the robotics community and ECHORD about expected future trends in robotics. These future trends will be examined in the following section.

The structured dialogue was designed as an iterative process of information gathering and consensus finding between all parties. This approach is well suited to the structurally diverse and interdisciplinary field of robotics, with many potentially interesting directions. Based on a collection of ideas gathered in polls, web consultations and expert meetings, an initial set of ideas will be profiled, redistributed with specific questions for discussion (filtered through an economic, scientific, technology perspective). Many results can be found in Röhrbein et al. [8].

7.1 Introduction

This contribution is going to summarize the results of our efforts to study industry-academia collaborations. We will report on research topics and foci.

The goal is to identify both current concerns of the respective communities and emerging trends. The method for indentifying such emerging developments is often a quantitative look at the literature in the respective field [5]. For this purpose two instruments were used:

1. We are going to look at publications in journals and at conference papers to see which topics are currently being addressed by research and development efforts.
2. We distributed questionnaires at different venues and are going to report on those results.

The results were compared to similar studies which were conducted by other parties. We will argue that our results are comparable to those found by the EUCog questionnaire which was distributed among its members and the results of a questionnaire which was distributed at IEEE/RSJ International Conference on Intelligent Robots and Systems (IROS) 2012[5].

We hope to encourage further academia-industry collaboration and help to avoid the pitfalls which threaten to hamper such endeavours. One factor that one needs to keep in mind is the fact that ECHORD focused mainly on technology-transfer and industry-academia collaborations. Therefore, one needs to contextualize the findings in order to make them usable for this purpose.

7.2 Are Robots Here?

In 2004 Rodney Brooks [2] predicted that robots would be as pervasive as electronic mail and the world wide web by approximately 2019. In 2013, one can

[5] `http://www.iros2012.org/` [accessed: 08/20/2013]

already see a trend towards this vision becoming a reality as robots have become part of a consumer entertainment market (for example the Sony AIBO [1]). They are also being deployed in hospitals, museums [6] and in households [3]. The largest market for robotic applications is, however, still in manufacturing. According to the IFR world robotics statistics[6], the market value for industrial robots is estimated to be US$ 8.5 billion "without cost of software, peripherals and systems engineering." Including these costs, the market value is estimated to be US$ 25.5 billion. In contrast, the professional service robotics market is estimated to be US$ 3.6 billion and the service robotics market for personal and domestic use is estimated to be US$ 636 million, according to the statistics of the IFR world robotics[7]. If robots are to become even more prominent in all these areas, there are a number of developments which need to accompany the growth in market share. The market pull and the research push need to accompany each other. Steels [11] notes that in the early 1980s one could notice the advent of a change in artificial intelligence with the dawn of an application driven research agenda which was accompanied by huge conferences, venture capital and the increased founding of spin-off companies.

This justifies increased research efforts in a technology related field which are sometimes questioned due to the double-boom pattern [10], which accounts for the lag between a science-push and a market-pull in technological fields. One of the main goals of ECHORD is to support a decrease in this lag by facilitating a productive exchange between industry and academia.

7.3 Investigating Current Research Topics

In order to encourage increased activities in R&D in the field of robotics, one first has to address the issue of what research topics are currently being pursued in the field of robotics. For this purpose, the most suitable method is the analysis of journal papers and conference contributions.

All conference contributions that were submitted in the years 2011 and 2012 to both ICRA and IROS have been analyzed. These are the most pertinent venues for presenting results to the robotics community[8]. We analyzed the keywords that are used by the authors for paper submission in PaperPlaza. IROS and ICRA use the same list of keywords, i.e., the one permanently available at `http://www.ieee-ras.org/ceb/areas.html[accessed:03/15/2013]`. The RAS Conference Editorial Board is in charge of refining and updating this list for ICRAs and this list has been very stable over the past couple of years.

Overall, there are 9,726 associations with the 143 keywords, which is a number about three times higher than the number of papers. This is due to the fact that typically three keywords are assigned to each paper.

[6] `http://http://www.worldrobotics.org/uploads/media/Executive_Summary_WR_2012.pdf` [accessed: 08/20/2013]

[7] `http://www.ifr.org/service-robots/statistics/` [accessed: 08/20/2013]

[8] See for example: `http://www.ias.tu-darmstadt.de/Miscellaneous/ConferenceQuality` [accessed: 03/15/2013]

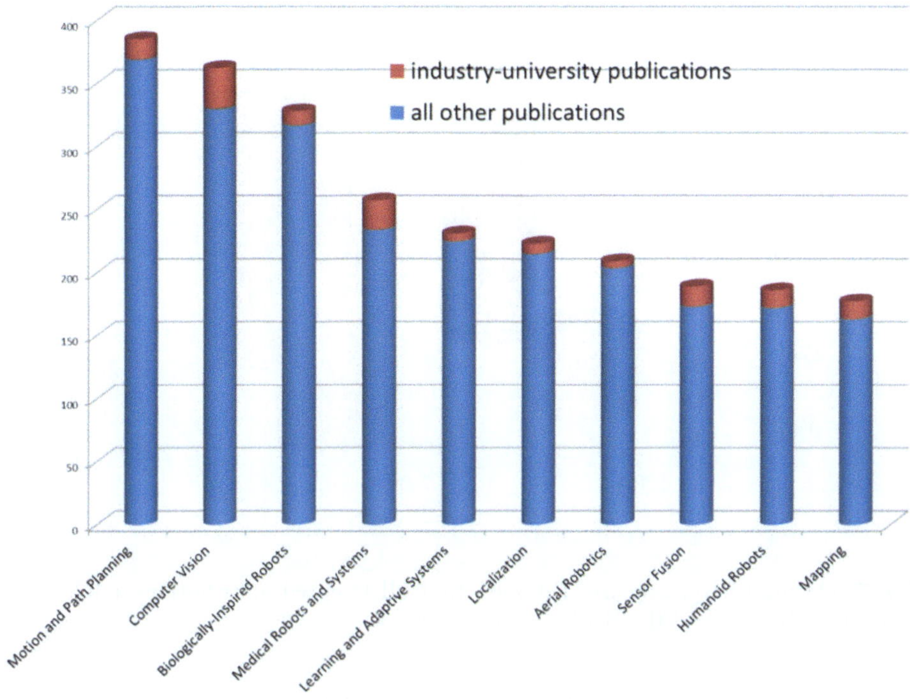

Fig. 6. The most frequently used keywords in all accepted contributions to ICRA 2011, ICRA 2012, IROS 2011 and IROS 2012.

Fig. 6 below shows the most frequently used keywords from all accepted contributions to ICRA-11, ICRA-12, IROS-11 and IROS-12. One can see the number of papers for each keyword. Though the graphs also differentiate between papers that resulted from an academia-industry collaboration (in red) or from research which was not based on such a collaboration (in blue) by assessing the affiliations of all authors. In total, 594 assignments stem from industry-academia papers.

This list of top research topics remains very stable with regard to conference (IROS, ICRA) and year (2011, 2012). The 7 highest-ranked topics from Fig. 6 above are in the individual top 10 lists of all four conferences (with only one exception)[9].

For these 7 research topics, there is a clear increase in the number of academia-industry collaborations: At IROS, the share increased from 5.5% to 7.3%. The share more than doubled at ICRA (see Fig. 7). The numbers are averages based on the topics Aerial Robotics, Biologically-Inspired Robots, Computer Vision, Learning and Adaptive Systems, Localization, Medical Robots and Systems, Motion and Path Planning.

[9] The topic "Learning and Adaptive Systems" was only ranked number 11 at ICRA 2011.

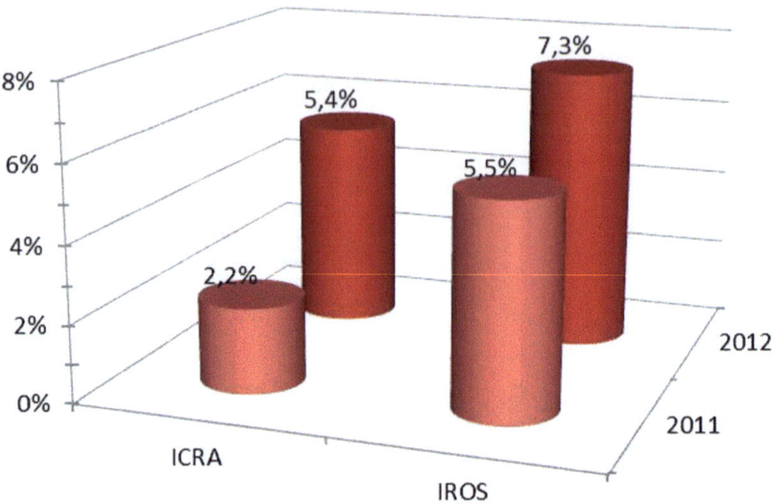

Fig. 7. The most frequently used keywords from all accepted contributions to ICRA-11, ICRA-12, IROS-11 and IROS-12

Next, we had a closer look at all topics with a high share of industry-academia collaborations. Fig. 7 below displays all those keywords that were attached to conference papers whose share in industry-academia was above 10%.

For journal papers, a similar keyword analysis is complicated by the fact that no predefined list of keywords exists from which the authors can select. As a consequence, a far larger set of keywords was used and they are overlapping etc. Here we would need an ontology. We have already begun work here but it has not been completed.

Nevertheless, there is a high overlap in keywords to be found in papers produced in collaboration between industrial partners and academic research institutes published in journals and those accepted for conferences. This is illustrated in the following Table 3 which lists the 10 most often used keywords for industry-academia collaborations.

How can one connect industry and academia effectively? To find an answer to this question, we decided to study the publications further. What leads to a successful, productive collaboration?

As part of the structured dialogue, we interviewed a number of experts with a record of successful technology transfer projects. The following quote is from an interview with Prof. Brooks, which illustrates what can be done to better connect industry and academia [9]:

> "I think that in industrial robots and manufacturing in general, we haven't really seen the impact of information technology in the same way we've seen it in the office and information spaces. [. . .] It is the

Table 3. Topics in conferences and journals

	Conferences	Journals
1.	Mechanism Design of Manipulators	Computer Vision
2.	Medical Robots and Systems	Mobile Robots
3.	Computer Vision	Medical Robots and Systems
4.	Intelligent Transportation Systems	Mobile Distributed Robotics
5.	Collision Avoidance	Sensor Fusion
6.	Motion and Path Planning	Motion Planning
7.	Recognition	Path Planning
8.	Localization	Localization
9.	Personal Robots	Distributed Systems
10.	Distributed Robot Systems	Force Control

simplicity of use which then leads to high adoption and high rate of adoption. So, I think we haven't seen that penetration. So, how to make the things easy for ordinary people to use, instead of making the people adapt to the technologies. Adapt them to people, not the other way around. So that's where I think the big payoff is going to be."

Rodney Brooks's statement suggests that a higher market penetration is desired by the industry. For this purpose, the technology needs to be developed further. Brooks mainly points toward human-machine interaction. He claims that this is the area where most work is required. However, we chose to look at future topics and emerging research trends in the field of robotics.

Our method used for investigating future topics and emerging trends is comparing results from our own literature survey (see [8]) with results from firstly, a recent poll conducted by *The European Network for the Advancement of Artificial Cognitive Systems, Interaction and Robotics (EUCog)*[10] and secondly, a survey organized by the IROS-2012 organization committee.

Main topics from the ECHORD literature survey were:

- Autonomy
- Bio-inspiration
- User interface, human robot interaction
- Vision & Recognition
- Sensor technology
- Language and emotion
- Advanced control
- Automatic path / motion planning
- Modular robotics & multi-agent systems
- Advanced cognition
- Safety and Security
- Test and Validation

[10] `http://www.eucognition.org/` [accessed: 08/20/2013]

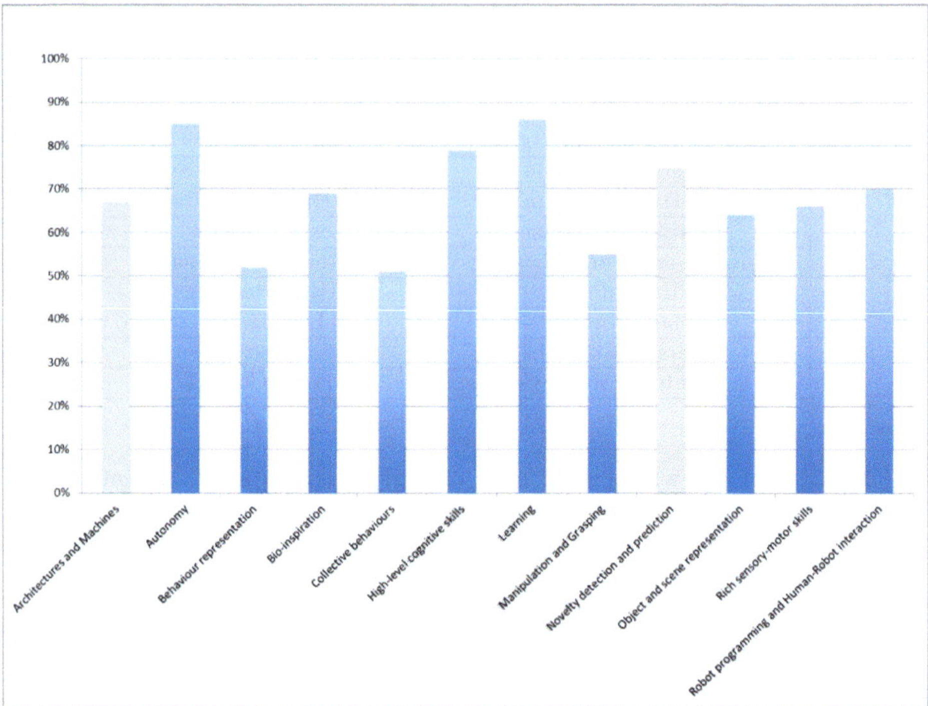

Fig. 8. Research topics to the EUCog questionnaire in comparison to the ECHORD questionnaire. The blue columns indicate research topics mentioned in the ECHORD replies and grey columns indicate research topics only mentioned in the EUCog replies.

EUCogII (the second phase of the network as an EU FP7 coordination action) conducted a survey among its members which is similar to those surveys conducted by the ECHORD team, but with a broader focus on future research topics in cognitive systems & robotics. In the EUCog survey, a list of research topics was given to the participants and their task was to rate them on a scale from 1 (not important) to 5 (very important). Fig. 8 briefly summarizes the main results by focusing on those topics that received a rating of 4 or higher by at least 50% from all 211 participants.

It is evident that most topics are similar to those identified previously (blue columns), but there are also some noteworthy differences.

Topics only mentioned in the EUCog list (grey columns):

- architectures and machines
- novelty detection and prediction

Topics only mentioned in the ECHORD list (blue columns):

- Language and Emotion
- Advanced Control
- Automatic path motion planning

- Modular robotics & multi-agent systems
- Safety and Security
- Test and Validation

Mainly due to a different focus with regard to robots, in our survey the results differ slightly. Our results echo the Brooks statement above in determining interactive capabilities (Language and Emotion, Safety and Security) more than the EUCog participants, who seem more focused on the reasoning of an individual agent, though both groups also mentioned Human-Robot Interaction as one possible topic for further research.

The other items in our list, which were not mentioned frequently by the EUCog members, are all related to robot motion apart from Test and Validation, which is, of course, needed in practical applications. The terms in the ECHORD list are more strongly oriented towards the need of the market and practical applications whereas many (but by no means all) EUCog members are more interested in more fundamental research questions.

In comparison to the IROS-2012 survey 116 forms were collected. The participants (the IROS attendees) were able to select up to three areas of research for future research for each question asked. Here we only present the results regarding the answers to two out of the four questions:

1. for effectively tackling such grand societal challenges, research should mainly focus on
2. in the next decade robotics R&D should focus mainly on the following grand research avenues

The IROS questionnaire identifies four topics which more than 10% of the participants suggested future research should focus on. The most freuqent answer given was improved control schemes and AI methods. This is related to the items Advance Cognition and Advanced Control, which were also frequently named by the participants in our study. The second most frequent response in the IROS answers relates to sensors which were also named frequently our study. The third most frequent reply to the IROS questionnaire relates to human-machine interaction which is also an important topic to the people who were we asked. The final topic which was named by more than 10% of respondents for the IROS questionnaire relates to system integration. This is a topic which the participants questioned in our study did not mention explicitly.

It is very likely that this difference arises from the fact that IROS is a conference which attracts system integrators and therefore this topic is assigned a high value. Also, there is quite a lot of overlap between the 10% of research avenues which the IROS 2012 participants named and the topics for future research which our participants named. Medical robots and prosthetics is the most frequent answer for the IROS 2012 participants. This is hard to map on to the responses to our questions and therefore, constitutes a difference. However, the other answers above 10% – Embodied Intelligence, Cognitive Vehicles and Cognitive Robotics – all relate to the topics Advanced Control, Advanced Cognition and Autonomy in our topic set. Therefore, a large overlap between the answers can again be seen.

7.4 Conclusions Regarding Current Research Trends

We found a strong agreement between the topics named by the EUCog members, the IROS 2012 participants and our ECHORD study. Many answers relate to autonomy and cognition, which have already been identified as an important topic for the future [7]. These are, of course, issues which relate to the core interests of all three communities.

Future trends will be decided by both a science-push and a market-pull. Increasing activities which bring together academia and industry will ensure that both parties decide jointly which topics are relevant. Our investigation of research topics does, however, suggest that the topics and interests are already converging to a certain degree.

References

1. Bekey, G.A.: Autonomous Robots. MIT Press, Cambridge (2005)
2. Brooks, R.A.: The robots are here. Technology Review 107, 30 (2004)
3. Forlizzi, J., DiSalvo, C.: Service robots in the domestic environment: a study of the roomba vacuum in the home. In: Proceedings of the 1st ACM/IEEE International Conference on Human-Robot Interaction (HRI), pp. 258–265. IEEE Press, New York (2006)
4. Knoll, A., Siciliano, B., Pires, N., Lafrenz, R.: ECHORD-The new face of academia-industry collaboration in European robotics. IEEE Robotics & Automation Magazine 17(4), 21–22 (2010)
5. Leydesdorff, L., Cozzens, S., Van Den Besselaar, P.: Tracking Areas of Strategic Importance Using Scientometric Journal Mappings. Research Policy 23, 217–229 (1994)
6. Pitsch, K., Wrede, S., Seele, J.C., Süssenbach, L.: Attitude of German Museum Visitors towards an Interactive Art Guide Robot. In: Proceedings of the 8th ACM/IEEE International Conference on Human-Robot Interaction (HRI), IEEE Press, New York (2011)
7. Müller, V.C.: Autonomous cognitive systems in real-world environments: Less control, more flexibility and better interaction. Cognitive Computation 4(3), 212–215 (2012)
8. Röhrbein, F., Griffiths, S., Voss, L.: On Industry-Academia Collaborations in Robotics. Technical Report TUM-I1338 (2013)
9. Röhrbein, F.: Academic and industry collaboration (Part 1): Interview with Rodney Brooks, `http://robohub.org/academic-and-industry-collaboration-part-1-interview-with-rodney-brooks/` (accessed: October 09, 2013)
10. Schmoch, U.: Double-boom cycles and the comeback of science-push and market-pull. Research Policy 36, 1000–1015 (2007)
11. Steels, L.: Fifty years of AI: From symbols to embodiment-and back. In: Lungarella, M., Iida, F., Bongard, J.C., Pfeifer, R. (eds.) 50 Years of Aritficial Intelligence. LNCS (LNAI), vol. 4850, pp. 18–28. Springer, Heidelberg (2007)
12. Veiga, G., Silva, C., Araújo, R., Pires, N., Siciliano, B.: The ECHORD project proposals analysis Research profiles, collaboration patterns and research topic trends. Elsevier Expert Systems with Applications 40(17), 7132–7140 (2013)

Part I
Future Industrial Robotics

Part I, Future Industrial Robotics, summarizes the results of five experiments funded by ECHORD focused on the challenges of industrial robotics. Industrial robots represent nowadays the vast majority of the operational robots worldwide but their use is still dominated by some traditional applications, like welding, assembly or handling. Furthermore, the majority of these installations still present low levels of autonomy/ flexibility, relying heavily on programming tasks, with reduced human-robot interaction during operation.

The human robot co-worker scenario is one of the key enablers for future robotics applications. The removal of fences or hard safety measures will in the future allow robots to perform more complex relying on close cooperation with humans. One of the crucial challenges for the robot co-worker scenario is the development of sensing systems that permit close and safe human robot interaction. In this scope the experiment EXECELL describes a novel projection-based sensor system that is intrinsically safe, can dynamically adapt the safety spaces and provides soft safety features such as the visibility of the space boundaries and visualization of intended robot movements.

The time required for robot installation and (re)programming represents a major technical and economical barrier for a broader use of industrial robots. With respect to industrial manipulators, programming is currently almost exclusively made via teach pendant which is not only time consuming but also economically not viable for many applications, since it is performed by trained technicians and requires the work cell to be stopped during programming. With respect to mobile logistics robots, or Autonomous Guided Vehicles AGV's, installation and programming is still performed by engineering experts in a case by case procedure, limiting in this way even more the number of companies that can benefit from it. Therefore, some of the most significant research challenges for industrial robotics are related with increased levels of autonomy/flexibility and advanced human robot interfaces and programming. In the following part, two chapters are focused on the development of programming by demonstration techniques. In the experiments FREE and dimROB, In-Situ Robotic Fabrication: Advanced Digital Manufacturing Beyond the Laboratory, the use of human movement capture systems allows the robot to be programmed explicitly. In this scenario, the operator uses a tool to describe the trajectory that is afterwards automatically processed to produce valid robot programs. In the experiment MoFTaG, Kinesthetic teaching using assisted gravity compensation for model-free trajectory generation in confined spaces the operator can perform kinesthetic programming, i.e. through physical guidance of a robot with active compliance enabled by impedance-based control.

The needs for frequent reprogramming and reengineering of robotics systems call for increased levels of autonomy. The use of techniques such as machine learning, automatic planning or model driven programming brings industrial robotics autonomy, and consequently applicability, to a new level. In the experiment MoFTaG, the use of an extreme learning machine (neural network) and implicit scene modeling allows the transfer of the user's implicit knowledge via kinesthetic programming. In the experiment TRAFCON models from the AGV traffic control problem are used to design a novel traffic manager that is more efficient and flexible for different industrial setups and above all can dynamically change the paths robots are following.

Finally, the Future of Industrial Robotics also builds on top of new applications. Among the areas where robotics presence is weak or null the construction industry case is one of the most outstanding due to its economic size and importance. The experiment dimROB comes from a robotic laboratory that has a long tradition in digital manufacturing. In this experiment, the use of a mobile manipulator opens great perspectives in terms of large scale digitally generated architectural pieces.

Below, a brief synopsis of each experiment in Part I is presented.

The experiment EXECELL presents the application of a novel projection-based safety system capable of ensuring hard safety in human-robot collaboration. It presents the enhancement of the development of newly developed projection-based safety system to utilize the current state of the robot to obtain optimizes safety spaces and describes the machine vision algorithms required for robust violation detection. Furthermore presents an in depth discussion about the compliance of the system with the IEC standards

The experiment FREE presents a flexible and safe interactive human-robot environment, achievable through a combination of standard commercial robot with the state of the art safety and control technologies. The experiment presents a control loop, the Superior Hierarchical control that interfaces the human and the robot with human position detection and operator work recording for task learning. The efficient robot learning procedure allows reducing or eliminating the need of accurate positioning of the workpiece with jigs and fixtures.

The experiment dimRob, In-Situ Robotic Fabrication: Advanced Digital Manufacturing Beyond the Laboratory fosters a non-standard digital robotic fabrication using a mobile manipulator that can be directly applied on the construction site and it is easily scalable. The experiment deals with challenges in robotics, like tolerance handling and human robot interaction but also with the architectural implications of the integration of the findings that resulted from experimentation at the earliest stages of design. The results show mobile manipulator that is capable of performing on situ digital fabrication of complex structures using advanced sensing for tolerance handling, perform self-localization and also being programmed explicitly by the user.

The experiment TRAFCON proposes a novel traffic manager that capable of efficiently control the coordinated motion of the AGVs and dynamically adapt the paths the robots are following. TRAFCON traffic manager was experimentally validated on simulated real plants and on a small-scale automatic warehouse and shown the significant reductions in the completion time through the introduction of dynamic routing.

The experiment MoFTaG, Kinesthetic teaching using assisted gravity compensation for model-free trajectory generation in confined spaces, presents approaches programming of redundant robot manipulator in a co-worker scenario from a user-centered point of view. In contrast with other kinesthetic approaches, in the MoFTag experiment the operator rather configures the robot by providing training data in different areas of the robot's workspace for the learning algorithm to infer an appropriate redundancy resolution. In a subsequent stage, the system provides assistance to the operator about a learned or configured redundancy solution.

Experimental Evaluation of Advanced Sensor-Based Supervision and Work Cell Integration Strategies - EXECELL -

Christian Vogel, Christoph Walter, and Norbert Elkmann

Fraunhofer Institute for Factory Operation and Automation IFF,
Magdeburg, Germany
{christian.vogel,christoph.walter,
norbert.elkmann}@iff.fraunhofer.de

Abstract. This paper presents the application of a novel projection-based safety system for ensuring hard safety in human-robot collaboration. We adapted the proposed sensor system to incorporate the joint positions of a collaborative robot, thus offering the opportunity to establish minimal and well-shaped safety spaces around the robot at any time. In this contribution we explain in detail main challenges and their solutions for generating and monitoring such safety spaces. Furthermore, we build up a collaborative workplace and evaluate the sensor system concerning its behavior and detection capabilities under operational conditions.

Keywords: Human-Robot Collaboration, Collision Avoidance, Safety.

1 Introduction

When developing or deploying machinery it is required to carry out a risk assessment specific to the apparatus and the custom environment setting. This machinery also includes robots and robotic systems. Special care must be taken when considering human-robot collaborative workplaces in order to avoid injuries of the human workers. Typical risks when considering robots in workspaces shared with humans are for example the crushing of human body parts by robot structure or in combination with environment structure or injury of a person as a result of a collision with the robot. These types of hazards are plausibly counteracted by means of power and force limiting either by design or by control. In general we can differentiate between approaches that prevent humans from potential collisions with the robot and approaches that deal with different reaction strategies at collision time. Here, especially developments concerning intrinsic safe robots like the DLR KUKA LWR or touch-sensitive sensors [1] [2] used as an artificial skin that encloses parts of the robot are well known. Other risks are more difficult to handle by collision reaction techniques alone. Such risks include for example hazards due to the handled part or medium or injuries due to dropping a gripped object with object falling down on persons. Here, separation must be maintained between robot and human worker in order to counteract the risk of injuries. Maintaining separation can be implemented by

F. Röhrbein et al. (eds.), *Gearing Up and Accelerating Cross-Fertilization between Academic and Industrial Robotics Research in Europe*, Springer Tracts in Advanced Robotics 94,
DOI: 10.1007/978-3-319-03838-4_2, © Springer International Publishing Switzerland 2014

means of monitoring separation distance and performing a protective stop if separation cannot be maintained. In [3] [4] sensors measuring the distance to unknown objects are applied to the robot. While these systems are applicable for small or mid-sized industrial robots, collision avoidance by external optical sensor systems is more general. Multi-camera setups in various configurations and combinations, i.e. multiple time-of-flight cameras [5], time-of-flight and stereo cameras [6], multi cameras [7], are used to estimate the minimal distance between human and robot, thus leading to a defined reaction of the robot to avoid potential collisions.

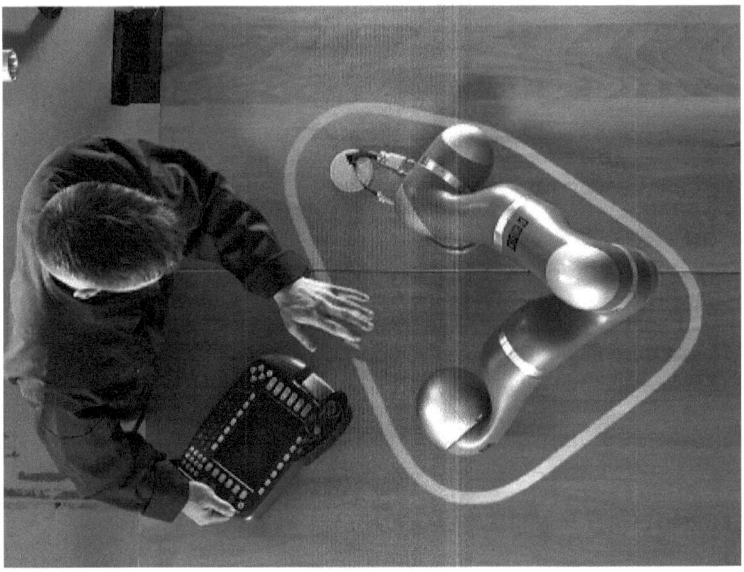

Fig. 1. Dynamically established safety space (white line) enclosing an industrial robot

In [8] we presented our novel projection-based sensor system and discussed the usage of them as a safety system in human-robot collaboration scenarios. We argued that our system has main advantages against other current research activities because of its intrinsically safety, dynamic adaption of safety spaces and a reduced influenceability by changing environmental conditions. Beside the aspect of "Hard-Safety", the system concerns also additional requirements regarding availability by its "Soft-Safety" features like visibility of safety space boundaries and visualization of intended robot movements. Furthermore, the system is capable of supporting the user by additional functionalities like information display and augmented reality for situation awareness, interaction, as well as 3D scene capturing [9].

The objective of this contribution is to demonstrate safe interaction between robots and humans in a shared workspace environment by using the aforementioned newly developed projection-based safety system in conjunction with one of the most current robots aiming at human-robot collaboration scenarios, the KUKA LWR 4+. We extended this safety system to utilize the current state of the robot thus leading to the capability of generating safety spaces dynamically at operation, as seen Fig. 1. On basis of the current joint angles and velocities of the robot the system is able to establish safety spaces that on the one hand keep the required separation distance and on the other hand limit the size of safety spaces to a minimum offering the user to use at most as possible of the workspace. Beside the process of online generation of safety spaces another main task of the experiment concerns the fast and robust detection of violations of them.

In the following sections we describe the overall setup of our collaborative workplace and sensor system accordingly. The focus of this contribution is on explaining the methods and algorithms for dynamically generating safety spaces on basis of the current state of the robot and detecting violations of the established safety spaces. We further elaborated various test criteria based on the Technical Report 61496-4 and its subparts 2 and 3 to evaluate the proposed sensor system according to the detection capabilities, system behavior and limitations.

2 System Setup

For testing and demonstrating usability and practicability of the proposed safety system we build up a prototypical collaborative human-robot workplace, as schematically depicted in Fig. 2. This also allows the validation and evaluation of the developed algorithms under operational conditions.

The overall setup consists of a system carrier that is used to mount the sensor system and additional hardware and electronics. In detail, this includes the four monitoring cameras, which are installed in the four corners of the carrier at a height of 2.50 m. Every camera has a resolution of 640 * 512 pixels and acquires images with a frame rate of about 50 fps. For trigger purposes and synchronization we connected the cameras to the projector by additional electronics.

While the original optic of the monochrome DLP-projector (resolution: 800*600 pixels) was not able to establish required large volumes of the safety spaces, we had to replace the wide angle lens by a lens with longer focal length, which leads to an increasing distance between projector and projection plane. We overcome the issue of a very high positioned projector by using a 45 degrees angled mirror on the top of the carrier. So, the emitted light of the projector, which was horizontally mounted on an arm outside of the system carrier, is reflected by the mirror and establishes the safety spaces on the workbenches, i.e. the projection plane. We further equipped the two workbenches that are placed under the carrier with an industrial lightweight robot (KUKA LWR 4+).

The four cameras and the projector are calibrated intrinsically and extrinsically relative to a common coordinate system positioned at the surface of the workbenches. Additionally, we measured and further refined by calibration the position and orientation of the robot.

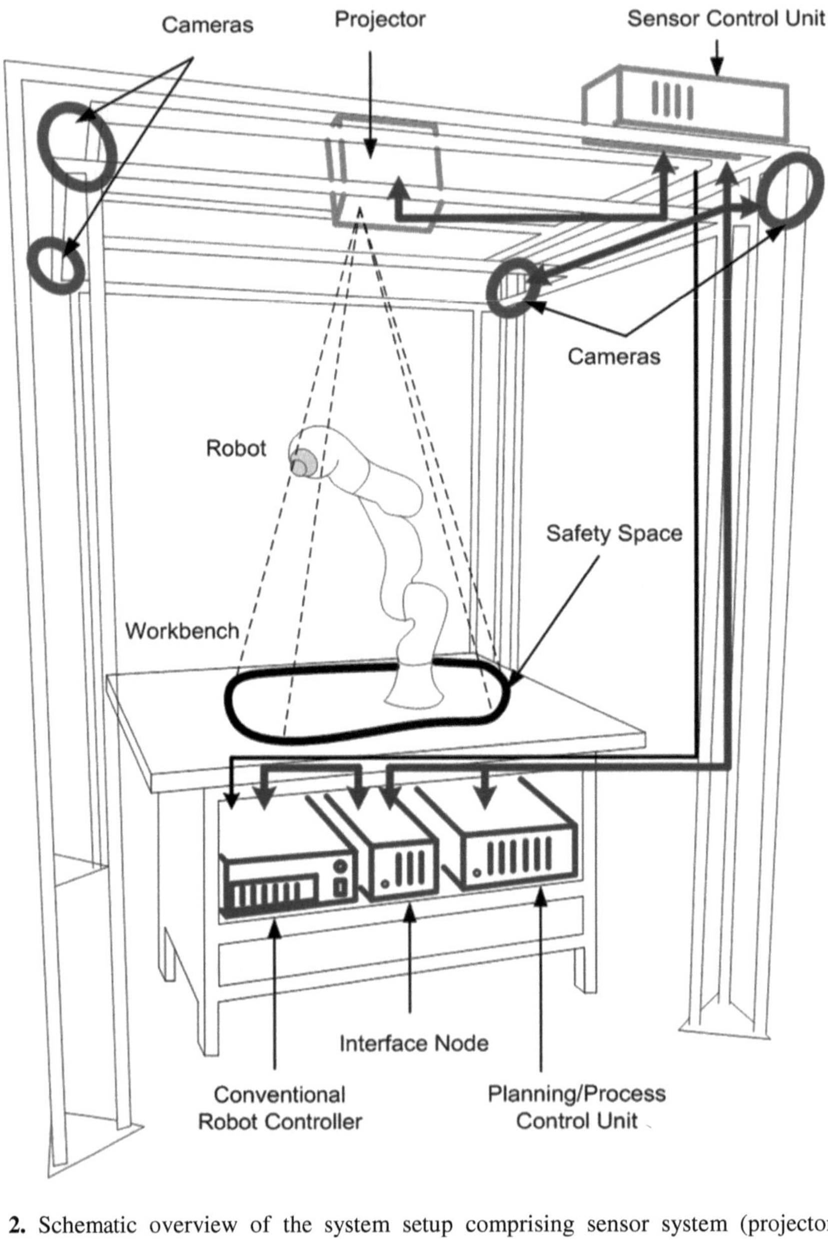

Fig. 2. Schematic overview of the system setup comprising sensor system (projector and cameras), robot, additional hardware and electronics

3 Planning and Intrusion Detection Algorithms

In this section we want to describe on the one hand the planning algorithms used to generate appropriate safety spaces based on the current state of the robot, and on the other hand the collision algorithms used to detect intrusions of the established safety space. Therefore, we begin by explaining the underlying internal robot model, its update methods and its purpose at safety space generation and collision test. Further on we describe in section 3.2 the generation process of safety spaces in detail and explain in section 3.3 how we detect collisions of these established safety spaces.

Fig. 3. Visualization of internal robot model consisting of several primitives like spheres and cylinders (wired)

3.1 Internal Model Representation

The internal representation of the robot, as well as the representation of additional objects in the environment depends on the sensor-system specific operation principle. While the interface between robot and safety system provides the exchange of an annotated robot model of the robot, the safety system adapts this model to its own optimized representation. So, the algorithms for safety space generation and collision detection can efficiently use this internal model and thus providing real-time capabilities. This means that the specific internal model is a main component of the respective sensor-system.

Since we use in our scenario a KUKA LWR 4+, the internal model consists of simple primitives like spheres and cylinders as depicted in Fig. 3. The pictures illustrate the view frustum and camera image of one of the four monitoring cameras as well as the current state of the internal robot model. Here, we updated the internal model by adapting the position of every joint with the current data of the real robot. As the size of the potentially generated safety space even depends on both the used tool and optional workpiece on it, we also have to consider them in the internal robot model. Instead of defining a certain primitive for every tool and workpiece it may be useful to identify a worst-case primitive that comprises all possible tools respectively workpieces. However, all primitives can be dynamically added to or removed from the internal representation at operation, thus influencing the resulting established safety space.

3.2 Safety Space Generation

For generating safety spaces we can differentiate between two approaches. The first one is based on the current robot state, i.e. the current joint angles and velocities, thus dynamically encloses the robot in a minimal manner at any time, as seen in Fig. 4 a) – c). The second approach is the semi-static one. Here, we use a sequence of robot states for a planned trajectory to generate an overall safety space. Such a trajectory defines the movement of the robot between tow motionless positions. In Fig. 4 d) we depicted such a safety space that was generated before the robot starts moving.

Fig. 4. a) – c) While the robot moves the safety space is continuously updated. d) Before the robot moves, the safety space is generated once for a planned trajectory.

In the following we describe the single steps of the online process of safety space generation in detail. The semi-static approach is also based on this sequence.

Update the Internal Robot Model. Since the generation of the safety space is based on the internal model we firstly have to update them. Therefore we adapt the positions of the 3-dimensional primitives of the internal robot model by the current data of the robot controller, finally representing the current state of the real robot.

At this point it is also of utmost importance to consider the current used tool and potentially workpiece on it. While in most cases the tool (e.g. gripper) is static, its size, shape and position at the robot can be approximated by another primitive at the internal robot model. Concerning the workpiece, this may be different. Here, the current used workpiece may be dependent by the single process steps and can differ in size, shape and position at the tool. To overcome this issue we identified the possible work pieces, approximated them by suitable primitives and added them dynamically to the internal robot model regarding the current process step. While the entire work process of the robot was implemented at the robot controller, the signal for adding or removing a workpiece to the internal robot model as well as the appropriate type of workpiece are provided by the robot controller.

Beside the adaptation of the internal robot model, at this point it is also possible to manage further objects, which should have an influence to the safety space generation process. At the end of this step we got an updated internal world representation that forms the basis for the safety space generation process.

Computation of Projector Image. The 3-dimensional internal robot model states an abstract representation of the real robot at certain times. As we sized the shapes of the primitives greater than the real sizes of the single parts of the robot, we can consider the internal robot model as a closure of the real robot. More precisely, it can be assumed as a "safety-hull". So, by simply perspective transforming these 3-dimensional primitives to the 2-dimensional image plane of the projector, we got an image that contains the boundaries of all primitives represented as white lines on a black background. The resulting projector image is determined by computing the overall contour of these boundaries. We further emit this projector image into the environment and established the corresponding safety space around the robot. Here, we identified a major drawback at certain robot configurations. In unfortunate circumstances i.e. particular joint angles, it was possible that no camera has the ability to observe single parts of this safety space because of the occlusion by the robot itself.

For avoiding this issue we implemented a second approach that computes the convex hull of the contours in the projector image. The resulting safety space is indeed in most cases greater than the safety space of the first approach but here the projected lines of the safety space are always visible to at least one camera. In Fig. 5 (right), we exemplarily depicted such a projector image.

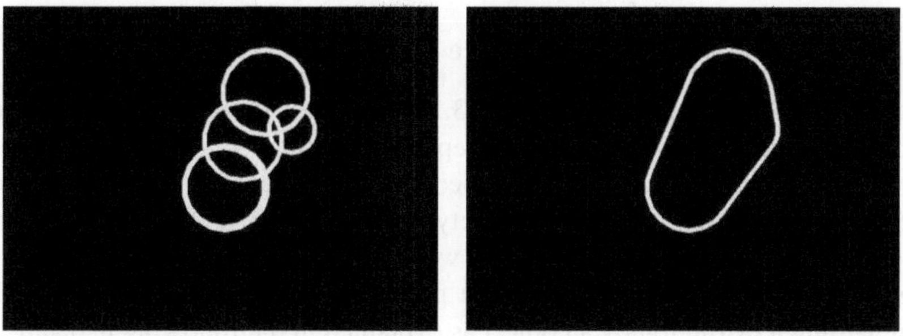

Fig. 5. Intermediate (left) and final projector image (right) after perspective transforming the 3D internal robot model to 2D image plane of the projector

As a prerequisite of the perspective transformation we need the updated internal world representation as well as the intrinsics and extrinsics of the projector. We further improved this transformation process by simplifying the internal robot model. As the spheres at the robot joints always have a greater radius than the adjacent cylinders, there is no need to incorporate these cylinders to the safety space generation process, because they never influence them. In Fig. 5 (left), we depicted such an intermediate projector image with transformed spheres of the internal robot model.

Safety Space Configuration. The projector image defines the safety space from the perspective of the projector. At this point there is no information about the shape, position or size of the safety space in world. But this is necessary for computing the virtual reference images i.e. expected state mask for every camera. For this, we compute the 3-dimensional representation of the safety space by using the intrinsics and extrinsics of the projector as well as the internal world representation. In detail, we determine for every pixel in the projector image that belongs to the safety space the corresponding light ray by using the intrinsics and extrinsics of the projector and intersect them with the projection plane defined by the internal world representation. The total of all these intersection points defines the safety space in world and will be stored as a list of 3D coordinates in the appropriate safety space configuration. Here, we also define the method of interpreting these coordinates to compound the safety space, e.g. as single dots or line strip, and specify additional information like line width.

Actually, we generate safety spaces that are based on the robot's state i.e. joint angles at a certain time. As we wanted to incorporate also the velocities of the robot joints, we extended the safety space generation by adapting the computation of the projector image to a two-stage process. After transforming the internal robot model representing the current robot's state to the 2D image plane, we secondly update the internal robot model by joint angles according to the current joint velocities of the robot and transform the appropriate primitives additionally to the projector image. The final projector image is further determined by computing the convex hull of all these primitive boundaries.

3.3 Reliable Detection of Safety Space Violations

As we described in the previous section 3.2 the process of generating safety spaces, we explain in the following the single steps of detecting violations of them. Such a safety space violation occurs, if an object disrupts the emitted light rays of the projector, which are representing the safety space border. As described in section 3, we use 4 cameras that are monitoring the workplace by different perspectives as seen in Fig. 6. The cameras are observing the projected safety spaces and are analyzing them regarding a disruption by an object as explained in the following in detail.

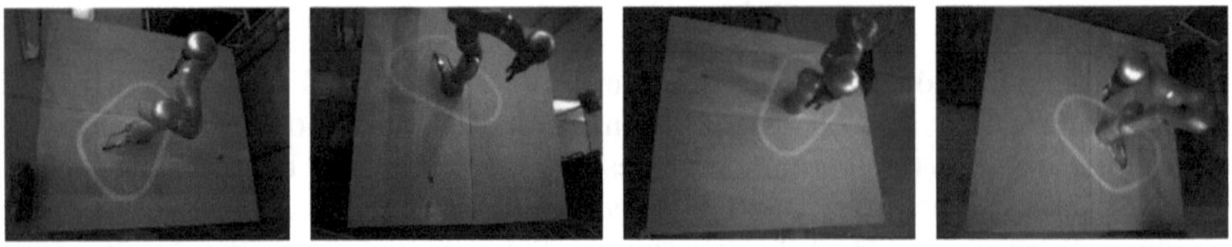

Fig. 6. Images of the four monitoring cameras

Concerning the position and orientation of the cameras they are capable of monitoring the whole workplace or only single parts of them. Beside this setup constraint it is also possible that a camera cannot see single parts of the workplace because of occlusions (e.g. by the robot) at certain times, as can be seen in Fig. 6. Generally, it is of utmost importance that the composition of all camera images leads to an area-wide monitoring of the entire workplace at any time. The overlapping of single observation parts by several cameras additionally increases the safety capabilities of the system.

Image Preprocessing. The cameras are mounted in the upper corners of the system carrier and were adjusted in a way that they can observe almost the entire work space (see Fig. 6). Obviously, projected lines closer to the camera appear in the camera image with a wider line width than projected lines far away. As we further process the images by several morphological operations for extracting these lines, the varying line widths can lead to weak results. Therefore, we resample the camera images from perspective of a virtual camera positioned in the center above the workplace. We defined a resolution of 800 * 800 pixels for this virtual camera, because this is a suitable trade-off between accuracy and processing time of the following algorithms. The resampled camera images of all four monitoring cameras are further used for the processing of collision detection and are depicted in Fig. 7.

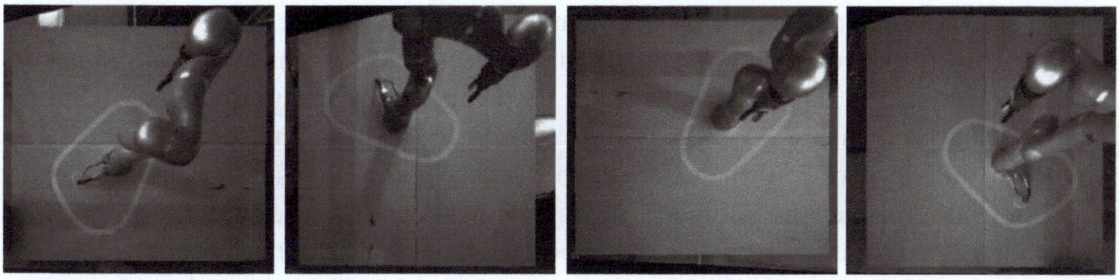

Fig. 7. Resampled images from perspective of a virtual camera positioned in the center above the workplace

Extraction of Current-State Mask. The first step for detecting a disruption of the safety-relevant emitted light-rays concerns the extraction of them in the camera image. Here, we pixel-wise compute the absolute differences of consecutive camera images. While the image acquisition is synced with the light emission of the projector, consecutive images differentiate remarkably at pixel positions representing safety spaces. This difference image forms the basis for applying some additional morphological operations and final thresholding. Any pixel position in the resulting binary current-state mask that represents a safety space has a value of 1, the others 0. Fig. 8 shows exemplarily the extraction process for one camera.

Actually, the cameras are triggered by a frequency of about 50 Hz for alternate image acquisition of a light (image with emitted light rays) and dark image (image without emitted light rays). The short time lag between imaging of these two images reduces the influence of changing light conditions to the extraction process.

Fig. 8. Computation of current-state mask (right) by differentiation of light (left) and dark image (middle)

Generation of Expected-State Mask. While the current-state mask represents the pixel positions of the actual safety spaces in camera image, it is not possible to decide if there exists a disrupted light ray or not. So, we need a reference camera image that determines the pixel positions of the safety-relevant light rays in an undisrupted state. This binary virtual reference image is computed on basis of the intrinsics and extrinsics of the corresponding camera and projector and is further called expected-state mask. So, at all pixel positions in the camera image a light ray of the projected safety space is expected, the pixel gets a value of 1, the others 0.

For determining the expected-state mask we take the current safety space configuration into account and perspective transform the 3-dimensional coordinates of the safety space to 2D image plane of the camera by using the corresponding intrinsics and extrinsics. As the safety space configuration specifies just the parameters and properties of the safety space, at this point the resulting expected-state mask does not consider the geometry of objects in the workplace like the robot. This means, under certain circumstances the expected-state mask defines pixel positions of light rays the camera is not able to see because of the occlusion by the robot. This depends on the position of the camera and the current kinematics of the robot. In Fig. 9 such an expected-state mask and corresponding current-state mask are depicted. It can be seen, that at some pixel positions light rays are expected, which are actually not visible for the appropriate camera.

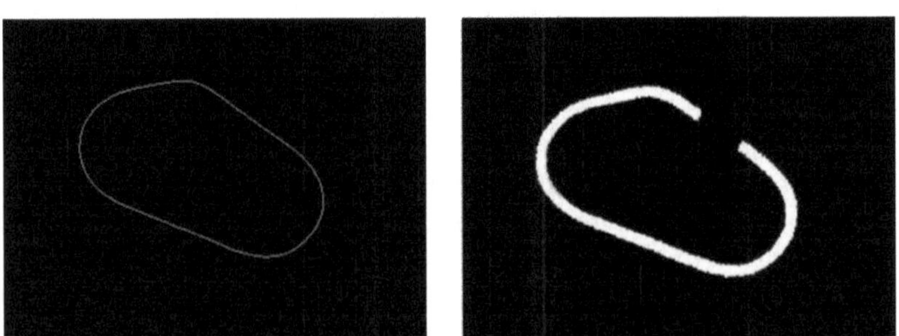

Fig. 9. Erroneous expected-state mask (left), appropriate current-state mask (right)

So, for determining the final expected-state mask we also have to consider the internal robot model, which is used to eliminate such pixel positions of expected light rays in the expected-state mask that are not visible for the camera. Similar to the generation of safety spaces in section 3.2 we transform the updated primitives of the model to the 2D image plane of the camera and compute a hull around the shapes. So, this hull represents the area in the camera image, at which the camera observes the real robot. The final expected-state mask is generated by erasing this area from the intermediate expected-state mask.

Determine Safety Violation. For detecting a safety violation it is necessary to compare the expected-state mask against the current-state mask (see Fig. 10). So, at every pixel position in the expected-state mask that contains a value of 1, the corresponding pixel in the current-state mask has to coincide. If there is a mismatch at one position, a safety violation is signalled.

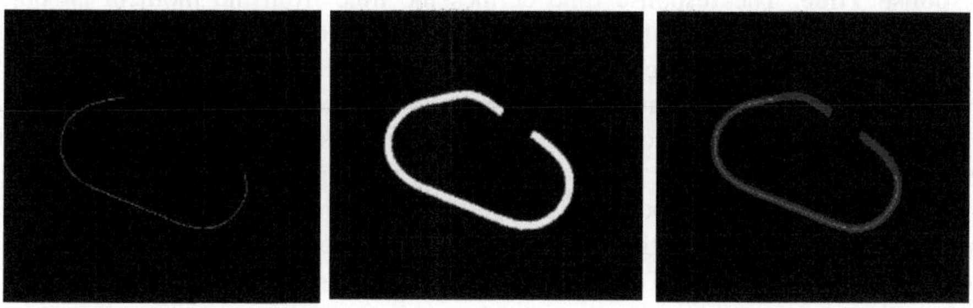

Fig. 10. Expected-state mask (left), current-state mask (middle), resulting match of both (right)

4 System Evaluation

In this section we present remarkable results and findings of the system evaluation. Here, the aim was on the one hand to analyze the system behavior and on the other hand to identify the limitations of the detection capabilities of the system. Therefore, we elaborated various test criteria on basis of the Technical Report IEC 61496-4 "Safety of machinery – Electro-sensitive protective equipment Part 4: Particular requirements for equipment using vision based protective devices (VBPD)" and its corresponding subparts -2 and -3. These subparts are specializations and focus on "Additional requirements when using reference pattern techniques" (Part 2) and "Additional requirements when using stereo vision techniques (VBPDST)" (Part 3). As our system can be seen as a composition of both, as it consists of similar hardware and functionality, we elaborated the test criteria accordingly. These papers define and specify functional, design and environmental requirements of electro-sensitive protective equipment designed specifically to ensure the safety of humans. The evaluation on basis of these test criteria represents first insights to a future safety certification and its feasibility.

In detail, we determined and analyzed the response time of the system at safety violation and identified the time-consuming system components and modules accordingly. This gives us the opportunity to effectively and precisely optimize and improve this system property. Another main part of the evaluation process covers the detection capabilities of the system. Here, especially the examination of object properties and their influences to the detection capabilities were involved. Different colors, minimal sizes and transparency are of special interest. Another aspect of the evaluation process concerns the environmental conditions and their influences to the detection capabilities. Here, it is of utmost importance that the system does not fail to maintain safety even in the presence of interference from extraneous light, dust or mechanical influences like vibrations.

4.1 Test Criteria and Results

Response Time. The response time defines the time from the moment an object intrudes the safety space to the moment the system generates the violation signal. In this period of time several computation tasks are executed. So the main task is to identify the relevant factors and its appropriate time periods. The hardware of our test system consists of 2 Intel Xeon E5645 2.40 GHz processors and 12 GB Memory. The determination of the safety space violation signal comprises several steps. Here, image acquisition, image processing and collision test are identified as the substantial parts of the entire response time. The identified and measured computation times of the single steps are listed in Table 1.

Table 1. Significant computation times and latencies

Module / Processing Step	Time [in ms]
Image acquisition	40
Image processing	15
Collision detection	10
Additional latencies	5
Total	***70***

Object Size. The minimal detectable size of objects is an important property of the detection capabilities of the system. Regarding the safety aspect in human robot scenarios this object size is a main fact of the applicability of the system. Therefore, we have to identify the minimal detectable object size and have to analyze the main factors this minimal object size depends on. As our system is developed to detect objects in the size of fingers, we use test pieces with cylindrical shape as seen in Fig. 11 (left).

The cylindrical test pieces have diameters of 25mm, 16mm and 12mm. However, as the system just detects the disruption of the safety space at the projection plane (i.e. the shadow of the object) as seen in Fig. 11 (middle, right), we can generalize the detection issue of different object sizes in various distances above the projection plane to the task of detecting the smallest possible shadow width. Here, we determined a minimal width of the disruption at the projection plane (workbench) of at least 15mm.

So, the system is capable of detecting the 25mm cylindrical test piece positioned at the surface of the projection plane, the 16mm test piece at about 15mm above the projection plane and the 12mm test piece at about 20mm above the projection plane. Thinner objects are also detected by increasing the distance between object and projection plane thus resulting in an appropriate disruption width at the projection plane.

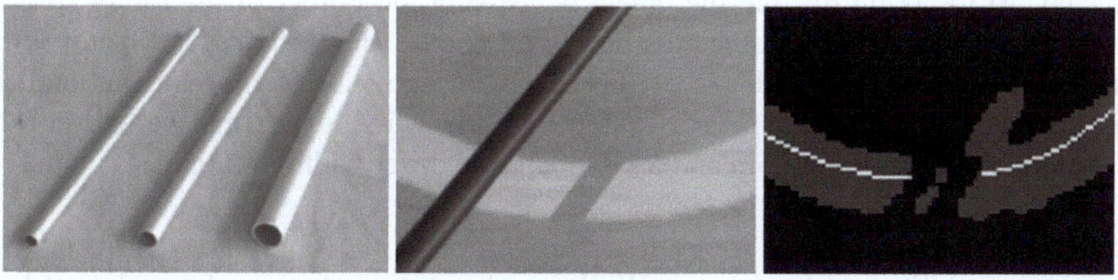

Fig. 11. Left: Cylindrical test pieces with diameters of 25mm, 16mm and 12mm. Middle: Black cylinder and disrupted light rays of the safety space. Right: Overlay of current state mask (grey pixel) and expected state mask (white pixel).

Furthermore, we analyzed the main factors this minimal width depends on. Here, the fundamental factor is the resolution of the real cameras, as well as the resolution of the virtual cameras. Higher resolutions of the cameras (real and virtual) result in a larger number of pixels representing the disruption at the projection plane. This is further affecting the subsequent execution of morphological operations that are used on the one hand to eliminate noise and on the other hand to improve the extracted safety-related projection light. Here, the amount and sequence of execution of these operations can further affect the detection capability.

Object Position. In this section we want to analyze the system's behavior by placing an unknown object at different positions in the workplace. This position can be on the one hand static or on the other hand dynamic.

Static Object Inside or Outside Safety Space. In general, the safety space is defined by an enclosing projected line that defines the outer border of the safety space. Thus, objects located entirely outside or inside the safety space are not detected. Objects outside the safety space are not safety-relevant and do not negatively affect the detection capabilities of the system. Unknown Objects inside the safety space are also not safety-relevant but may influence the availability of the system.

Static Object at Safety Space Border. An object is located at the border of the safety space that results in a disruption of the safety-relevant emitted lights. The object is detected and a positive safety violation signal is generated. This is the common case of detecting humans or parts of them that are potentially colliding with the robot thus disrupting the projected border of the safety space.

Dynamic Object Inside or Outside Safety Space. Moving objects inside or outside the safety space are not detected. As well as static objects placed inside or outside the safety space, moving objects which are not crossing the safety space border are even not safety-relevant. They do not affect the detection capabilities but may influence the availability of the system.

Dynamic Object Into or Out of Safety Space. Objects that are moving from outside to inside (or from inside to outside) of the safety space are safety-relevant and have to be detected. At the point of time the object is crossing the border of the safety space it can be assumed as a static object as described above. Here, the crucial factor is the velocity of the object. On basis of the current frame rate of the cameras and the operational principle of the sensor system, we identified a maximal velocity of 2.5 m/s for an object size of 0.05m and a safety space border width of 0.05m.

Object Surface Property. The surface properties of the object may influence the detection capabilities of the system. While an object disrupts the safety-relevant light of the safety space, it may be possible that some surface properties of the object can lead to a non detection of this safety violation. So, here we analyze the following properties and clarify their influence to the detection capabilities of the system:

Colors. In the previous test criteria 'Object Size' we analyzed that at safety violation not the object itself, but the shadow of the object at the projection plane is detected. Generally for all opaque objects, its shadow is not affected by the color. So, we can constitute that the color property of opaque objects do not affect the detection capability of the system.

Textures. As a texture can be assumed as a combination of different colors, it leads to the same result as the aforementioned 'Colors' property. Therefore, it is also of no importance if the object is textured like the projection plane.

Transparency. As these objects do not cast a shadow the system is not able to detect them, which results in a false negative violation signal. This drawback of the safety system does not necessarily reduce the safety of humans as they are still safely detected. Furthermore, to avoid this issue during operation the usage of transparent objects should be banned.

Reflectance. Objects with low reflectance do not affect the detection capabilities, whereas Objects with high reflectance may influence the detection capabilities. In worst case such an object can be a mirror, as seen in Fig. 12. Here, we positioned a mirror in a way that the projected light is directly reflected to one camera. As a result, in the corresponding current state mask of the camera image, the pixel values of the whole area around the mirror are assumed as projected lights. Disruptions of the safety space in this area are not recognized. More precisely, safety cannot be guaranteed.

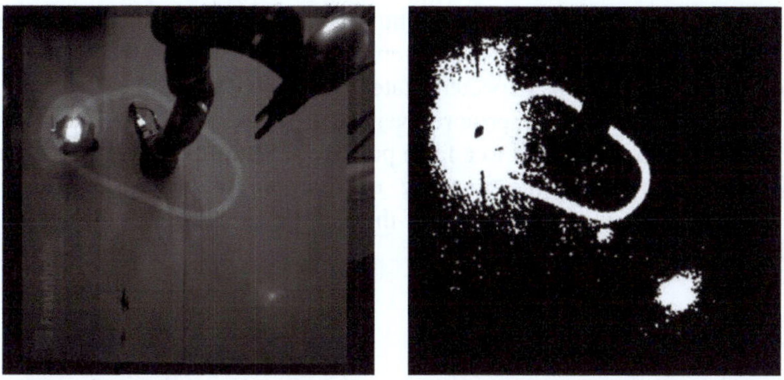

Fig. 12. A mirror reflects the emitted light of the projector directly to a camera. Left: Camera image of the illuminated camera. Right: Extracted current state mask of the camera image.

Projection Plane Surface Property. The projection plane is a fundamental part of the entire optical monitoring system. Here, the light emitted by the projector has to be reflected by the projection plane in a way the monitoring cameras can recognize them. As this does not affect the safety, the surface properties are essential for the availability of the system. While we use in our system setup light wooden workbenches the emitted light of the projector is well recognized by the cameras while specular reflections are suppressed. In general, there must be a trade-off between diffuse and specular reflection of the projection plane surface. Colors and textures can also influence the reflectance locally or globally of the projection plane.

Environmental Conditions. In general, optical systems are influenced by environmental conditions. Here, especially camera based systems are highly dependent from light conditions. Beside several tests concerning different light conditions and light changes we additionally test the system concerning mechanical influences like bumps and vibrations.

Mechanical Interference. Firstly, we tested the system against slow vibrations by softly pushing the system carrier and the system further works fine. At faster and stronger vibrations the system signals a positive safety violation. The reason for this is that the cameras and projector become decalibrated which results in a mismatch of expected-state mask and current-state mask. By bumping the entire system carrier, or a single camera, or the projector we got the same behavior. Generally, mechanical influences to the system do not affect the safety but decreases the availability of the system.

Light Interference. Here, we tested the system by using different light sources and light changes. In general, the illumination intensity dictates the camera settings like exposure, gain and brightness. So, if the illumination intensity changes too much, these camera settings have to be adapted. Otherwise the camera images are on the one hand too dark or on the other hand too light to observe a difference between projected light and background. As a result, the system generates a weak current state mask that does not corresponds to the expected state mask and that further leads to a false positive violation signal. Furthermore, very fast changes of the illumination like stroboscopic lights can also lead to a false positive violation signal. If the illumination intensity changes between the imaging of light and dark image, the resulting difference image is very weak because of the significant change of all pixel values in one of the camera images.

4.2 Conclusion

In this section we described several elaborated test criteria to analyze our system concerning safety and availability. These test criteria are based on the technical report IEC 61496-4 and its subparts -2 and -3. Here, we focused on tests that are realizable concerning its technical equipment and feasibility; especially the tests concerning different temperatures, humidity or aging of components were not realizable. These steps are clearly within the scope of product development, not research.

The detection capabilities and especially the tests the system fails to danger were of special interest. Here, we got remarkably insights to the system behavior concerning reflecting object surfaces and transparent objects. However, beside these exceptions the system's behavior remains in a safe state at varying environmental conditions and offers potential for real industrial applications.

5 Summary

In this contribution we presented a projection-based sensor system for monitoring human-robot collaborative workplaces. At first, we described the setup of our exemplary collaborative workplace, which was equipped by our novel sensor system that was adapted to communicate with the robot controller for receiving the current state of the robot. We explained in detail the process of generating optimally shaped safety spaces enclosing the robot in a minimal manner by incorporating the current joint positions and velocities of the robot. Beside safety space generation, the focus was also on collision detection. Here, we described the algorithms of processing and analyzing the camera images to robustly detect violations of the established safety spaces. We further evaluated the system on basis of test criteria elaborated on the technical report IEC 61496-4 and its subparts 2 and 3. The achieved results demonstrated the feasibility of our approach to robust workspace monitoring for applications that require spatially close human-robot coexistence while maintaining separation of robot structure and operator.

Acknowledgements. This work was supported by the EU Project EC FP7-ICT-231143 ECHORD.

References

1. Heiligensetzer, P., Wörn, H.: Safeguarded human-robot interaction in production. In: 33rd International Symposium on Robotics (ISR), Stockholm, Sweden (2002)
2. Fritzsche, M., Elkmann, N., Schulenburg, E.: Tactile sensing: A key technology for safe physical human robot interaction. In: 6th ACM/IEEE International Conference on Human-Robot Interaction, Lausanne, Switzerland, pp. 139–140 (2011)
3. Ostermann, B., Huelke, M., Kahl, A.: Freed from fences - Safeguarding industrial robots with ultrasound. In: 6th Working on Safety Conference (2012)
4. Feddema, J.T., Novak, J.L.: Whole Arm Obstacle Avoidance for Teleoperated Robots. In: IEEE Robotics and Automation Proceedings, San Diego, USA, pp. 3303–3309 (1994)
5. Fischer, M., Henrich, D.: 3D Collision Detection for Industrial Robots and Unknown Obstacles using Multiple Depth Images. In: German Workshop on Robotics (GWR), Braunschweig, Germany (2009)
6. Walter, C., Vogel, C., Elkmann, N.: A Stationary Sensor-System supporting Manipulators at Safe Human-Robot interaction. In: 41st International Symposium on Robotics (ISR) and 6th German Conference on Robotics (ROBOTIK), Munich, Germany, pp. 1–6 (2010)
7. Henrich, D., Gecks, T.: Multi-camera collision detection between known and unknown objects. In: 2nd ACM/IEEE International Conference on Distributed Smart Cameras (ICDSC), Stanford, USA (2008)
8. Vogel, C., Poggendorf, M., Walter, C., Elkmann, N.: Towards Safe Physical Human-Robot Collaboration: A Projection-based Safety System. In: IEEE/RSJ International Conference on Intelligent Robots and Systems (IROS), San Francisco, USA, pp. 3355–3360 (2011)
9. Vogel, C., Walter, C., Elkmann, N.: Exploring the possibilities of supporting robot-assisted work places using projection-based sensor system. In: IEEE International Symposium on Robotic and Sensors Environments (ROSE), Magdeburg, Germany, pp. 67–72 (2012)

FREE: Flexible and Safe Interactive Human-Robot Environment for Small Batch Exacting Applications

Dario Antonelli[1], Sergey Astanin[1], Gabriella Caporaletti[2], and Francesco Donati[2]

[1] Department of Production Systems and Economics, Politecnico di Torino, Italy
{dario.antonelli,sergey.astanin}@polito.it
[2] EICAS Automazione, Torino, Italy
{g.caporaletti,donati}@eicas.it

Abstract. The FREE experiment addresses the field of small-batch production where handwork is still the main manufacturing option since automation is more expensive and lacks the prescribed flexibility. FREE aims at addressing this situation by introducing a flexible and safe interactive human-robot environment, achievable through a combination of standard commercial robot equipments with the state of the art safety and control technologies. The core idea is to add to the system a further control loop operating at a level hierarchically superior with respect to the standard robot controller. Such a control loop, defined "*Superior Hierarchical Control*", is the interface between the robot and the human operator through a variety of sensors providing contact-less human position detection for safety and human work recording for task learning .

Keywords: Training by demonstration, automated welding, human-robot interaction.

1 Introduction

Modern production is facing tough price competition. At the same time, the requirements for flexibility are increasing. Product life cycles are getting shorter, the variety of products is increasing, while the demand is steady. This necessitates the reduction of batch size. These changes require more flexibility in automation and simultaneous engagement of human workers and robots. More than 228,000 manufacturing Small Medium Enterprises (SMEs) in the European Union (EU) are a crucial factor in Europe's competitiveness, wealth creation, quality of life and employment. To increase the competitive assets of EU companies with respect to the new developing regions in the world, the Commission emphasized research efforts aimed at strengthening knowledge-based manufacturing in SMEs as agreed at the Lisbon Summit and as pointed out in the roadmap for robotics in Europe by CARE [1].

Significant efforts are requested to overcome the barriers that prevent SMEs from automating their processes because automation often appears cumbersome and costly. Robotic systems, while classified as a flexible production technology, are basically

F. Röhrbein et al. (eds.), *Gearing Up and Accelerating Cross-Fertilization between Academic and Industrial Robotics Research in Europe*, Springer Tracts in Advanced Robotics 94,
DOI: 10.1007/978-3-319-03838-4_3,

focused on high volume productions. The main hindrance in installing robots for small-batch production is the amount of product-specific costs. Each work phase for each product has to be programmed, and auxiliary equipments must be manufactured based on part-specific geometry. If the product volumes are small it is necessary to enhance robot flexibility in terms of part family variety and work phases.

Assembly processes are composed of two different stages: handling and joining of the parts to be assembled. A common joining technique is arc welding that is usually executed by robots. Robot requirements in order to perform handling and welding are different in terms of accuracy, payload and work area. For sake of cost reduction, the trend leads to a single robot for both tasks, changing end-effectors during the work.

Nevertheless some robots must be reserved offline, in order to test new work programs without slowing production activities. To install robotized assembly systems in small batch production it is necessary to address two issues: the learning phase should be made the shortest and robot setup and fine tuning of the tool path should be programmable right on the assembly station in the contemporary presence of human workers and without requiring dedicated personnel.

2 The Proposed Collaborative Man-Machine Assembly Cell

To comply with the issues presented in the previous section, we propose a collaborative working cell with both humans and robots working together. The collaboration between man and robot derive from the capability of the robot to learn in real time the tasks to execute by observing the gestures of the human co-worker.

This approach leads straightforward to Learning from Demonstration techniques (LfD). In [2] it is complained that a large effort has been done investigating 'how to imitate' and 'what to imitate' in applying LfD, while 'who to imitate' and 'when to imitate' remain unexplored. In other words, a inadequate attention is paid to the type and amount of experience the teacher should have and to the attention to the amount of time spent for the demonstration. Trying to apply LfD to actual industrial processes the importance of solving these last two issues becomes apparent. The robot should be trained to execute a welding task by observing a human welder with no experience on robot programming. The demonstration phase should waste a minimum amount of time, considering that welding is a fast process even when manually operated.

Following [3] we can classify our problem as derivation of a policy by a mapping function approach that apply regression on a dataset built by the external observation of the demonstration through remote sensors.

We developed our approach using the framework proposed in [4] consisting of five phases: demonstration presentation, recognition, analysis, policy generation and execution. We paid particular attention to the recognition phase where the movements of the operator have to be recorded accurately and the welding tasks must be identified and taken apart from the complete demonstration. As there is a consistent literature on LfD we could make reference to [5] and [6] to build a continuous trajectory starting from the observed raw poses, to [7] for hints about how to extract the features we are interested in (the welding operations) and to [8] for the selection of the optimal policy. First results of the research were presented in [9] and [10].

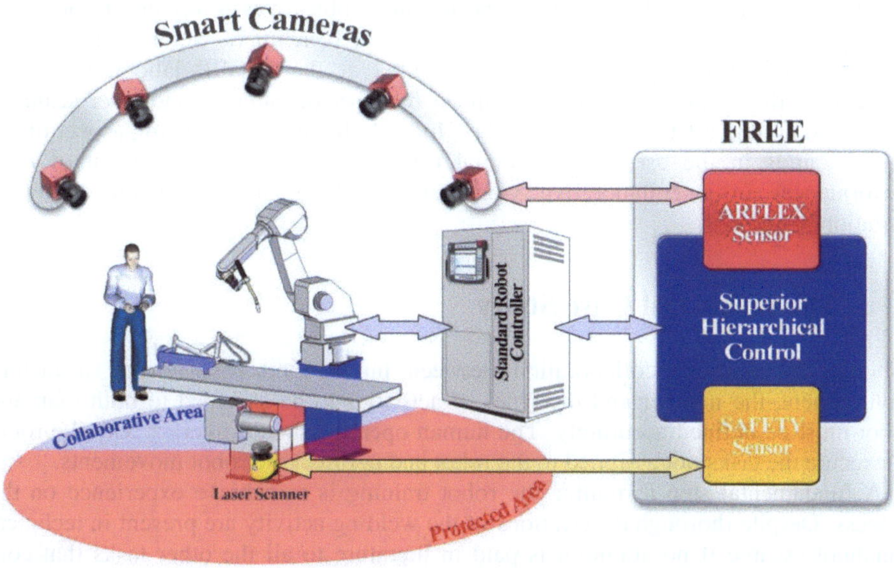

Fig. 1. Schematic overview of the FREE project showing the use of both measuring sensors to teach movements to the robot and safety sensor to avoid the operator entering the protected area

In order to be able to employ a standard industrial robot, we introduce a further control loop operating at a level hierarchically superior than standard robot control – denoted as "Superior Hierarchical Control" (SHC) – equipped with dedicated sensors, interfaced with the human operator and able to direct the robot standard controller (Fig.1). SHC receives measurement data from sensors located in the working area and, in the human co-worker scenario, it receives commands directly from the human co-worker. The sensor set includes the ARFLEX sensors [11] [12] for real-time contact-free measurement of the pose of human operated devices and laser scanners for detection of the presence of the human co-worker in the protected area reserved to robot working. SHC performs the following tasks:

1. "Safety Manager": it operates at the higher hierarchical level, aiming at guaranteeing the human worker safety by reducing the robot velocity or stopping it, if necessary. The application of the FDIR (Fault Detection Isolation and Recovery) methodology in the "Safety Manager" conception and design is a guaranty for a safe human-robot cooperation.
2. "Enhanced Programming by Demonstration": SHC works by interacting with the human operator through the observation of a pointer handled by the operator. The ARFLEX sensor is use to detect the pointer's pose with a minimum impact to the work, due to its contactless capabilities. Robot programs are automatically generated, assessed and refined through an iterative procedure.
3. "Flexibility and accuracy improvement": the ARFLEX sensor can get real time measurements of the pose of both robot tool and workpiece. Such measurement data can be used by SHC to increase robot accuracy.

The achievement of the scientific and technical objectives is verified through the experiment, consisting in the execution of a set of benchmark welding tasks in the assembly cell. The experiment is implemented at the DIGEP laboratory of the Politecnico di Torino. The cell is equipped by a set of cameras and laser scanners positioned around a COMAU NS16 robot. Indeed, the most challenging aspect of the experiment is in the use of commercial robots and in the enhancement of their performances in a collaborative workstation without any significant hardware variation.

3 The Industrial Case Study

In order to establish collaboration between human and robot inside a factory environment, the number and the types of activities to be assigned to both man and robot must be defined accurately. The human operator should interact with the robot to execute the tasks not assigned to the robot and to direct the robot movements.

A fundamental step to realize the robot training is to achieve experience on the process. Despite thorough descriptions of the welding activity are present in technical handbook, scarce if no attention is paid in literature to all the other tasks that com plement the direct execution of the welding.

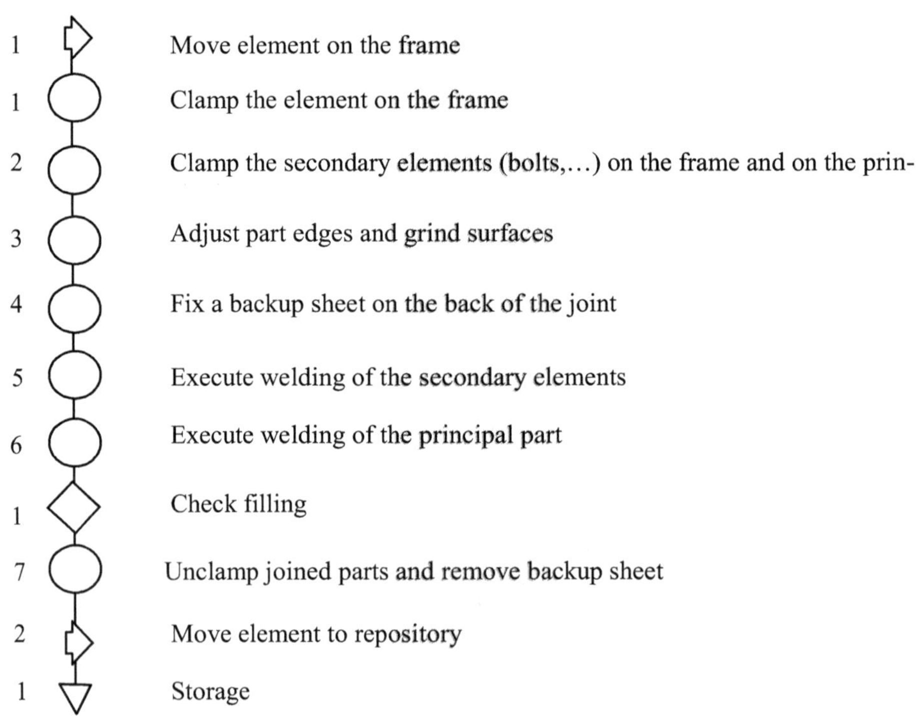

Fig. 2. Process flow diagram of a complete manual welding process (ASME standard)

To gain firsthand knowledge about the details of the process we observed and recorded it on the field thanks to the 'friend' factory, Eurodies. We drew up a detailed Process Flow Diagram (PFD) according to ASME standards [13].

Fig.2 reports the main welding operations. Circle stands for 'operation', diamond for 'inspection', arrow for 'transport' and triangle for 'storage'.

The process is composed of the following main steps. The first step moves the part from a storage to the specific frame. Then operators prepare the edges to be joined and maintain them in a proper position by the use of clamps and fixture. In a second time they adjust part edges and grind surfaces to facilitate the welding. Then they fix backup sheets on the back of large weld beads to prevent the spillage of molten metal. The fourth step is actual welding. After the welding the operators unclamp joined parts and remove backup sheets, then the assembly is moved to another frame for tests and eventually is moved to the storage.

4 The Experiment Description

The problem that present research confronts is the enabling of a human-robot collaborative process organized as follows:

- The human operator fixes the workpiece on a standard frame (the position of the workpiece relative to a reference coordinate system changes with every new part).
- The human operator teaches the robot the welding tasks.
- The human operator moves to a safe position, out of the welding zone, but not necessarily out of the reach of the robot (to go beyond the fences requires far too much time).
- The robot moves the torch to the points indicated and executes the welding.
- The robot return to a safe position so that the operator can approach the workpiece to check of the welding quality.

The problem can be subdivided in several sub-problems that have to be separately solved:

- Guarantee the *safety* to human inside robot workspace by avoiding that the robot enter inside a given safe area surrounding the operator (exclusion space).
- Implement a *machine learning* system based on multi-cameras observation of a pointer trajectory, handed by a human welder and detected through markers mounted on it.
- Implement a *user interaction* system to make the human operator able to give orders and indicate points on the workpiece.
- Put together the solutions of each sub-problem in a comprehensive control system that supervises all the robot actions (the SHC).

In the following sections we separately deal with the solution of all sub-problems.

5 The Safety Manager

The considered robot multi-hierarchical control framework is shown in Figure 3. The lowest level is the robot conventional control, generally based on the encoder measurements. The higher levels typically receive inputs from smart sensors mounted on the robot or in the working environment and provide the corrections to the lower control levels in order to meet the system requirements.

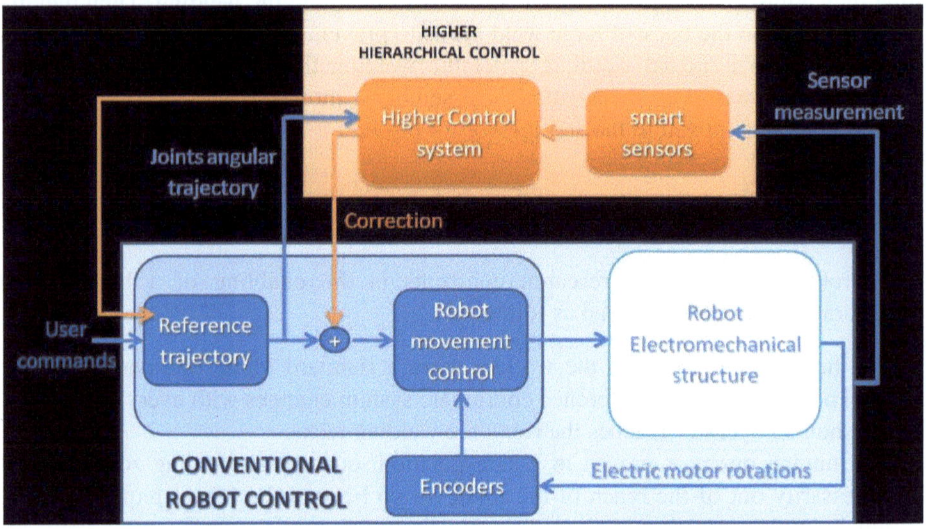

Fig. 3. Multi-level hierarchical control architecture

The robot part program and the trajectory is known and the safety control can give to the lower levels of control all the commands necessary to achieve [14]:

- full utilization of robot flexibility in relation to the safety requirements through the modifications of the operations sequence, but allowing the robot to carry out its job;
- otherwise a reduction of the working speed, up to the complete stop of the robot, disclaiming the work done for sake of human safety.

The above mentioned levels of control are designed as a discrete-event dynamic system (DEDS). The system is equipped with SICK laser scanners to identify the occurrence of events that modify the robot working safety conditions. Therefore, when event are identified, the system can actuate the control strategy able to keep the level of safety within the prefixed limits.

The control system must operate carrying out the following steps [15]:

1. Detection. A change in the safety conditions is detected, like a human or an object (foreign body) entering in the safety areas.

2. Identification. The type and the characteristics of the foreign body that have influence on the safety are identified. The occupied volume is identified.
3. Prevision. With reference to the prefixed temporal horizon, a forecast of the trajectory that the robot will execute is computed, together with the prevision of the possible movement of the volume occupied by the foreign body in the safety area. In the case in which the crash probability is higher than a threshold the next step (Correction) is executed.
4. Correction. The safety control system acts on the robot movement control system according to the specific detected dangerous condition, both at level of flexibility (variation of trajectory) and at level of variation of the working speed (leaving unchanged the programmed trajectory).

6 The ARFLEX Sensor

The ARFLEX sensor is an instrument for the contact-less measurement of the pose of moving rigid bodies. Two Cartesian reference frames are defined. The first one is the *user reference frame* and it is the reference frame respect to which the user wants to measure the moving object pose. The second one is the *body reference frame* and it is the reference frame associated with the moving rigid body, the pose of which has to be measured.

The instrument outputs, for each sampling step, the following output data:

- the moving body position measurement, expressed through the value of the three user reference frame coordinates of the body reference frame origin;
- the moving body attitude measurement, expressed by the three-dimensional rotation to be applied to the user reference frame in order that its axes take the same direction of the body reference frame axes;
- the current estimate of measurement error covariance matrix.

The ARFLEX sensor presents the following innovative aspects with respect to the current state of art in pose measurement instruments for industrial applications.

- It provides directly as an output – fully contact-less - the *pose* measurement of the moving *body reference frame* with respect to the *user reference frame* together with the updated current estimate of the measure error covariance matrix.
- When the sensor is installed, the user should not take care to detect each camera pose with respect to *user reference frame* or to the other cameras. The camera pose is automatically detected during the instrument start-up phase and it is continuously updated while the instrument is normally operating.
- Calibration is not required before installing cameras. Camera calibration is automatically carried out during the instrument start-up and it is continuously updated while the instrument is normally operating.
- It is equipped with FDIR (Fault Detection Identification and Recovery) functions, which - without interrupting the normal operations - allow to tolerate faults, that might accidentally occur, such as a partial or total occultation of a camera and displacements of a single marker from its nominal position.

The hardware and firmware structure of the smart camera are provided respectively in Fig. 4 and Fig. 5.

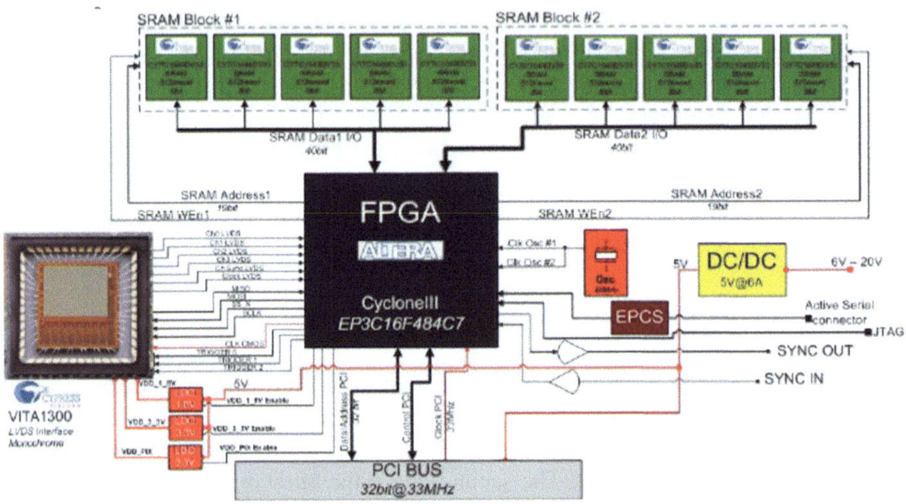

Fig. 4. The hardware architecture of the ARFLEX camera

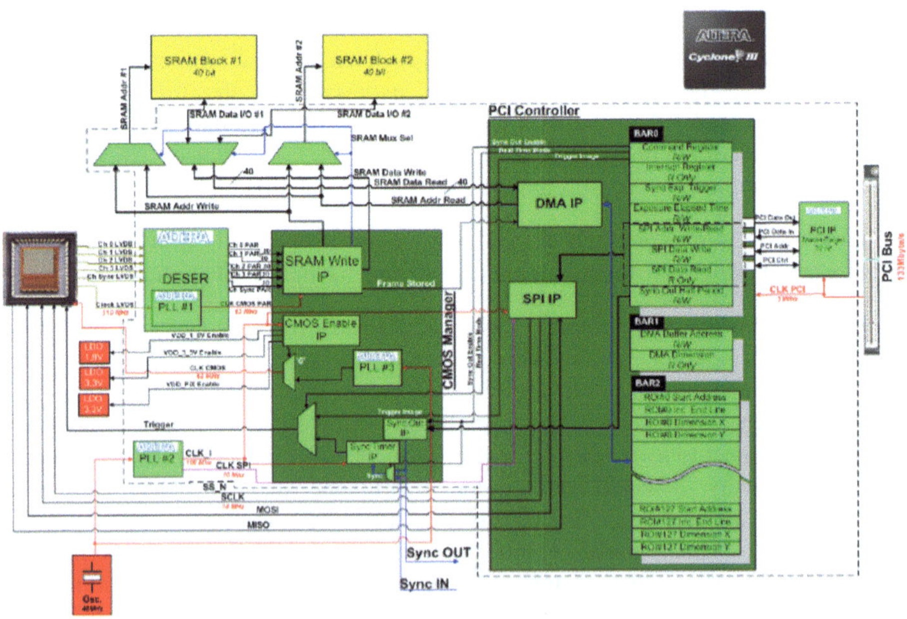

Fig. 5. The firmware architecture of the ARFLEX camera

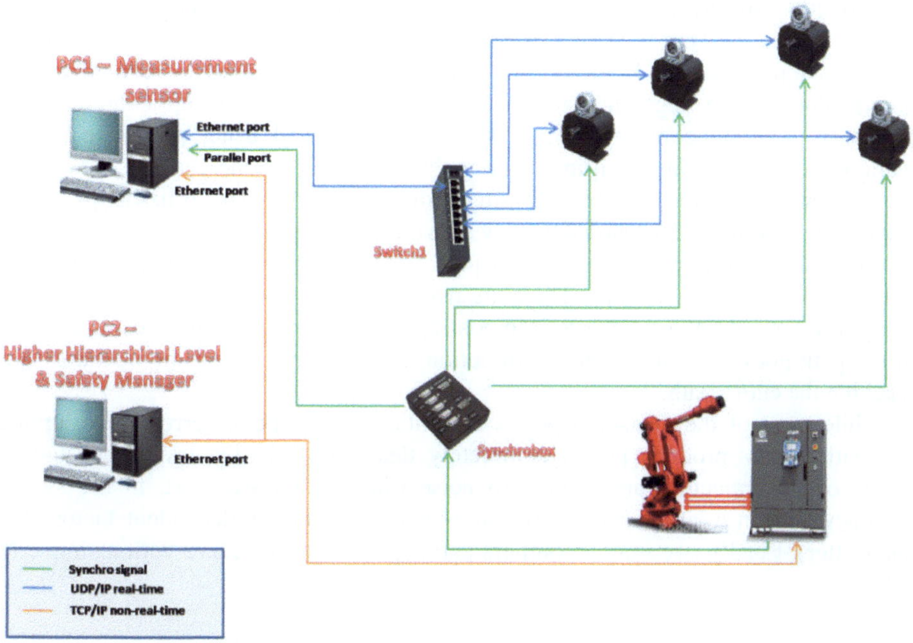

Fig. 6. The hardware architecture of the multi-level hierarchical control

Both cameras and markers are redundant, in order to guarantee simultaneously accuracy and reliability. In the experiment four cameras and six markers were employed (considering only the ones on the pointer). The smart cameras uses the configuration of Fig.6. Each camera is connected to the "Measurement sensor" PC and to a synchrobox to replicate to each camera the external synchronization signal provided by the robot.

Each camera receives it, through a serial port. The "Measurement sensor" PC1 is able to receive the same external signal, replicated by the synchrobox, through a parallel port. The external synchronization signal is generated by the robot at a frequency of 8 KHz that is way higher of the camera frame rate (50-100fps). The communication between the "Measuring sensor" PC and the smart cameras is carried out through a real-time gigabit Ethernet connection using the UDP/IP protocol. The real-time communication is made possible thanks to the RT-net driver, running on the real-time Linux operating system, patched with RTAI.

7 Training by Demonstration: Trajectory Reconstruction

In the path learning phase the operator moves a lightweight tool (a *pointer*) along a given trajectory. The action is recorded and reconstructed by the FREE measuring sensor that produces a list of pointer poses. The list is used to produce a smooth and continuous approximation of the trajectory that provides the input data for a PDL2 program governing the robot along its working task.

Derivation of the path from the measured points is not trivial [5] [16]. The dataflow of the process is shown in Fig.7. It consists of splitting the original record into non-stop segments. Velocity data and their history is used to recognize stop-points. Only those segments are taken into consideration which contain actual trajectory. Automatic recognition of stop points allowed using a passive pointer without electrical wiring. Multiple overlapping segments are merged and their points are properly reordered to obtain a single directed curve. The timestamps of the original data points are discarded as they aren't needed anymore.

Capability of the system to build a path from multiple overlapping segments is critical to real life applications. There are many circumstances when indicating a path using just one continuous gesture is impossible or undesirable: 1) long paths 2) a part of the path goes in an inconvenient direction, 3) obstacles prevent the operator from reaching the entire path.

While most of the measured points lay within an acceptable error range, optical acquisition may produce points completely detached from the trajectory, and all points of the trajectory are subject to noise (due, in a lesser part, to the system accuracy and, in a larger part, as we show below, to human dependent factors like hand jitter). Finally, we remove outliers and approximate the curve with a simple B-spline.

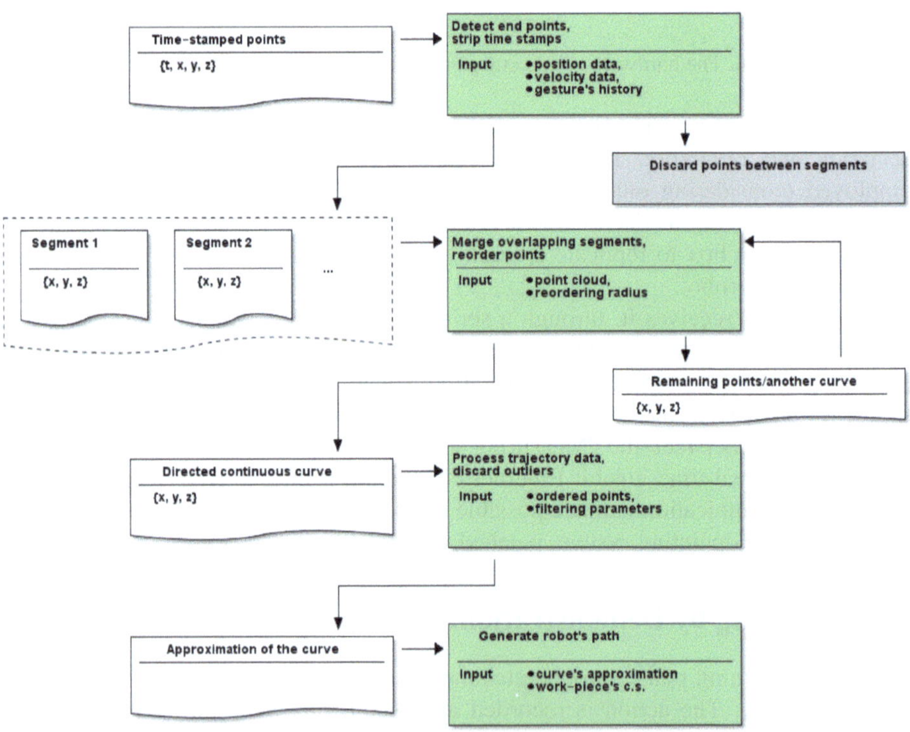

Fig. 7. Dataflow of the trajectory learning process

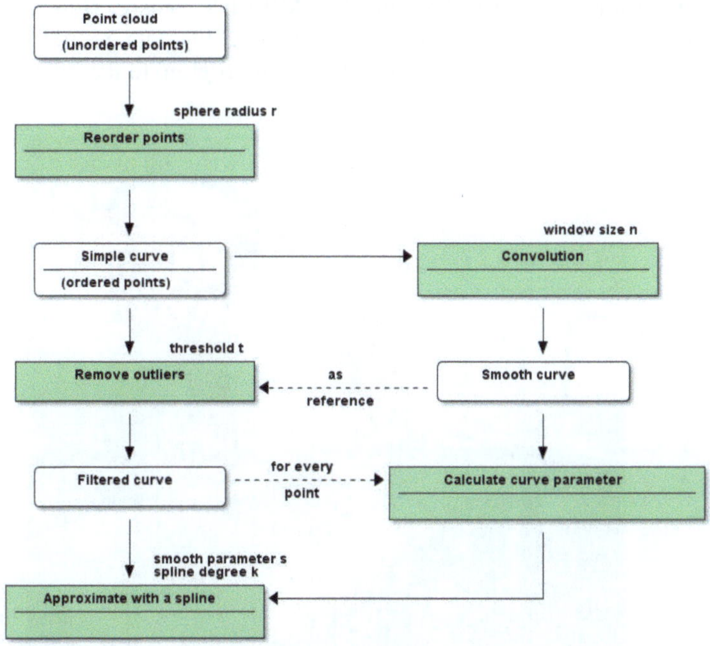

Fig. 8. Trajectory processing dataflow and algorithm parameters

Assumptions about curve properties are: the curve is simple (no self-intersections), regular (no backtracks), curvature radius is bounded from below. These restrictions do not affect most of the welding applications.

To build an approximating vector spline we need to provide some algorithm parameters. We observed that the best results are reliably achieved when we apply a smooth filter to the data points, and calculate the curve parameter from the smoothed curve. In our implementation we use a convolution with Hann window function separately for all three sequences x_i, y_i, z_i. This should work well for curves with uniform distribution of points. The smoothed curve is used to calculate curve parameter for a B-spline and eliminate outliers. For details see [17].

8 Experiment

The experiment is arranged in an area properly fenced and alarmed, as required by present security laws. There is also a verification system based on a mobile CMM. Fig.9 shows the elements of the workstation. The robot executes the welding process and is equipped with a welding gun or a marking pen. Two laser scanners detect the human's location. Whenever a laser scanner registers human presence in proximity to the collaborative safety area, the robot slows down as far as the emergency stop if the predicted human trajectory bring to a collision. The operator uses the pointer equipped with retro-reflective markers to indicate the workpiece's reference system and the trajectory (Fig.10) through the ARFLEX sensor (Fig.11). The FREE algorithm extracts trajectory segments from the tracking data, reorders the points and smooth the curve, to eventually generate the input program to the robot (Fig.12).

Fig. 9. Robot and CMM in the experimental area. There is a benchmark plate with a Z-shaped groove on the desk.

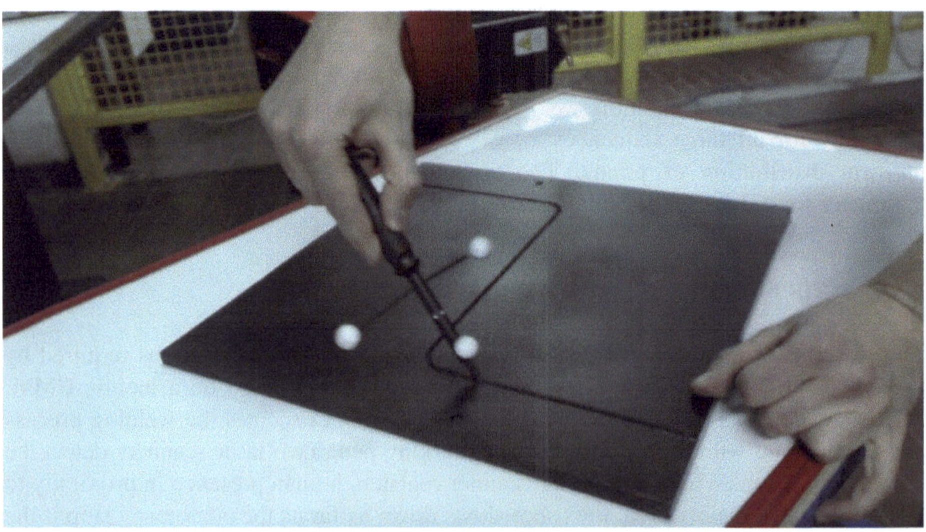

Fig. 10. A pointer with retro-reflective markers is used to indicate path on the Z benchmark

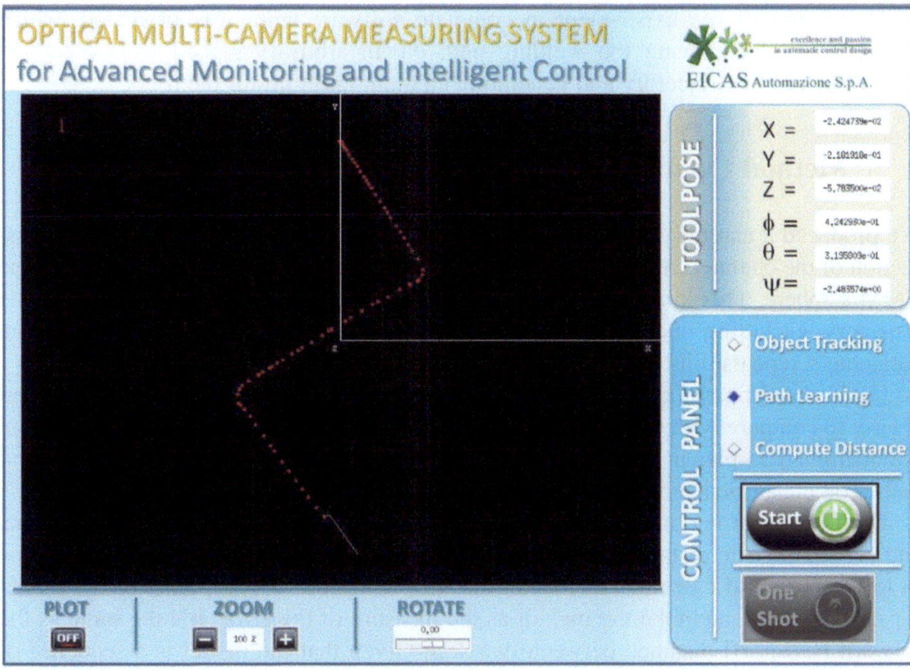

Fig. 11. Screenshot of the capturing software

Fig. 12. A robot repeats the Z-shaped path (welding simulation)

At the end of the experiment, a complete software suite was produced, which allowed for 1) capturing the welding path on an arbitrary placed work-piece, 2) reconstructing the trajectory and generating an input program for the robot.

9 Method Qualification and Parameter Selection

We produced a carefully studied benchmark plate with a Z-shaped channel. The cross-section of the channel can accommodate the end tip of the hand-held tool, as well as the CMM's sphere. This kind of benchmark plate allowed to repeatedly indicate the same trajectory, excluding operator errors, highlighting the other factors such as algorithm parameters, repeatability of the recording, and of the workpiece positioning.

The Z-shaped path (a model) represents a common element for a welding path and is easy to compare with the executed trajectory. We'd like to underline that the system is not limited to planar or regular shapes, and may reproduce an arbitrary free-hand path. The planar benchmark was chosen only for sake of method qualification.

The form error is obtained by aligning two curves, the real and the measured one, and calculating the distance between them. We sample N points on the output spline (a test curve), and calculate Euclidean distance di to the closest points of the model for each of them, and take the max di as the measure of the error. It's the same as the forward Hausdorff distance between two sets. Given that the model is a sequence of few straight line segments and circular lines, calculating only the forward distance is much faster (O(N)) than the complete Hausdorff distance (O(N 2 log N)), and the result is not much different in practice.

To align the test curve and the model, we find a proper rigid transformation to minimize the above mentioned distance. No matter which method is used to align the curves, it needs a good initial approximation. Most registration methods concentrate on matching geometric elements as in [18]. In our case we know in advance that the model and the output curve should overlap completely. This allows us to sample the same number of points on the model and on the spline uniformly, and postulate one-to-one correspondence. Then we may find the best rotation between two point clouds using quaternion method [19]. The calculated quaternion may be converted to arbitrary Euler or Tait-Bryan angles as in [20].

To determine the optimal method parameters, we studied the distribution of the minimal form error across multiple scans as a function of every method parameter (Fig.13).

In general the method is robust with respect to the selection of the parameters (it consistently produces similar results in the wide range of parameters' values). Threshold parameter t and the smoothness parameter s are not independent: the greater is t (less points are removed as outliers), the greater is optimal s (more smoothness has to be permitted in the spline). In our case the relation between the optimal t and the optimal s appears to be almost linear, but may depend on the properties of the acquisition system.

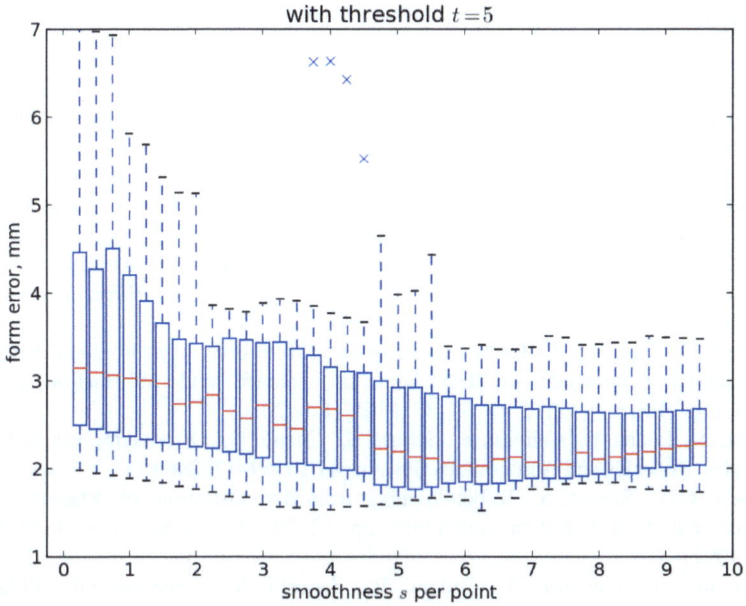

Fig. 13. Distribution of the form errors as a function of spline smoothness parameter s

10 Conclusions

FREE aims at eliminating the barriers against robot application in small batch production. The focus on SMEs is strategic to enable the EU to become more competitive, both because technology manufacturers need to find new markets and because SMEs have a considerable weight in terms of GDP and employment. SMEs lack the engineering skills necessary for installation, running and maintenance of new manufacturing technologies.

Future robotics needs easy and standardized interfaces with built in diagnostic devices to ease service and support. Following current trends it is apparent that robot operations will increasingly depend on information generated by sensors, worker instructions and/or CAD product data.

In the project, the efficient robot learning procedure allows to reduce or eliminate the need of accurate positioning of the workpiece with jigs and fixtures. Application of the SHC opens new possibilities in configuration of the work area. FREE capitalizes on successful results achieved in the previous ARFLEX EU project. FREE can contribute to new products/services in a reasonable time frame, and opens new ways to reduce costs and increase quality in small batch production.

Acknowledgements. This work was supported by the EU Project EC FP7-ICT-231143 ECHORD.

References

1. An executive summary of the strategic research agenda for robotics in Europe. Coordination Action for Robotics in Europe (CARE), FP6 IST-045058 (June 2008)
2. Argall, B.D., Chernova, S., Veloso, M., Browning, B.: A survey of robot learning from demonstration. Robotics and Autonomous Systems 57(5), 469–483 (2009)
3. Billard, A., Calinon, S., Dillmann, R., Schaal, S.: Handbook of Robotics Chapter 59: Robot Programming by Demonstration (2007)
4. Kuniyoshi, Y., Inaba, M., Inoue, H.: Learning by watching: Extracting reusable task knowledge from visual observation of human performance. IEEE Transactions on Robotics and Automation 10(6), 799–822 (1994)
5. Fenghua, G., Caiming, Z.: Curves reconstruction from ordered point cloud data. In: 2010 International Conference on Environmental Science and Information Application Technology (ESIAT), vol. 2, pp. 84–87. IEEE (July 2010)
6. Billard, A., Epars, Y., Calinon, S., Schaal, S., Cheng, G.: Discovering optimal imitation strategies. Robotics and Autonomous Systems 47(2), 69–77 (2004)
7. Atkeson, C.G., Schaal, S.: Robot learning from demonstration. In: Machine Learning-International Workshop then Conference, pp. 12–20. Morgan Kaufmann Publishers, Inc. (July 1997)
8. Gasparetto, A., Lanzutti, A., Vidoni, R., Zanotto, V.: Experimental validation and comparative analysis of optimal time-jerk algorithms for trajectory planning. Robotics and Computer-Integrated Manufacturing 28(2), 164–181 (2012)
9. Corfiati, M., Antonelli, D.: A Semi-Automated Welding Station Exploiting Human-Robot Interaction. In: AMST 2011, Mali Losinj, HR (2011)
10. Antonelli, D., Astanin, S., Galetto, M., Mastrogiacomo, L.: Training by Demonstration a Welding Robot. In: ICME 2012, Ischia, Italy (2012)
11. ARFLEX (Adaptive Robots for Flexible Manufacturing Systems), FP6 EU project EC-NMP2-CT-2005-016680, `http://www.arflexproject.eu`
12. Caporaletti, G.: The ARFLEX project: Adaptive robots for flexible manufacturing systems. Intelligent Manufacturing Systems 8(1), 253–258 (2007)
13. `http://syque.com/quality_tools/toolbook/Flowproc/flowproc.htm` (last accessed March 4, 2013)
14. Koskinen, J., Heikkila, T., Pulkkinen, T.: Monitoring of co-operative assembly tasks: functional, safety and quality aspects. In: IEEE International Symposium on Assembly and Manufacturing, ISAM 2009, pp. 310–315. IEEE (November 2009)
15. Zaeh, M., Roesel, W.: Safety aspects in a human-robot interaction scenario: a human worker is co-operating with an industrial robot. In: Kim, J.-H., et al. (eds.) Progress in Robotics. CCIS, vol. 44, pp. 53–62. Springer, Heidelberg (2009)
16. Ude, A.: Trajectory generation from noisy positions of object features for teaching robot paths. Robotics and Autonomous Systems 11(2), 113–127 (1993)
17. Dierckx, P.: Curve and surface fitting with splines. Oxford University Press (1995)
18. Salvi, J., Matabosch, C., Fofi, D., Forest, J.: A review of recent range image registration methods with accuracy evaluation. Image and Vision Computing 25(5), 578–596 (2007)
19. Horn, B.K.: Closed-form solution of absolute orientation using unit quaternions. JOSA A 4(4), 629–642 (1987)
20. Shoemake, K.: Euler angle conversion. Graphics Gems IV, 222–229 (1994)

In-Situ Robotic Fabrication: Advanced Digital Manufacturing Beyond the Laboratory

Volker Helm, Jan Willmann, Fabio Gramazio, and Matthias Kohler

Architecture and Digital Fabrication, DFAB,
at Swiss Federal Institute of Technology (ETHZ),
Zurich, Switzerland
{helm,willmann,gramazio,kohler}@arch.ethz.ch

Abstract. This paper takes an important step in characterizing a novel field of architectural research where a robotic system moves on a construction site and positions building components in-situ. Developed by the research group of Gramazio & Kohler at ETH Zurich, this approach offers unique advantages over traditional building technology: it fosters non-standard building processes, it can be directly applied on the construction site and it is easily scalable and it offers digital integration and informational oversight across the entire design and building process. Featuring a comprehensive new approach to architecture and technology, this paper considers 1) research parameters and components of in-situ robotic fabrication (such as tolerance handling, man-machine cooperation and localisation), 2) experimentation and building prototypes at full architectural scale, and 3) the architectural implications of integrating these findings into a systemic, unifying process at the earliest stages of design. As a result, in-situ robotic fabrication opens up entirely new possibilities of automated construction that are not limited by the constraints of prefabrication; the most evident and radical consequences of in-situ robotic fabrication are the ability to digitally oversee and control a large number of aspects of design and fabrication within an efficient and flexible building process.

Keywords: In-situ, automated construction, additive fabrication, robotic manufacturing, computational design, non-standard architectural structures, digital integration, tolerance handling, human-machine-interaction.

1 Introduction

Industrial robots are becoming extremely significant to the field of architecture [1]. Not only can they lead to significant time and cost savings in architectural production, but their ability to connect digital design data directly to the fabrication process enables the efficient construction of very complex building

F. Röhrbein et al. (eds.), *Gearing Up and Accelerating Cross-Fertilization between Academic and Industrial Robotics Research in Europe*, Springer Tracts in Advanced Robotics 94,
DOI: 10.1007/978-3-319-03838-4_4,

components at full scale. Yet fixed robotic systems have predefined working areas that limit their scale of action and thus constrain the size of the work-piece they act upon [2]. As such, in-situ robotic fabrication presents a radically different approach to building prefabrication: First, in-situ robotic fabrication enables direct operations on the construction site and the build up of architectural structures according to specific geometric configurations simply because the utilized robotic system moves and mounts construction parts directly to their required position; Second, because the construction process is continuous, in-situ robotic fabrication allows the assemblage of a great number a small building components and therefore reduces the segmentation of large building components [3]. Consequently, not only the need for costly and unsustainable transportation of large structural elements, but also for manual measurement or auxiliary construction devices such as formwork is removed; Third, in-situ robotic fabrication operates under the explicit guidance of a digital architectural design [4], and can place and manipulate material according to a precise digital blueprint. Here, the robotic system has a virtual overview of the construction-relevant data, leading to highly flexible and adaptable [5] building processes, even when the design and construction information are highly complex [6]. As such, the information additionally gained from the building environment for instance the actual dimensions of the building space can be directly incorporated into the logic and assembly of architectural structures, fostering the combination of digital and physical data towards the important factor of real-time information at building time; Fourth, the concept of in-situ robotic fabrication can be applied either individually or cooperatively, using a multiple of mobile construction robots or performed in direct cooperation with human workers. This allows to scale the productivity of the system according to specific requirements and building situations.

These unique features, that no other computer controlled fabrication system has today in architecture [7], makes in-situ robotic fabrication apt to yield a variety of construction processes and building applications. In contrast to conventional construction machinery such as stationary cranes, in-situ robotic fabrication allows the implementation of many different manufacturing scenarios, where the use of robots is regarded as an open system [8] that can be customized by various tools [9]. These features allow this research to pave the way for a multitude of novel building processes that are not currently possible with stationary building construction machinery.

However, the investigation of in-situ robotic fabrication is still in its infancy, and presents many theoretical, practical and methodological obstacles [10]. Obvious challenges are wide-ranging and include the need for particular construction systems and tolerance handling, man-machine interaction and the integration of advanced localisation techniques. In order to develop a schema for addressing these challenges, the research group of Gramazio & Kohler at ETH Zurich has created a first experimental setup for in-situ robotic fabrication that stems from the unique combination of an industrial robotic arm, a caterpillar carriage and an additional sensor system [11]. This approach – advanced by the ECHORD project[1] within the Seventh Framework Programme of the European Union – resulted in several building experiments and required many innovations, including the exploration of new robotic assembly processes of brick and timber elements, the in-depth investigation into the handling of building tolerances and the characterisation of robust localisation techniques[2].

In the following text (see section 2) the context of the work is presented from both an architectural and a construction perspective. Section 3 suggests essential parameters for the investigation of in-situ robotic fabrication, including construction methodologies, features of mobile robotic systems, as well as feedback and localisation techniques. Section 4 discusses an experimental setup that is based on the aforementioned research components, and Section 5 presents a detailed description of three robotic building experiments. In Section 6, the challenges of in-situ robotic fabrication and strategies for addressing them are summarised. The conclusions are presented in Section 7.

2 Context

In-situ robotic fabrication in architecture is a novel area of research, and only a few applications currently exist. Research on robotic construction in architecture dates back to the early 1990s [12]. There have been several attempts to develop mobile bricklaying and construction robots for the usage on site, the most advanced of them being the ROCCO [13] and the BRONCO [14] projects. The motivation behind this research was to improve the productivity and economy of building construction, mainly by utilizing the machines ability to handle an increased payload in contrast to humans. Although highly advanced, these developments for robotic prefabrication did not find access into the market since they were not flexible enough to adapt to diverse material and constructive systems, and to react to different designs and spatial situations [15]. In the course of the recent shift towards digital technologies in architecture, the Gramazio & Kohler group has set up the first industrial robotic fabrication laboratory for non-standard architectural fabrication processes in 2005 at ETH Zurich. This was followed by universities such as Harvard GSD (2008) [16], Carnegie Mellon (2009) [17], University of Michigan (2009) [18] and University of Stuttgart (2010) [19]. Parallel to ETH Zurich, they have fostered promising architectural case-studies and prototypical structural elements, elevating robotic assembly processes to the role of constitutive design and construction tool. Such novel technologies now motivate new approaches to the design of architectural structures and advanced fabrication systems that show that the use of robotic technology in architecture (see figure 1) is clearly feasible [20].

---

[1] European Clearing House for Open Robotics Development (ECHORD), `http://www.echord.info/wikis/website/home`

[2] Gramazio & Kohler, ETH Zurich, `http://www.dfab.arch.ethz.ch/web/e/forschung/198.html`

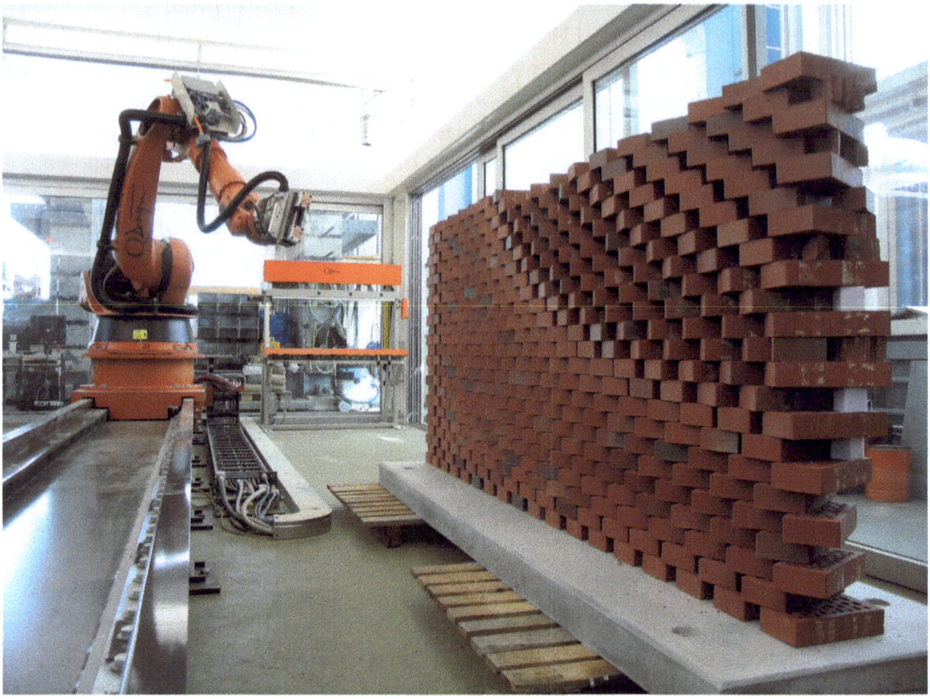

Fig. 1. Robotic fabrication of a non-standard brick wall at ETH Zurich (Gramazio & Kohler, ETH Zurich, 2006)

Despite the potential advantages using robotic technologies, in comparison to other industries the construction sector has been slow to adopt innovative digital fabrication methods; most construction work on site is still carried out using manual methods. The main reasons for this are, firstly, that existing machinery to automate building construction (such as material-handling robots) were for a long time not able to compete with manual labour, as they are, in most cases, too slow and for the majority of building companies simply not affordable [21]. And secondly, automated construction machinery was mainly developed towards repetitive building tasks, neglecting the fact that every building task and construction site is different [22].

In the course of the recent shift towards digital fabrication technologies in architecture, however, the exploration of automated construction has returned, in which generic and flexible industrial robotic machinery is used to perform a range of non-standard assembly tasks. In particular, the Gramazio & Kohler group has deployed promising experimentation [23], resulting in robust, highly adaptable and sustainable construction systems for both prefabrication and on-site fabrication. For example, in 2008 the group developed the transportable fabrication unit R-O-B[3], which consists of a flexible robot cell where an industrial robotic arm is mounted on a linear axis and housed in a modified freight container (see figure 2, left). The container can be moved directly onto the construction site and can perform on-site prefabrication tasks. As such, this approach combines the advantages of prefabrication (such as precision and consistent work quality) with the advantages of short transportation routes and just-in-time production on the building site. Furthermore, the mobile fabrication unit allows to use local materials and is not restricted to a predefined manufacturing process or a particular building material, but allows for manufacturing building elements with highly specific forms, which would be difficult and expensive to transport otherwise [24].

Fig. 2. left: On-site prefabrication performed with the R-O-B fabrication unit; right: Resulting in a highly articulated non-standard brick system at the 11th Venice Architectural Biennale (Gramazio & Kohler, ETH Zurich, 2008)

This exemplifies that the gradual advancement and availability of industrial robots has enabled the creation of novel robotic systems that have led to impressive progress in architectural fabrication. However, despite the fact that industrial robotic systems are becoming a mature technology in the field of architecture, novel architectural applications for in-situ fabrication are not existing yet. Ideally such technologies should integrate seamlessly with humans in the same environment or in other places that are not specially designed *a priori* for robots, such as construction sites.

3 Research Components

Research into in-situ robotic fabrication (as defined in section 1) is based on specific components and strategies to manage and perform complex construction

[3] Gramazio & Kohler, ETH Zurich, and Keller Ziegeleien AG, `http://www.keller-systeme.ch/de/nachhaltigkeit_und_innovationen/digitale_fabrikation/`

tasks. Three general categories of research have been identified: 1) non-standard construction systems; 2) mobile robotic machinery; and 3) advanced feedback and localisation techniques. This can be outlined as follows:

3.1 Non-standard Construction Systems

Digitally controlled assembly processes in architecture are by nature "additive" [25], they are flexible and can incorporate complex information about the individual elements to physically position them in a geometrically differentiated way, which allows to accommodate economic and programmatic criteria [26]. Because non-standard construction systems are also determined from robotic handling considerations, these must be designed according to specific building capabilities of the specific robotic arm. In the context of in-situ robotic fabrication, one important key is to develop construction systems that are suitable to direct implementation on the construction site and are architecturally lean [27] (see figure 3).

Fig. 3. An 8-metre-long modular wall, digitally fabricated in the parking garage of the Department of Architecture at ETH Zurich using an additive assembly method, for which the robot had to reposition itself several times (Gramazio & Kohler, ETH Zurich, 2012)

The building components, for example, must allow complex assemblies with multiple degrees of freedom. As long as they can be physically placed in a stable manner, they can be individually positioned and rotated by the robotic arm. This creates a set of rules and parameters that informs a varied design for the overall structure, in terms of both the horizontal and the vertical aggregation. These rules and parameters, together with usability and structural requirements, provide a comprehensive framework not only to additive digital fabrication processes in general, but specifically to the design of in-situ aggregated building structures. The consequence is a design that is never monotonous or repetitive, but rather specific and adaptable to different architectural, material and site characteristics. It is both comprehensive and versatile all at once [28]. As a further consequence, the utilized construction technology must encompass tolerances of both the buildup material and the robotic placement of it. It must also account for discrepancies[4] in material properties, such as weight and friction, rigidity and the connection system. This fosters investigation into adaptive material systems [29], specifically the efficiency with which particular levels of formal complexity and construction performance (such as stability) can be achieved.

This "information" logic between dynamic contingencies and physical characteristics – such as the requirements of robotic positioning and the physical constraints of the material – must be seen as integral; consequently, the construction system must be allocated a very high priority in in-situ robotic fabrication. Overall, this thinking fosters the creation of architecture that profoundly reinvents its construction repertoire. In-situ robotic fabrication could thus lead to profound changes in the design, performance and expressive language of architecture.

3.2 Mobile Robotic Machinery

The focus of in-situ robotic fabrication is not only to precisely examine the impact of novel constructive systems, but also to develop essential criteria of machinery and tools. In fact, this encourages the exploration of mobile robotic systems (see figure 4) that are flexible and adaptable, so that singular building elements can be precisely handled and accumulated in space where needed, enabling the implementation of additive in-situ fabrication on an architectural scale.

As such, mobile robotic systems are nowadays a research topic for many research groups in the field of robotics, and their abilities have drastically improved in recent years[5], but full robustness for outdoor capability and versatility – as is required for in-situ construction work – is still missing. Further, mobile robotic systems will have to seamlessly integrate into the digital fabrication process, whereas the investigation of extensive data transfer [30], both offline and online, will be crucial. Within the field of in-situ fabrication, the development of

[4] For more information, please refer to:
`http://www.dfab.arch.ethz.ch/web/e/lehre/131.htm`

[5] For more information, please refer to: Autonomous Systems Lab (ASL), Prof. Roland Siegwart, `http://www.asl.ethz.ch/`

Fig. 4. The experimental setup: An industrial robot, ABB IRB 4600, mounted on a compact mobile track system that is sized to fit through a standard door frame on a construction site (Gramazio & Kohler, ETH Zurich, 2011)

mobile robotic machinery also involves tight interaction with architects, robotic specialists, mechanical and civil engineers, CAD and computation specialists; this interaction will inform design and construction decisions of mobile robotic machinery to match the requirements of a real world in-situ fabrication process in architecture.

3.3 Advanced Feedback and Localisation Techniques

The development of mobile robotic machinery for in-situ fabrication is completed by the elaboration and integration of advanced feedback techniques. In contrast to a controlled industrial environment (such as a factory), a key challenge is that a typical construction site contains not only insufficient information about the exact geometry of the building space but also significant uncertainty, such as a changing environment, imprecise knowledge about the robots position, as well as uncertainty about the position of the humans or other robots assisting it in a task. Further, this includes uncertain robot dynamics, poorly known loads and contact forces. This necessitates the development and implementation of advanced feedback systems and will require the deployment of scanning and sensor systems to handle the significant level of uncertainty that characterises such building sites.

Until today, however, interactions between a robot and an uncertain environment are avoided in automated construction work as much as possible, to prevent breaking the robot and/or elements of the environment or, more dramatically, causing harm to human workers. Solving this problem involves the engagement with three main topics: perception, planning, control. These must seamlessly integrate and work robustly in a challenging environment. While there has been significant progress in all these fields, solutions for in-situ robotic fabrication are not yet at hand.

Because in-situ robotic fabrication is intended to be employed directly on a construction site, it needs a localisation system to orient itself while recognizing obstacles or people working in its surroundings. As the construction site is continuously expanding, with the work being focused at different locations, it becomes crucial to exploit the information available in the digital fabrication work flow and to develop a localisation system that allows a number of mobile robots to cooperate. As a concrete example, one can imagine a number of mobile fabrication units, moving between perching points in the space to provide an optimal site-wide fabrication system. As such, most systems still rely on external localisation methods, but advances in sensor capabilities (cameras, laser range finders, etc.) and their growing miniaturization drastically improves the possibility to use them for in-situ robotic fabrication and, as a further step, to develop semi-autonomous or fully autonomous robotic mobility on the construction site [31].

4 Experimental Setup

As a consequence of the identification of the above-described challenges, the requirements for an experimental setup are determined, so that account can be taken of the overall capabilities and limitations regarding its physical building performance. For instance, the setup is sized to fit through a standard door frame on a construction site in its folded position and each individual component was integrated into the system, enabling the mobile robotic unit to connect to different objects and building materials, steadily move them to a target point in space and place them with a given orientation. As such, the experimental setup[6] was developed in three stages: First, initial system design and evaluation of single components (see figure 5), second, the integration of the components and prototypical building processes, and third, man-machine integration and on-site testing. The components of the experimental setup can be outlined as follows: 1) an ABB IRB 4600 which is relatively lightweight, has a suitable load capacity, and provides a wide operational range (and a 6 DOF), and was tested in a building construction environment; 2) a mobile track system was engineered, supported by side-hinged telescoping outriggers with integrated raising jacks and an attached diesel engine; 3) a 2D-line scanner (Sick LMS 500) for detection of dimensional tolerances and a 3D scanner (see section 5.2) were installed to the robotic arm, enabling the detection of objects or obstacles in the workspace and facilitating man-machine interaction; 4) two vacuum grippers were added to the end-effector, enabling the unit to grip bricks from different sides. This includes a vacuum pump installed at the back of the fabrication unit.

[6] The experimental setup was developed in cooperation with the Bachmann Engineering AG (Zofingen, CH),
`http://www.bachmann-ag.com`

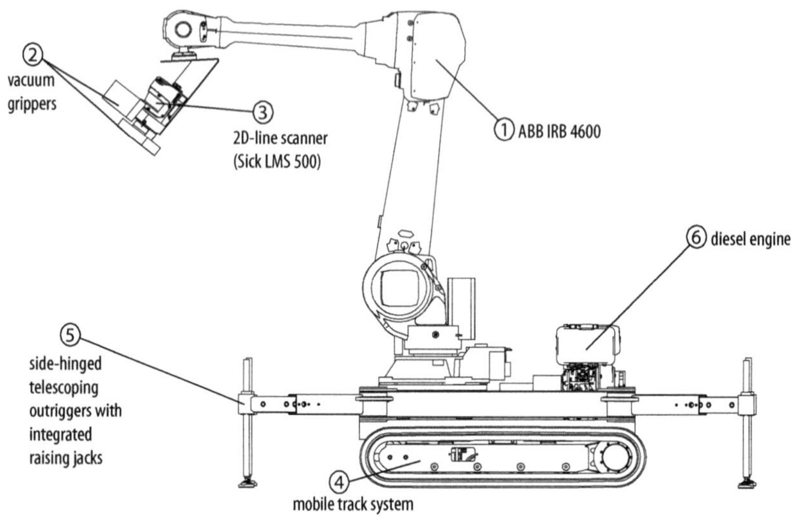

Fig. 5. The experimental setup and the single components of the mobile unit (Gramazio & Kohler, ETH Zurich, 2011)

5 Experimentation at Full Architectural Scale

In fact, in-situ fabrication represents a challenging approach to building construction with regard to the complexity of the robotic fabrication process and the overall localisation and tolerance handling performance [32]. Additionally, when working with robotic fabrication processes, where discrete elements are assembled to form a greater whole, the robustness of each fabrication step must be guaranteed. However, in-situ fabrication processes are of high interest, as they make efficient use of material (i.e. material is directly placed where it is needed) and allow for structurally optimized constructions that use less material and can be assembled without additional scaffolding or measuring devices. Consequently, this results in a considerable reduction of transportation costs for material and highly specific architectural designs. If the precision of a robot is sufficient to deal with unique and specific situations, and localisation systems are enabled an in-situ robotic system will be capable of handling the physical conditions of any given construction site through the introduction of responsiveness to the process, ensuring the necessary adaptability. In the following section three experiments are presented where novel in-situ fabrication methodologies and tools are based on the coherence between parameters used in computational design and the physical degrees of freedom available to robots on the construction site.

5.1 Tolerance Handling

The objective of the first in-situ robotic fabrication experiment was to explore a robust assembly technique to validate material and construction tolerances for the implementation of in-situ robotic fabrication at full scale. On that scope, a material setup was developed which featured discrete wooden blocks of varying sizes and a geometrically differentiated circular buildup to explore the tolerances of the buildup and the material, and the tolerances of the robotic placement, as well as material properties such as weight and friction. A first demonstration was given at the FABRICATE 2011 conference in London[7]. This test structure, circular in plan, was assembled from 1,330 timber building blocks of three different thicknesses, simulating the range of dimensional tolerances that would be faced on a construction site.

During the fabrication process itself, the fabrication unit maps the indeterminacy (see figure 6) between the fabrication data and the physical output by scanning each layer of blocks as they are laid. The mapped measurements are then sent back to the design/control software, and the robot arm re-orientates itself according to the new set of height and angle data. As a result, the assembly process with its specific material properties and unpredictable buildup tolerances had a direct impact on the structural expression of the materialised architectural artefact (see figure 7).

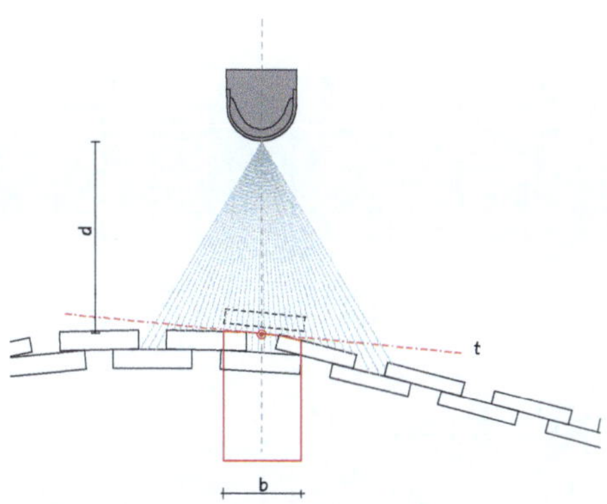

Fig. 6. After laying individual wooden building blocks of different thicknesses (3 cm, 4.5 cm, and 6 cm), the fabrication unit was mapping the indeterminacy (b=position; d= distance; t=inclination) by scanning each layer (Gramazio & Kohler, ETH Zurich, 2011)

[7] The project "Stratifications" was developed 2011 at the Fabricate Conference in London and realised with the support of Bachmann Engineering AG (project leader: Andrea Kondziela; team: Volker Helm, Ralph Bärtschi, Dominik Weber). For more information, please refer to
`http://www.dfab.arch.ethz.ch/web/e/forschung/206.html`

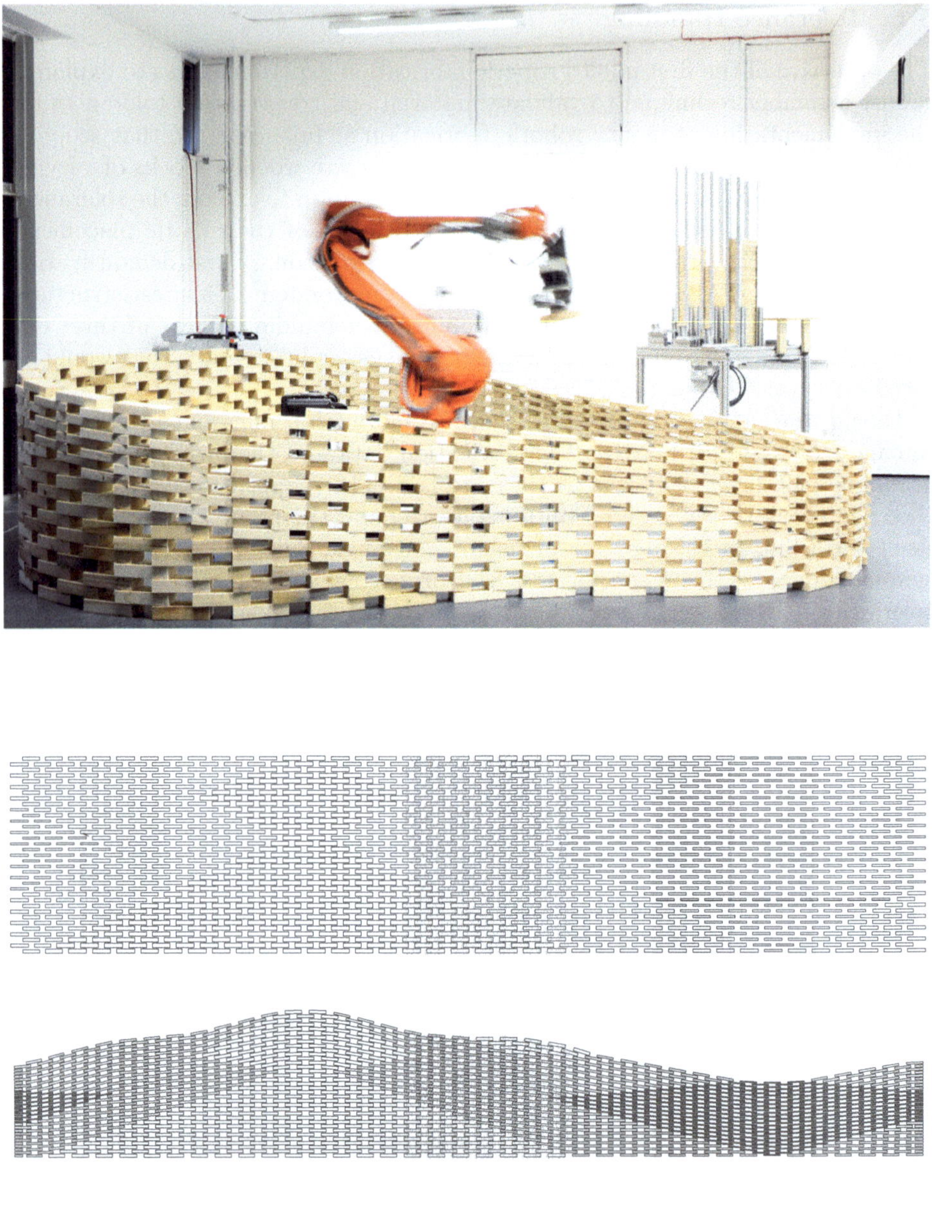

Fig. 7. top: The 1:1 installation "Stratifications" at the FABRICATE 2011 conference, performing an experimental demonstration of the handling of building tolerances; middle: The initial configuration of the installation's structure in which differently sized timber blocks are positioned according to a pixel image; bottom: The actual configuration after the blocks were robotically assembled, aggregated freely on top of each other (Gramazio & Kohler, ETH Zurich, 2011).

5.2 Man-Machine Interaction

As for the handling of tolerances 2D scanning technologies have been successfully integrated, the while for the recognition of the workspace, it is the innate orienting skills of the human counterpart that are called upon to lead the supervision of the robot. For this, a 3D scanning technology is integrated into the mobile fabrication system in order to map the hand movements of the human collaborator, which are then processed by the fabrication unit, allowing it to build in an unknown context such as a construction site. In this way, the system reacts to visual instructions and accordingly applies an additive assembly strategy (see figure 8).

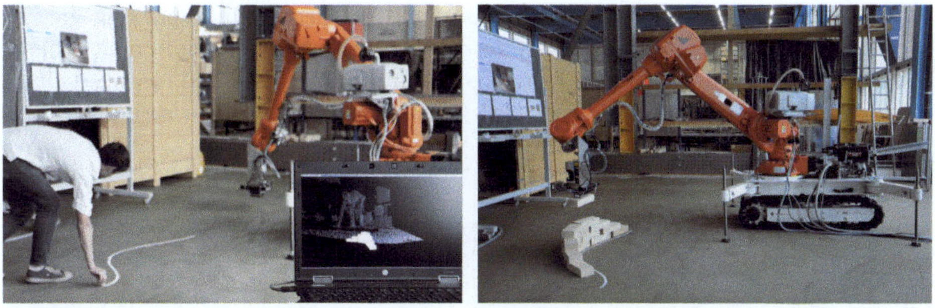

Fig. 8. left: User interaction with the mobile robotic unit whereas the hand gesture is scanned and imported into the CAD software as a line segment; right: the configuration of the wall structure is built accordingly (Gramazio & Kohler, ETH Zurich, 2011).

A demonstration at 1:1 scale of a fabrication cycle applying this interactive process was accordingly made at the 2011 Scientifica Exhibition in Zurich[8], in which audience participation was effectively employed as the human supervision. In this interactive fabrication process, the 3D scanning technology detects and processes hand movements, so that the human collaborator is able to "show" the robot its working area and building zone with reference to its relative position. With the implementation of this recognition technology, the human counterpart, working in collaboration with the robot, leads the building process through freely-made gestural instructions. On the software side, a feedback loop is programmed that provides continuous communication between the robot controller and the scanning system. This continuous exchange of information feeds into the fabrication process, with hand movements scanned and imported as CAD data into the design software. This data is then converted into robot code, with

[8] Gramazio & Kohler, "Scientifica - Zurich research days 2011" (project leader: Andrea Kondziela; team: Volker Helm, Dr. Ralph Bärtschi, Ryan Luke Johns, Dominik Weber). For more information, please refer to
`http://www.dfab.arch.ethz.ch/web/e/forschung/216.html`

which the fabrication unit is continuously informed in real time. The CAD software acts as an intermediary in the communication between the fabrication unit and human counterpart.

5.3 Robotic Localisation

In-situ robotic fabrication necessitates the development of a localisation systems where the robot and its peripheral system must permanently know their position in relation to the available material and the specific place of construction. Traditional ground robots (such as fixed industrial robots or CNC machines) have predefined working areas that limit their scale of action and thus constrain the size of the work-piece they act upon [33]. However, this research aims at mobile robotic units that can operate in space and assemble architectural structures at different scales. This experiment was deployed directly in a parking garage because the spatial situation closely resembles that of a construction site; here, a slanted floor, columns, and a restricted ceiling height provided essential characteristics for using a mobile robotic system and adapting the structure to a specific surrounding (see figure 9).

Fig. 9. In-situ robotic assembly of 1,000 wood elements without additional fasteners, using a customized localisation technique (Gramazio & Kohler, ETH Zurich, 2012)

Here it was possible on one hand to recognize the environment and its geometric deviations compared to the idealized counterpart described by the computer model. On the other hand, the data of the real physical situation could be immediately entered into the building process [34]. This principle of synchronising digital design data and real world data during the assembly process yields future potential in digital fabrication; consequently, material-based procedures must be tolerant to incongruences between physical reality and digital model. As such, the aforementioned 3D scanner integrated into the fabrication unit for the recognition of the workspace by following hand movements is employed by the self-positioning system. With the help of two metal disks ("satellites"), whose center point is used as a local reference marker, the fabrication unit is able to reposition itself according to the scanned and mapped coordinates of these reference points. The scanning device finds/measures the center points of the "satellites", then it is set as the origin of the coordinate system that is defined in the CAD model for the current position (see figure 10). After working in a sequence of positions, a fragile wall structure was fabricated in-situ[9]. Here, the ideal position of the robotic unit as well as the partitioning of the building structure in different steps has informed the design.

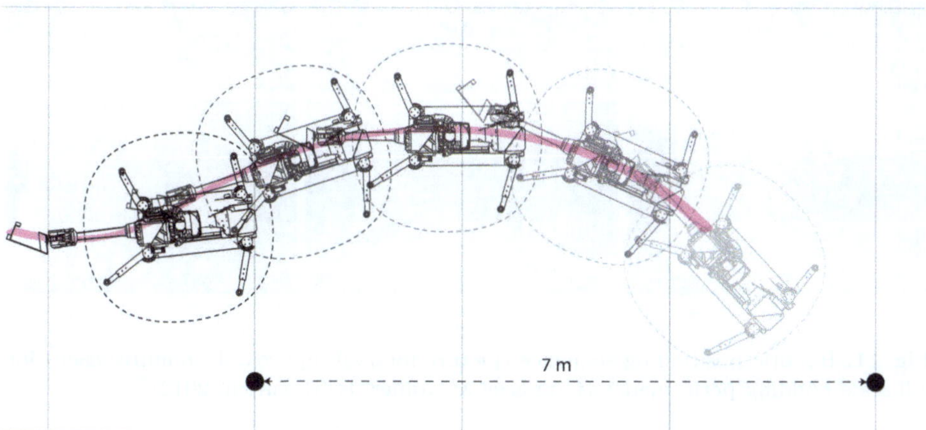

Fig. 10. The scanning mechanism developed for finding the center point of a metal disk and setting it as the origin of the plane defined in the CAD model for each new position of the robot (Gramazio & Kohler, ETH Zurich, 2011)

[9] The project "The Fragile Structure" was developed 2012 at the ETH Zurich and realised with the support of Schilliger Holz AG (project leader: Luka Piskorec; team: Volker Helm, Selen Ercan, Thomas Cadalbert; students: Leyla Ilman, David Jenny, Michi Keller, Beat Lüdi). For more information, please refer to `http://www.dfab.arch.ethz.ch/web/e/lehre/225.html`

The complexity of the building structure arises from the superposition of algorithmic rules which cannot be applied easily in a manual fabrication process while several aspects of the design are driven by the selected robotic fabrication method (see figure 11). The scale and modularity of the structure were directly derived from the surrounding spatial conditions, one, for instance, being the ceiling height. These are used as key design parameters and simulated on the software side. Through a collaborative process in design scripting and offline programming, the fabrication data from the CAD model of the structure is exported to the robot controller.

Fig. 11. Robotic positioning sequence (picture interval: approx. 40 minutes each) for full scale building performance (Gramazio & Kohler, ETH Zurich, 2012)

In sum, the experimental combination of robotic logics and strong computational and material orientation has proven to be an essential concept of in-situ robotic fabrication. Despite the complexity of the different experiments, it was purposely chosen to examine in depth the specific building construction characteristics of this combination, in order to unlock a new and interdisciplinary research direction for architecture.

6 Future Challenges of In-Situ Fabrication

The consequences of these experiments are indeed multiple and intertwined, and cannot be reduced to a simple perspective. It is therefore difficult to obtain a universal principle model of in-situ fabrication in architecture. It is possible, however, to isolate important potentials from the experiments discussed in this paper, evaluating future challenges of in-situ robotic fabrication.

6.1 The Integration of Architecture and Robotics

Indeed, tremendous advances of digital technologies in the field of architecture have come from the close interrelation of computational design and digital manufacturing. Industrial robots enable the implementation of complex material systems on an architectural scale [35]. It is possible to design a robots specific set of "manual skills", to determine its movements, and to assign it particular types of assembly sequences. When freed from operating within the predefined parameters of specialized machinery (e.g. automotive industry), robots can allow a wide range of manufacturing possibilities. However, when exploring these links, the differing characteristics of architectural material systems and in-situ robotic fabrication are becoming a prominent issue, such as the handling of tolerances or man-machine cooperation (see section 5.1, 5.2). As a consequence, three principal factors determine this research: 1) adaptive design and construction systems; 2) advanced robotic fabrication routines; and 3) enhanced interaction with the user. In fact, the different research areas are faced with many potentials, considerations, and conflicting issues, as is to be expected when pioneering a new field of research.

6.2 Determination of Parameters and Scale

In-situ robotic fabrication is a multi-variant field of research with no *a priori* set of parameters. Given the degree of its constructive differences, and difficulties relating to its fidelity for direct implementation on a typical construction site, specific experiences for this research are not presently available. Thus, experimental research in in-situ robotic fabrication is challenged by the determination of particular experimental setups and the selection of specific tasks that can be transferred to practical applications. The basic idea of the research approach presented here relates to performing research with the most realistic impact possible; this includes components such as a mobile robotic unit, different advanced material and constructive systems, and not least, control and localisation strategies to address full-scale implementation. On that scope it can be accurately stated that in-situ robotic fabrication can be transferred to specific building tasks and environments, and that these capabilities are definitely necessary when conducting this research.

6.3 Cooperation and Control

The experiments in this paper (see section 5) have shown that a mobile robotic system is able to fulfil in-situ construction tasks, e.g. the assembly of discrete building elements within an uncontrolled environment. However, such mobile robotic systems can perform a desired building action also collectively and deploy an intelligent dynamic construction cooperation. In addition to direct collaboration, their work capacity is also to a large degree scalable, a trait that digitally controlled robots as presented in this paper share with many other digitally driven technologies. In fact, these have potential to cooperate in many ways: as mentioned before, they can collaborate to lift heavy loads and react to human intervention. For example, two mobile robotic units can carry two building parts (such as modules or bars) while another one helps them with the assembly. Consequently, multi-robotic cooperation is a central theme in in-situ robotic fabrication and will allow the development of scalable fabrication processes, but requires the investigation of the possibilities of collaboration between vehicles when defining the material system and assembly strategy at the very beginning of the construction design. For this reason, the manufacture of adaptive construction systems using mobile robotic systems will require an advanced control and feedback systems, and a dynamic structuring of space.

7 Conclusion

To date, construction automation has already reached a sufficient level and led to impressive prefabrication routines and standardized building solutions. However, the building industry continues to be strongly driven by manual on-site construction processes, and cannot yet efficiently perform advanced non-standard building task. As a result, it is considerably lacking in flexible and efficient construction processes, is subject to high costs (both material and human resources) and is plagued by inconsistent workmanship. As such, accompanied by changing technical, cultural and economic conditions, and a different conceptual approach of construction robotics, there are signs of this being overturned and to make a complementary link to overcome the boundaries between efficient automation and flexible on-site operations, therefore using standard components and to realise the assembly on site. This will require flexible construction robots that can be effectively utilized for various constructions tasks and are able to plan and control, like their human counterparts, autonomous building operations within an unstructured and changing environment. In this respect, in-situ robotic fabrication contributes significantly to the field of constructions robotics, ranging from theoretical findings to sophisticated computational design techniques, from innovative robotic building processes to smart material designs and innovative construction systems. As such, it represents a fundamentally new paradigm and forms a long-term vision to substantially improve the flexibility and environmental impact of the building construction on different scales, and to foster in-depth research into sophisticated control systems, feedback mechanism and advanced robotic machinery.

Acknowledgments. This work was supported by the EU Project EC FP7-ICT-231143 ECHORD. The authors thank their team for their pioneering efforts, particularly Dr. Ralph Baertschi, Tobias Bonwetsch, Selen Ercan, Ryan Luke Johns and Dominik Weber. The authors would also like to thank the involved industrial partner, Bachmann Engineering AG, for the precious collaboration and for the engineering of the experimental setup. We are also grateful to ABB for its generous project support. An extra thanks goes to Lauren Vasey for her efforts on the editorial revision of this paper.

References

1. Gramazio, F., Kohler, M.: Digital Materiality in Architecture, pp. 8–11. Lars Müller Publishers, Baden (2008)
2. Bonwetsch, T., Gramazio, F., Kohler, M.: R-O-B: Towards a Bespoke Building Process. In: Sheil, B. (ed.) Manufacturing the Bespoke: Making and Prototyping Architecture, pp. 78–87. John Wiley & Sons, London (2012)
3. Bonwetsch, T.: The Informed Wall: Applying additive digital fabrication techniques on architecture. In: ACADIA 2006: Synthetic Landscapes. Proceedings of the 25th Annual Conference for the Association for Computer-Aided Design in Architecture, Louisville, p. 489 (2006)
4. Burry, J.: Philosophy of Mathematics for Computational Design: Spatial Intuition Versus Logic. In: Menges, A., Ahlquist, S. (eds.) Computational Design Thinking, pp. 168–178. John Wiley & Sons, West Sussex (2011)
5. Bonwetsch, T., Gramazio, F., Kohler, M.: Digitales Handwerk. In: GAM 2006, Nonstandard Structures, Graz, pp. 172–179 (2010)
6. Willmann, J., Kohler, M., Gramazio, F.: Die Operationalitt von Daten und Material im Digitalen Zeitalter. In: Hofmeister, S., Hellstern, C. (eds.) Positionen zur Zukunft des Bauens. Edition DETAIL/Institut fuer int, pp. 10–11. Architektur-Dokumentation, Munich (2011)
7. Helm, V., Ercan, S., Gramazio, F., Kohler, M.: In-Situ Robotic Construction: Extending the Digital Fabrication Chain in Architecture. In: Synthetic Digital Ecologies: Proceedings of the 32nd Annual Conference of the Association for Computer Aided Design in Architecture, San Francisco, pp. 169–176 (2012)
8. Gramazio, F., Kohler, M.: Digital Materiality in Architecture, p. 9. Lars Müller Publishers, Baden (2008)
9. Willmann, J., Kohler, M., Gramazio, F.: Roboterbasiertes Bauen, Architektur und digitale Fabrikation. Deutsche Bauzeitschrift 11(60), 56–59 (2012)
10. Willmann, J., Kohler, M., Gramazio, F.: Digital by Material: Towards an Extended Material Performance in Architecture. In: Robotic Fabrication in Architecture, Art and Design, RobArch 2012, pp. 12–27. Springer, Vienna (2012)
11. Helm, V., Ercan, S., Gramazio, F., Kohler, M.: Mobile Robotic Fabrication on Construction Sites. In: IEEE/RSJ International Conference on Intelligent Robots and Systems, Vilamoura, pp. 4335–4341 (2012)
12. Gramazio, F., Kohler, M.: Digital Materiality in Architecture: Bridging the Realms of the Virtual and Physical. In: Geiser, R. (ed.) Explorations in Architecture, pp. 179–199. Birkhäuser, Basel (2008)
13. Andres, J., Bock, T., Gebhart, F.: First results of the development of the masonry robot system ROCCO. In: Proceedings of the 11th International Symposium on Automation and Robotics in Construction, Brighton, pp. 87–93 (1994)

14. Dalacker, M.: Schriftenreihe Planung, Technologie, Management und Automatisierung im Bauwesen. In: Bock, T. (ed.) Entwurf und Erprobung Eines Mobilen Roboters zur Automatisierten Erstellung von Mauerwerk auf der Baustelle, Band 1, Stuttgart (1997)
15. Bonwetsch, T., Gramazio, F., Kohler, M.: R-O-B: Towards a Bespoke Building Process. In: Sheil, B. (ed.) Manufacturing the Bespoke: Making and Prototyping Architecture, pp. 80–83. John Wiley & Sons, London (2012)
16. Schodek, D., Bechthold, M., Griggs, K., Kao, K., Steinberg, M.: Digital Design and Manufacturing: CAD/CAM Applications in Architecture. John Wiley & Sons, New York (2004)
17. Ficca, J.: Inclusion of Performative Surfaces material and fabrication research. In: Iwamoto, L. (ed.) Digital Fabrications: Architectural and Material Techniques. Princeton Architectural Press, New York (2009)
18. Pigram, D., McGee, W.: Formation embedded design: A methodology for the integration of fabrication constraints into architectural design. In: Proceedings of the 31st Annual Conference of the ACADIA, Calgary, pp. 122–131 (2011)
19. Menges, A.: Integrative Design Computation: Integrating Material Behaviour and Robotic Manufacturing Processes in Computational Design for Performative Wood Constructions. In: Proceedings of the 31th Conference of the Association For Computer Aided Design In Architecture (ACADIA), Banff, Canada, pp. 72–81 (2011)
20. Dunn, N.: Digital Fabrication in Architecture, pp. 26–39. Lawrence King Publishers, London (2012)
21. Von Both, P.: Industrialisierung versus Individualisierung: Neue Methoden und Technologien. In: Hofmeister, S., Hellstern, C. (eds.) Positionen zur Zukunft des Bauens. Edition DETAIL/Institut für int, pp. 20–23. Architektur-Dokumentation, Munich (2011)
22. Obayashi, S.: Current Status of Automation and Robotics in Construction in Japan. In: 9th International Symposium, Automation and Robotics in Construction, Tokyo (1992)
23. Willmann, J., Kohler, M., Gramazio, F.: Die Operationalität von Daten und Material im Digitalen Zeitalter. In: Hofmeister, S., Hellstern, C. (eds.) Positionen zur Zukunft des Bauens. Edition DETAIL/Institut für int, pp. 6–19. Architektur-Dokumentation, Munich (2011)
24. Geiser, R. (ed.): Explorations in Architecture, pp. 210–216. Birkhäuser, Basel (2008)
25. Soar, R.: Additive Manufacturing technologies for the Construction Industry. In: Hopkins, N., Hague, R., Dickens, P. (eds.) Rapid Manufacturing: An Industrial Revolution for the Digital Age, pp. 249–273. John Wiley & Sons, London (2006)
26. Bärtschi, R., Knauss, M., Bonwetsch, T., Gramazio, F., Kohler, M.: The wiggled Brick Bond. In: Ceccato, C., Hesselgren, L., Pauly, M., Pottmann, H., Wallner, J. (eds.) Advances in Architectural Geometry, pp. 137–147. Springer, Wien (2010)
27. Gramazio, F., Kohler, M.: Die Digitale Materialität in der Architektur. In: Archplus, Issue 198/199, 43 Haus der Zukunft, Berlin, pp. 42–43 (2010)
28. Gramazio, F., Kohler, M.: Digital Materiality. In: Grobman, Y.J., Neuman, E. (eds.) Performalism: Form and Performance in Digital Architecture, pp. 160–169. Routledge, New York (2011)
29. Menges, A.: Material Computation: Higher Integration in Morphogenetic Design. In: Menges, A. (Guest ed.) Material Computation, Architectural Design (AD), vol. 82, pp. 14–21. John Wiley & Sons, London (2012)

30. Helm, V., Ercan, S., Gramazio, F., Kohler, M.: Mobile Robotic Fabrication on Construction Sites. In: IEEE/RSJ International Conference on Intelligent Robots and Systems, Vilamoura, pp. 4335–4341 (2012)
31. Lupashin, S., Schoellig, A., Sherback, M., Dandrea, R.: A simple learning strategy for high-speed quadrocopter multi-flips. In: IEEE International Conference on Robotics and Automation, Anchorage, pp. 1642–1648 (2010)
32. Willmann, J., Augugliaro, F., Cadalbert, T., Dandrea, R., Gramazio, F., Kohler, M.: Aerial Robotic Construction: Towards a New Field of Architectural Research. International Journal of Architectural Computing 10(3), 439–460 (2012)
33. Kohler, M.: Aerial Architecture. In: Log 25, Reclaim Resi[lience]stance, New York, pp. 23–30 (2012)
34. Helm, V., Ercan, S., Gramazio, F., Kohler, M.: Mobile Robotic Fabrication on Construction Sites. In: IEEE/RSJ International Conference on Intelligent Robots and Systems, Vilamoura, pp. 4335–4341 (2012)
35. Sheil, B., Glynn, R. (eds.): Fabricate: Making Digital Architecture. Riverside Architectural Press, Cambridge (2011)

TRAFCON – Traffic Control of AGVs in Automatic Warehouses

Cristian Secchi[1], Roberto Olmi[2], Cesare Fantuzzi[1], and Marco Casarini[2]

[1] DISMI - University of Modena and Reggio Emilia
via G. Amendola 2, 42122 Reggio Emilia, Italy
{cristian.secchi,cesare.fantuzzi}@unimore.it
[2] Elettric80 s.p.a., via G. Marconi 23, 42030 Viano, Italy
{olmi.r,casarini.m}@elettric80.it

Abstract. In this chapter we illustrate the main results of the TRAFCON experiment. We consider a real AGV based automatic warehouse system and we model the traffic control problem exploiting coordination diagrams and taking into account all the constraints holding in real plants. We propose a novel traffic manager that, besides efficiently controlling the coordinated motion of the AGVs, can dynamically change the paths the robots are following. The TRAFCON traffic manager is experimentally validated on simulated real plants and on a small-scale automatic warehouse.

Keywords: AGV, traffic management, coordination, automatic warehouse, dynamic routing.

1 Introduction

Because of their flexibility and efficiency, many Autonomous Guided Vehicles (AGVs) are used in production lines and today they are more and more exploited also at the end of the production line, namely in automatic warehouses where huge quantities of goods are continuously moved. Companies producing convenience goods, like those in the food field or in the paper field, have greatly invested and are still investing in the automation of the logistics. The possibility of covering 24 hours of work using a single AGV increased the market of AGV systems in those countries where the labor cost is high and where, therefore, the return of the investment is short.

Nowadays, automatic warehouses need to guarantee a higher and higher delivery rate (e.g. for loading trucks timely) and, in order to avoid congestion of goods in entrance and/or in exit, they need to never stop during working hours. For these reasons, the number of AGVs that need to be used is growing more and more and their motion need to be controlled in such a way that each AGV reaches its destination as quickly as possible and that the system never stops, even in case a fault on a vehicle or other unexpected events take place. Thus, the problem of coordinating the vehicles in an efficient way is getting more and more relevant. The problem of traffic control is recognized as one of the main issues for

F. Röhrbein et al. (eds.), *Gearing Up and Accelerating Cross-Fertilization between Academic and Industrial Robotics Research in Europe*, Springer Tracts in Advanced Robotics 94,
DOI: 10.1007/978-3-319-03838-4_5,

the development of an AGV system both by the industrial and by the scientific communities (see e.g. [1]). If this problem is not properly treated, congestion, deadlocks or even collisions can take place. These situations can block part of the system and they can require to stop the AGVs to allow the intervention of qualified personnel for a manual restart. The traffic control problem is further complicated by the presence of unpredictable events that can take place in automatic warehouses. For example, a pallet may fall during the transportation or an AGV can suddenly stop because of a fault. These events produce some unplanned static obstacles along the routes tracked by the vehicles. Furthermore, more and more "mixed" (i.e. automatic and manned) warehouse systems are present in the market. Because of economical or logistic issues, only a portion of a warehouse can be made automatic and part of the goods is still managed by human guided forklifts. In these situations, the AGVs are working in an environment populated by moving obstacles (the forklifts) they cannot absolutely collide with because of safety reasons. The human factor can be considered as an unpredictable disturbance to the motion of the AGVs. In summary, an efficient traffic management strategy has to consider both collisions avoidance issues and the optimization of the delivery time in an environment characterized by a strong, unpredictable variability. Nowadays, the approach adopted by most of the AGV systems providers for handling traffic is manual. A group of specialized personnel (mostly engineers), whose size depends on the size and on the complexity of the warehouse to be automated, is sent on site for manually designing an optimal set of traffic rules that is layout dependent and on the delivery pattern of the company. Furthermore, a set of exceptional rules, taking into account possible faults and/or unexpected obstacles, are developed and their efficiency is highly dependent on the experience of the personnel sent on-site. It can be easily guessed that, for AGV systems manufacturers, the design of a traffic control algorithm that is collision-free requires a lot of personnel time and it needs to be heavily re-adapted when installing the AGV system in another warehouse. Furthermore, unpredicted obstacles and/or faults often require to stop the system for manual recovery. Besides drastically reducing the performance of the system, manual interventions/restarts have a negative impact on the customer perception of the AGV system.

In the TRAFCON experiment the Elettric80, a global provider of AGV systems for automatic warehouses, and the ARScontrol group of the University of Modena and Reggio Emilia, an academic institution, have cooperated for developing an efficient, fault-tolerant traffic control strategy that can be successfully applied in AGV systems for automatic warehouses.

The main objective of the experiment is to develop the traffic management for the AGVs that is automatic and efficient and that satisfies all the constraints imposed by the architecture of an automatic warehouse. No tuning depending on the layout of the warehouse needs to be done and good performance have to be guaranteed also in presence of faults and mobile obstacles as human guided forklifts. This will drastically reduce the personnel time required for each installation and the number of required stops of the system leading to better performance, to a significant reduction of installation costs and to an increase of customers satisfaction.

As a common practice adopted by the AGV system providers, each time that a path is assigned to a vehicle (e.g. for reaching a delivery point), the AGV has to track it until the destination is reached and there is no possibility to replan the path. Thus, the traffic control problem is actually a coordination (e.g. tracking velocity control) problem.

The standard approach adopted by AGV systems provider to coordinate the motion of the vehicles is based on a set of traffic rules manually defined during the installation of the system (see e.g. [2]). This requires a lot of personnel work when an AGV system has to be deployed or modified since several exceptions have to be manually handled both for production and safety reasons.

In the literature many strategies for coordinating a group of mobile robots can be found. A Mixed Integer Non-Linear Programming problem is formulated in [3] in order to obtain the optimal velocity profile for each robot. Except from their very high complexity, this kind of approaches are non-robust to any contingencies arising during the system operation. They require to discard the plans every time the coordination problem must be updated (e.g., a new mission is assigned to a robot). Traffic problems have been also treated using distributed negotiation protocols (see e.g. [4]). This approach is successful if the environment is populated by cooperative robots that aim at solving traffic jams. Unfortunately, this is not the case in industrial AGV system. In fact a traffic problem can be caused by an unexpected obstacle or by a human guided forklift which are not cooperative entities. These situations can severely limit the efficiency of the system. It is also possible to model an AGV system as a Discrete Event System (DES) and the traffic control unit as a controller for this DES. Two relevant examples are [5] and [6]. The main drawback of DES-based approaches is that the main focus is the development of a collision and deadlock free routing for the AGVs. Performance of the fleet is not often taken into account and it can be rather low in some situations. Furthermore, it is not clear how to consider unexpected events that could block some vehicles.

The algorithm that we propose is based on the *coordination diagram* (CD) [7]. This tool allows to map a coordination problem into a planning problem. Previous works which apply the CD for coordinating the motion of mobile robots are [8], [9], [10] and [11]. All these approaches however suffer from a very high computational complexity so that they can not be applied on a real AGV system composed by many vehicles. Moreover these approaches are complete and, therefore, the coordination must be computed before the robots start moving and, if a vehicle deviates from the planned motion profile (e.g., due to an unexpected obstacle), the computation must be repeated from scratch.

The perfomance of a fleet of AGVs traveling in a shared environment can be defined as the total time necessary for delivering all the goods each AGV is transporting. It can be improved by making the system more flexible through a dynamic replanning of the paths the vehicles have to follow.

Multi-robot planning problems have been deeply investigated by the robotics community. Centralized planners that compute the best paths in the composite configuration space can produce an optimal solution [3,12] but they are computationally demanding and not suitable for AGV systems where up to 50 vehicles need to be handled. Furthermore, each time a path needs to be replanned (e.g. deviations, an AGVs is stuck for an emergency), the optimization needs to be executed again. Priority based approaches [13,14] are computationally efficient but require a rule for defining priority. Usually all the AGVs have the same priority. Even if a priority rule is forced, when a new mission is assigned to a vehicle or when a vehicle stops for safety reasons, it is not clear which priority the vehicle should assume. Finally several efficient approaches tackle the problem of planning by deforming paths assigned to the vehicles [15,16] but this approach is not implementable in AGV systems where the vehicles are constrained to travel along a fixed roadmap.

The TRAFCON traffic manager is based on novel coordination strategy coupled with a dynamic routing algorithm. The coordination strategy is based on an extended version of the coordination diagram tool proposed in [7] for multi-robot systems. A coordination diagram compliant with the constraints of an automatic warehouse is defined and a computationally efficient incremental coordination law is developed. The dynamic routing module evaluates the possibility of deviating from the nominal path for the AGVs on the basis of a risk factor that takes into account the likelihood of choosing a deviation that would take more time to be tracked than the original path because of coordination reasons. The traffic control strategy is tested and characterized both through simulations and in the arena, a small scale automatic warehouse built during the experiment, where all the problems and constraints arising in a real AGV system can be reproduced. Furthermore, a comparative analysis with the traffic control strategy currently used by the Elettric80 is done.

The outline of this chapter is the following: in Section 2 the architecture of an automatic warehouse system and the requirements for the traffic manager are illustrated. In Section 3 the TRAFCON traffic manager is presented and in Section 4 the arena, the small scale automatic warehouse where the traffic manager will be evaluated is shown. In Section 5 the experimental results are provided and, finally, in Section 6 some conclusions are drawn.

2 System Architecture and Requirements

2.1 Roadmaps

In automatic warehouse AGVs and manually guided forklifts are traveling for picking stored goods and for delivering them to trucks for external delivery. A typical layout of the Elettric80 automatic warehouse (the roof has not been drawn for clarity reasons) and an AGV are represented in Fig. 2.

Elettric80 AGVs are endowed with a navigation laser scanner and with a safety laser scanner. Several reflective surfaces (called *reflectors*) are properly distributed in the warehouse and each AGV localizes itself by integrating the

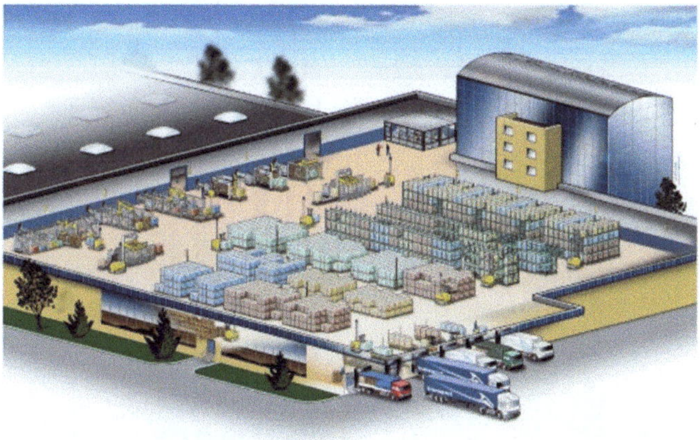

Fig. 1. The Elettric80 AGV system and an AGV

odometric data with the reflectors identified by the navigation laser scanner. Fake reflections (e.g. coming from pallets wraps) are properly filtered for avoiding inconsistent localizations. Laser triangulation is the localization system that is most used by AGV system producers [17]. Each AGV is endowed with a navigation controller that allows it to track desired setpoint trajectories. The safety laser scanner is set in front of each AGV and it stops the vehicle when it detects an obstacle at a certain tunable distance. For safety reasons, many AGV systems providers, as well as Elettric80, adopt the segmented roadmap paradigm, a particular version of the zoned controlled paradigm [1], for safely managing the motion of multiple AGVs in the warehouse. In each warehouse, a fixed (virtual) roadmap along which AGVs are constrained to move is designed. The roadmap is partitioned into segments (i.e. portions of curves) and each segment can be occupied only by one AGV. Furthermore, for safety reasons, AGVs cannot go backward in the warehouse.

2.2 Control Architecture

The AGV system control framework adopted by Elettric80 (but common to many AGV systems providers) is reported in Fig. 3. The higher layer, the plant level, is the interface between the ERP system of the company and the Elettric80 AGV system. Its main role is to receive a set of tasks and to generate a set of *missions* for the AGVs. A mission is nothing else than a set of points an AGV has to reach for picking or delivering some goods.

The Fleet level is run on a central unit and it is responsible of controlling the group of AGVs. It is run periodically and it is composed by two modules: the vehicle manager and the traffic manager system (TMS). The vehicle manager receives the mission queue and it checks the status of the fleet in order to decide which and how many tasks to execute and it sends this information to the TMS. The traffic manager assigns the missions to the AGVs and it computes the path

Fig. 2. The Elettric80 AGV system and an AGV

each AGV has to follow. Since paths are partitioned into segments it is necessary to compute how many segments to reserve (i.e. to allow to track) to each AGV for avoiding collisions and congestions. This is decided by the TMS considering the information about the position of the AGVs (i.e. the segment they are occupying) that is received by the vehicle manager module and by implementing some traffic management strategy. The lists of allocated segments per each AGV is sent to the vehicle manager that transmits them to the AGVs. Finally the AGV level is the low level control of each vehicle for tracking the segments allocated by the traffic manager and for handling safety issues.

In TRAFCON, the way missions are assigned to the AGVs has been left unchanged while the way the paths are computed and segments are allocated has been changed for improving the flexibility and the efficiency of the overall system.

2.3 Requirements

Currently, when a mission is assigned to an AGV, a path the vehicle has to follow is computed and the AGV is constrained to follow the path until the destination has been reached. The traffic of the AGVs is coordinated using a set of manually optimized traffic rules for allocating segments. Based on its experience on automatic warehouses, Elettric80 has established the main requirements the TRAFCON traffic manager has to satisfy for becoming the next generation traffic manager for AGV systems:

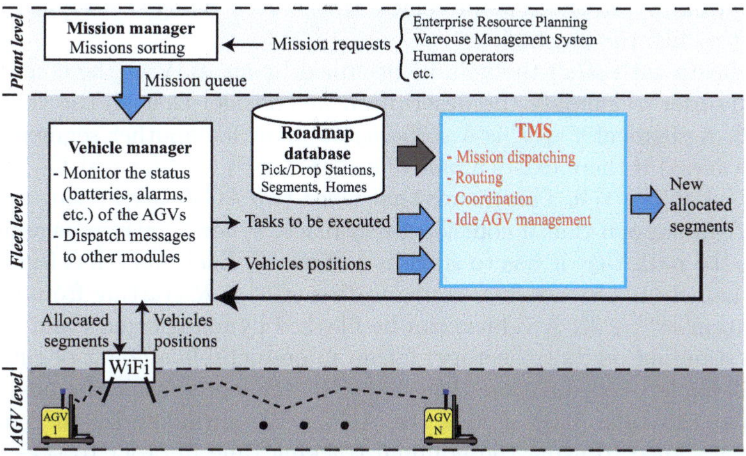

Fig. 3. The AGV system control framework

- **More Flexibility:** It has to be completely automatic and it has not to require any manual intervention (e.g. manual rules).
- **More Efficiency:** It has to improve the performance of the traffic manager currently adopted by Elettric80 both in terms of installation costs and in terms of missions per hour that can be executed.
- **Compatibility:** It has to be compliant with the Elettric80 architecture and with the safety constraints (e.g. segmented roadmaps).

3 The TRAFCON Traffic Manager

3.1 Mathematical Formalization of the Problem

We consider M AGVs that are moving in the same environment and that share the same configuration space $\mathcal{C}$ (e.g., $SE(2)$). In the application that we are considering, for a given plant to be served, a network of paths which the AGVs can follow is defined. We can model this network as a *roadmap* $\mathcal{R}$, that is a one-dimensional connected subset of $\mathcal{C}$. The roadmap is formed by a collection $\mathcal{T}$ of regular curves called *segments*. Each segment $\tau \in \mathcal{T}$ can be represented by a mapping $\tau : [0, l_\tau] \to \mathcal{R}$ (where, with a slight abuse of notation, the segments in $\mathcal{T}$ and the mappings used to represent them are indicated with the same symbol). A *path* p_i is defined as a sequence of n_i adjacent segments and it can be represented by the following mapping

$$p_i : k \to \mathcal{T}, \quad k = 1, 2, \ldots, n_i \tag{1}$$

$\mathcal{A}_i$ will denote the ith vehicle that is assigned to the path p_i. Each segment is characterized by a nominal velocity it needs to be tracked with. The velocity depends primarily on the type of vehicle that travels along the segment and on other factors like the payload.

We indicate with $\mathcal{A}(x)$ the volume occupied by an AGV at the configuration $x \in \mathcal{R}$ (in order to simplify the description we consider that all the vehicles are identical). A segment $\tau \in \mathcal{T}$ is a *colliding segment* for another segment $\tau' \in \mathcal{T}$ (and vice versa) if there exists a pair of scalars $(\alpha, \beta) \in [0, l_\tau] \times [0, l_{\tau'}]$ such that $\mathcal{A}(\tau(\alpha)) \cap \mathcal{A}(\tau'(\beta)) \neq \emptyset$. This means that when two AGVs are moving through τ and τ' it can happen that a collision takes place. When a mission is assigned to an AGV, the path that it has to track is computed. Each AGV has to execute a mission, namely to reach a final configuration $x_i^{\text{goal}} \in \mathcal{R}$ starting from its initial configuration $x_i^{\text{init}} \in \mathcal{R}$. A vehicle can be blocked by unexpected events (such as a person standing on its trajectory) for an unpredictable amount of time.

We will exploit coordination diagrams (CD) for representing the possible collisions that can take place among the AGVs. We parameterize the path each vehicle is assigned to using the nominal tracking time in order to take into account the minimum time necessary for tracking each segment. The real tracking time can be higher due to dynamic constraints and/or to unexpected events that could occur during the motion of the vehicle (e.g a vehicle stops to avoid a person on its way). For that reason the collisions between vehicles can be avoided only reserving at each vehicle a list of segments that are not colliding with other segments already reserved. Each path can be parameterized using the nominal tracking time as:

$$\pi_i : s_i \to \mathcal{R}, \quad s_i = \left[0, T^i\right] \tag{2}$$

where s_i is the nominal time necessary for tracking the segments composing the path. In particular $\pi_i(0)$ and $\pi_i(T^i)$ correspond to the initial and to the final configurations respectively. The Coordination Diagram is given by $\mathcal{S} = \left[0, T^1\right] \times \cdots \times \left[0, T^M\right]$. A point $s = (s_1, \ldots, s_M) \in \mathbb{R}^M$ in the CD represents a possible configuration of the robots along their paths.

Since each path can be split into a set of adjacent segments, the domain $\left[0, T^i\right]$ of its parameterization Eq.(2) can be partitioned into a sequence of adjacent intervals $\left(\Delta^i_{k_i}\right)$, with $k_i = 1, \ldots, n_i$. Each $\Delta^i_{k_i}$ corresponds to the time interval that the AGV $\mathcal{A}_i$ would take to travel along the segment k_i if traveling at its nominal velocity. When a new path is assigned to a vehicle, these intervals are computed according to the reference velocity by which each segment has to be covered by the vehicle. The path partition induced a partition of the CD into a set of *blocks* where a block is defined as $b_{k_1,\ldots,k_M} = \Delta^1_{k_1} \times \ldots \times \Delta^M_{k_M} \subseteq \mathcal{S}$.

A plane of the CD, denoted as $\mathcal{S}_{ij}$, is a coordination diagram as well and it represents the configurations of a pair of vehicles $(\mathcal{A}_i, \mathcal{A}_j)$. A block b_{k_i,k_j} identifies a rectangular region $\Delta^i_{k_i} \times \Delta^j_{k_j}$ of the plane $\mathcal{S}_{ij}$. Given two paths p_i and p_j, a block b_{k_i,k_j} corresponding to a pair of colliding segments $(p_i(k_i), p_j(k_j))$, is defined a *collision block*. An example of $\mathcal{S}_{ij}$ is reported in Fig. 4 in which the collision blocks are represented with black filled rectangles. The set of all collision blocks is called *collision region*. When a vehicle completes its mission, a new one is assigned to it. This implies that the CD must be modified in order to consider the possible collisions introduced by the new path. The a priori

Algorithm 1. Coordinator algorithm

function COORDINATOR(CD, $l_1, \ldots, l_M, X$)
  $\mathcal{F} \leftarrow$ UPDATE FORBIDDEN REGIONS(CD)
  while $X \neq \emptyset$ **do**
    $\mathbf{u}^* \leftarrow$ ACTION COMPUTATION($\mathcal{F}, X, l_1, \ldots, l_M$)
    for all $i \in X$ **do**
      $l_i' = l_i + u_i^*$
      if $u_i^* = 1$
        if $\exists\ k_j \in [c_j, \ldots, l_j]$ **:** b_{l_i',k_j} = collision block
          return l_i ▷ Return previous value
          $X = X \setminus i$
        end if
        if $l_i = n_i$ **or** $W_i(l_i) > \bar{W}$ ▷ Not starving
          return l_i'
          $X = X \setminus i$
        end if
      else ▷ If no segments are reserved
        return l_i'
        $X = X \setminus i$
      end if
    end for
  end while
end function

knowledge of the roadmap allows to simplify this process since the identification of the pairs of colliding segments (which requires the use of collision checking techniques) can be computed offline [18]. The definition of the CD allows to give a geometrical representation of the coordination problem. Through this representation all information about the possible collisions among the vehicles can be extracted in a way that is independent of the roadmap considered. A classification of the possible collisions that can take place between vehicles can be given by using the CD, [18]. In order to achieve a collision free coordination of the AGVs it is sufficient to find a sequence $(b_{k_1,\ldots,k_M})$ of blocks that joins the first block $b_{1,\ldots,1}$ (identified by the starting segments of the paths) to the block $b_{n_1,\ldots,n_M}$ (goal segments of the paths) and which does not contain any collision block.

3.2 Coordinator

In this section an incremental algorithm which determines the reserved segments (i.e., the segments that the AGV is allowed to track) for each vehicle in such a way that no collisions take place is presented.

The mapping defined in Eq.(1) specifies the sequence of segments that an AGV has to cover. In order to simplify the notation, the k-th segment of a path p_i, is indicated as $k_i \in \{1, \ldots, n_i\}$. Each vehicle $\mathcal{A}_i$ is associated to a list of adjacent segments, called *reserved segments*, that it is allowed to cover. The first and the last segments of the list are denoted with c_i and l_i respectively, where $i \in \{1, \ldots, n_i\}$, and they are the segment where the ith AGV is located on and the last segment that it can reach. The current segment c_i is automatically incremented each time that an AGV enters a new segment of its path. A vehicle covers all the reserved segments and stops at the end of the last. The list of reserved segments $[c_i, \ldots, l_i]$ can be updated by the coordinator with the addition of new segments (i.e. incrementing l_i). The coordinator is executed periodically and its period of execution is indicated with δ. The nominal time required by an AGV in order to reach the segment l_i is given by

$$W_i(l_i) = -\chi_i \|\Delta^i_{c_i}\| + \sum_{k_i=c_i+1}^{l_i} \|\Delta^i_{k_i}\| \tag{3}$$

where $\chi_i \in [0, 1]$ is the ratio between the distance covered by $\mathcal{A}_i$ on current segment and the total length of that segment and $\|\Delta^i_{k_i}\|$ indicates the length of $\Delta^i_{k_i}$. If the time $W_i(l_i)$ is less than a value δ the AGV $\mathcal{A}_i$ is defined *starving*. This condition indicates that a vehicle will reach the end of the segment l_i before the next execution of the algorithm and thus it will stop unless new segments are added to the reserved list. All the vehicles that are starving at the moment the coordinator is called are grouped into a set X.

The coordination algorithm is reported in Alg. 1. Each time, the algorithm updates the forbidden regions in the coordination diagram for taking into account variations due to new missions assigned to the AGVs. A loop (see line 3) where, for each starving vehicle, the last reserved segment l_i is updated is then started.

At each iteration of the loop the algorithm evaluates the last reserved segment l_i of each vehicle $\mathcal{A}_i$ (also those that are not in X) in order to choose how the reserved list of each vehicle has to be updated. This computation is done by Action Computation (Alg. 2) which returns an *action* $u_i \in \{0, 1\}$ for each AGV. If $u_i = 1$ a new segment is added to the reserved list (i.e. l_i is incremented by 1), see line 6. Otherwise, if $u_i = 0$ no segments are added and the $\mathcal{A}_i$ is removed from the set X. The block b_{l_i,l_j} identified by this set of segments is called *reference block*.

A vehicle $\mathcal{A}_i$ is removed from X (i.e., l_i is no more updated) if $W_i(l_i) \geq \delta$, which means that it is no more starving. An AGV that has to implement an emergency stop is automatically removed from X. In this way, the emergency handling is embedded online in the coordination controller. When the X becomes empty the loop terminates.

The algorithm Action Computation (see Alg. 2) executed at each iteration of the coordinator algorithm. For any pair of AGVs in which at least one of the two is in X, this function evaluates the position of the block b_{l_i,l_j} with respect to the collision regions of the CD $\mathcal{S}_{ij}$. The set X represents the AGVs for which the number of reserved segments has to be incremented. The function Action Computation returns a vector of actions $\mathbf{u} = (u_1, \ldots, u_N)$ (also called *coordinated action*) that will be used by the coordinator algorithm to update the

Algorithm 2. Action Computation

```
function ACTION COMPUTATION(F, X, l_1, ..., l_M)
    V = X,  E = ∅,  u_i = 0,  ∀i
    for all i ∈ X do
        for all j = 1, ..., M do
            c = FIND CONSTRAINTS(l_i, l_j, F)
            if c = "u_i + u_j ≤ 1"
                E = E ∪ (i, j)
            else if  c = "u_j = 0"
                V = V \ {j}
                u_j = 0
            else if c = "u_i = 0"
                V = V \ {i}
                u_i = 0
            else if c = "u_i,  uj = 0"
                V = V \ {i, j}
                u_i = 0,  u_j = 0
            end if
        end for
    end for
    S = MAXIMUM INDEPENDENT SET(V, E)
    return u_i = 1,  ∀i ∈ S
end function
```

reserved segments of the AGVs in X. We consider the best coordinated action $\mathbf{u}$ as the one which leads to the major advancement of the fleet.

For each collision region, the proposed algorithm builds a polygonal region, called *forbidden region* and indicated with $\mathcal{F}$, that encloses the collision blocks. The forbidden region is the polygon with light blue edges inFig. 4 The reference block b_{l_i,l_j} represents the configuration that the vehicles will reach, once that all reserved segments have been covered. When new segments have to be reserved, the position of this block is evaluated in order to avoid the possible forbidden regions. The blocks in which a plane $\mathcal{S}_{ij}$ is partitioned can be grouped according to their positions with respect to the forbidden region $\mathcal{F}$. At each iteration the algorithm evaluates the group the reference block b_{l_i,l_j} belongs to.

At each group of blocks, (that we call a *region*), a constraint that restricts the set of actions that can be chosen by the algorithm is associated. These constraints ensure that the two vehicles will reach their goals without colliding. Given a forbidden region (see e.g., Fig. 4) the groups of blocks and the associated constraints are:

- *Antumbra blocks:* As long as the reference block b_{l_i,l_j} is in this group, all the actions remain valid (i.e. $u_i,\ u_j \in \{0,1\}$).
- *Penumbra blocks:* To escape from this region the following constraint between the actions has to be satisfied:

$$\begin{aligned} u_i &= 0 \ \text{ if } \ b_{l_i,l_j} \ \text{in Penumbra left} \\ u_j &= 0 \ \text{ if } \ b_{l_i,l_j} \ \text{in Penumbra right} \end{aligned} \tag{4}$$

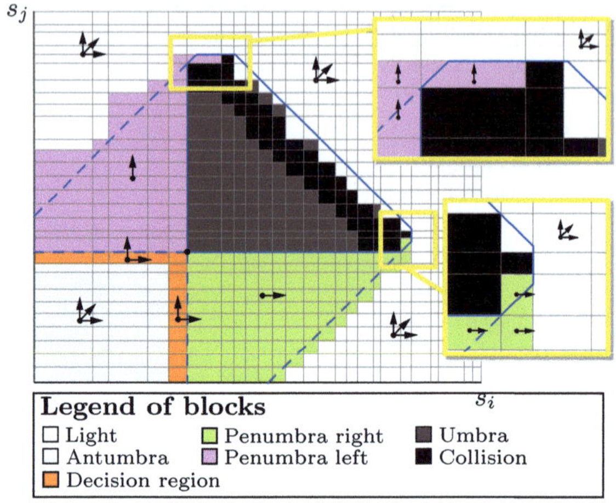

Fig. 4. The colors denote the block grouping. For each group the actions that are allowed are represented with black arrows.

- *Decision blocks:* When the reference block is in this region it has to be chosen which AGV has to stop in order to avoid the collision region. The following constraint is imposed:
$$u_i + u_j \leq 1 \tag{5}$$
- *Umbra blocks:* When b_{l_i,l_j} is in this region, no action are allowed to both the vehicles. Thanks to the constrains imposed the reference block b_{l_i,l_j} will never be in umbra group. However, a bad initial positioning of the fleet at start-up or a system failure could lead to this situation. In this case both the vehicles are stopped an error message is issued.
- *Light blocks:* All other blocks. There are no constrained actions.

On each plane $\mathcal{S}_{ij}$ in which a forbidden region is defined, the position of the block $b_{l_i^i,l_j^j}$ is evaluated. This can be easily done in constant time by geometric considerations on the coordination diagram. The problem of finding a coordinated action $\mathbf{u}^*$ which maximizes the advancement of the vehicles in X can be expressed as a Binary Integer Program (BIP) in which each constraint is either unary (Eq.(4)) or binary (Eq.(5)). The unary constraints are trivially satisfied since they simply require to put the related variable to 0 (see lines 10, 13 and 16 in Alg. 2). The remaining problem can be stated as follows.

$$\begin{aligned} &\text{Max } U = \sum u_i \\ &\text{Subject to } \begin{cases} u_i + u_j \leq 1 & \forall (i,j) \in E \\ u_i \in \{0,1\} & \forall i \in V \end{cases} \end{aligned} \tag{6}$$

Where V is the set of actions u_i that are not constrained by unary constraints and E are the pairs of actions that are constrained by binary constraints. This particular BIP can be represented as a undirected graph $G = (V, E)$ where V is the set of vertices and the E is the set of edges. This fact allows to transform this optimization problem into a *Maximum Independent Set* (MIS) problem. Given an undirected graph, the MIS problem consist of finding a maximum-cardinality subset S of vertices such that no two vertices in S have an edge between them. By solving the MIS for the graph $G = (V, E)$, the solution S indicates which variables have to be set to 1 in order to obtain a solution for the problem in Eq.(6) (see [19]). In other words, the set S corresponds to the maximum set of AGVs that can advance without violating any of the constraints.

In order to find the set S, the heuristic procedure developed in [19] is applied (see line 20 in Alg. 2). This algorithm incrementally removes the vertices which are not within S. At each iteration a vertex is selected and removed from the graph along with all its incident edges. The algorithm is based on two characteristics of the vertex: the *degree*, which is the number of neighbors of the vertex, and the *support*, which is the sum of the degree of the neighbors of the vertex. Among the vertices which maximize the support, the one with the minimum degree is selected. The process terminates when all edges of the graph have been removed. The set of remaining vertices is S. Note that, for our application, it is likely that there could be tie situations where more than one vertex exists with the same maximum support and minimum degree. In this case, one of these vertices can be indifferently selected. In summary, the algorithm COORDINATOR defined in this section (see Alg. 1) updates the lists of the segments that each vehicle is allowed to cover. The coordinated action $\mathbf{u}$, with respect to which the lists are updated, is computed by the algorithm ACTION COMPUTATION (see Alg. 2). This algorithm finds the coordinated action which maximizes the advancement of the fleet subject to a set of constraints.

As detailed in [20], the computational complexity of the coordination algorithm is polynomial. If M is the number of AGVs and n is the maximum number of segment a path can be composed of, then the complexity is given by $O(M^2n^3 + M^2n + M^3)$. In some plants, backward motion is allowed. An efficient coordination strategy for this case is described in [21,22].

3.3 Dynamic Routing

The coordination strategy presented in Section 3.2 allows to efficiently and automatically coordinate a group of AGVs traveling on preassigned paths. Nevertheless, unexpected stops due to manual forklifts, humans or other obstacles are common. After an unexpected stop, the time parameterization of the path of a vehicle (and, consequently the coordination diagram of the fleet) changes. Thus, the paths chosen may not be a good choice anymore and a re-planning would be helpful for avoiding traffic jams. The best choice would be to re-plan all the paths of the AGVs Unfortunately, in real plants where tens of AGVs are traveling, this would imply to stop the AGVs for the time needed to compute the new paths each time an unexpected event takes place. This is not acceptable for real implementations.

Thus, a re-planning algorithm, that is sub-optimal with respect to a global replanning but that can be implemented without stopping the fleet, has been developed. Each time that an AGV approaches to a crossroad or that it has the chance to change its path, it computes k alternative paths. Each alternative is a deviation and it has to conflue to the original one by T_{dev} seconds, where T_{dev} is a design parameter. This can be done very quickly exploiting the data structure available in the AGV system and the number of alternatives to be computed depends on the size and on the complexity of the roadmap. The choice of deviations rather that completely new paths is linked to the fact that each AGV computes the alternatives disregarding the presence of the others for keeping the computational burden low. If completely new paths may be chosen, then the traffic may significantly change and a good deviation for a vehicle may mean a catastrophic traffic jam for the rest of the AGVs. Deviations are "local" and their effect on the traffic is much more limited. The confluence time T_{dev} tunes the locality of the deviation.

Deviations need to be evaluated carefully. In fact, it is necessary to consider both their length and the traffic that can be encountered when tracking them. Paths are composed by segments and each AGV knows a *time window* during which it will be tracking a given segment. If a vehicle has to stop for coordination reasons, the time windows change and they have to be re-planned. Let ${}^jTW_m = [{}^jt^1_m, {}^jt^2_m]$ be the time window vehicle j plans to be on segment m.

Consider two AGVs i and j tracking paths p_i and p_j respectively and suppose that $p_i \cap p_j \neq \emptyset$ starting from a given instant t_d. Furthermore, let m_1 be the first segment that i needs to track after t_d and that is such that $p_j \cap m_1 = m_2 \neq \emptyset$[1].

We define $\Xi_i^j(t_d) \in \mathbb{R}^+$ as

$$\Xi_i^j(t_d) = |{}^jt^1_{m_2} - {}^it^2_{m_1}| \tag{7}$$

The quantity $\Xi_i^j(t_d)$ represents the closeness of a traffic situation involving AGVs i and j considering the planning of the robots at time t_d.

Suppose that at time t_c the AGV i, tracking the path p_i, approaches a point in the roadmap where it has the chance to change its path. It then computes D deviations $d_1 \dots d_D$. Let $\Gamma_{d_j}(t_c)$ be the set of paths assigned to other AGVs and intersecting with d_j. We associate to each deviation d_j the following *risk factor*:

$$\rho_j(t_c) = \begin{cases} \max_{p_k \in \Gamma_{d_j}(t_c)} \frac{t_j^{p_k} - t_c}{\Xi_j^{p_k}(t_c)} & \text{if} \quad \Gamma_{d_j}(t_c) \neq \emptyset \\ 0 & \text{otherwise} \end{cases} \tag{8}$$

where $t_j^{p_k}$ is the time at which deviation j plans to start tracking the first segment that is intersecting with p_k. The index $\rho_j(t_c)$ measures the risk of taking deviation d_j at time t_c, namely the possibility of having to stop for coordinating with other vehicles. The highest is the risk, the lower may be the benefit of taking the deviation.

If no other vehicle is intersecting d_j, the deviation is risk free otherwise the $\rho_j(t_c)$ depends on how close in time and AGV tracking d_j is contending a segment with another vehicle ($\Delta_j^{p_k}(t_c)$) and on the time it starts to track the common segment. The bigger is $\Delta_j^{p_k}(t_c)$, the less probable that there will be a need for

[1] It is possible that $m_2 = m_1$ in case a portion of the two paths coincide

coordination and, therefore, the lower the risk is. On the other hand, since a lot of unexpected events can take place in the warehouse, the risk grows with $t_j^{p_k} - t_c$. In fact, even if $\Delta_j^{p_k}(t_c)$ is small, the bigger $t_j^{p_k} - t_c$ becomes, the most likely the time planning of the segments can change in an unpredictable way (because of unexpected events) and lead to stops due to coordination.

Let $\bar{\rho} > 0$ be a *risk threshold*, a design parameter that depends on the particular warehouse and that represents a tolerated risk for taking a given deviation.

The dynamic routing algorithm works as follows.

If no deviations are possible at time t_c, AGV i continues to track its originally assigned path. Otherwise, for each deviation $d_j, j \in 1, \ldots, D$ the risk factor $\rho_j(t_c)$ is computed.

If some deviations are possible, the system looks ahead of T_{dev} seconds. in the path the AGV is tracking and it checks if some coordinating actions due to traffic are necessary (this can be done running for T_{dev} seconds the coordination algorithm reported in Section 3.2). If it is the case and if deviations free of traffic are available, then the AGV takes the shortest deviation d_k such that $\rho_j(t_c) < \bar{\rho}$.

If all the deviations involve some coordination or if their risk factors are greater than $\bar{\rho}$, the risk factor $\rho_i(t_c)$ of p_i is evaluated. If $\rho_i(t_c) < \bar{\rho}$ it means that it is safer to stay on p_i and the AGV doesn't change its path. Otherwise, the deviation d_j with the lowest risk factor is taken since it is probably more efficient to track the longer but free path rather than facing the traffic on the original path. In all other cases the AGV i stays on its original path.

A timeout parameter T_{block} is introduced for taking into account the situation in which a vehicle is stuck because of an obstacle on its path. If an AGV is stopped for a time greater than T_{block} then $\rho_i(t_c) = \infty$. This forces the AGV to take a deviation, if it exists.

4 The Arena

In order to evaluate the effectiveness of the TRAFCON traffic manager on a real industrial environment, a small scale automatic warehouse, called *arena*, has been set up in a facility of Elettric80.

The arena is an area of 400 m^2 and it is big enough to allow the circulation of up to three AGVs and it is small enough for avoiding dispersion of the vehicles. According to the on the field experience of Elettric80, the arena is appropriate for reproducing a realistic small scale warehouse where significant real-world traffic situations can be implemented. Modular shelves and racks have been used for building several physical layouts for the arena.

The localization system used for tracking the position of the AGVs is the same as that used in real warehouses. A set of reflective landmarks have been installed in the arena and their position has been precisely measured by specialized personnel. A laser scanner is mounted on top of each AGV. At each scan, the sensor detects a certain number of landmarks whose position is perfectly known and, using a triangulation strategy, the position of the AGV is recovered with a sufficient precision for navigation purposes.

Fig. 5. The arena is big enough for allowing the up to three AGVs circulating. Movable shelves allow to reproduce different layouts.

The IT infrastructure has been installed and the arena has been endowed of all the feature of Elettric80 automatic warehouse system. In particular a mission manager, that receives tasks to be executed by a management program or by the user, dispatches missions to the AGVs, Elettric80 CAD system can be used to design segmented roadmaps the AGVs can travel along in the arena.

In Fig. 5 a picture of the arena with two of the AGVs that have been used during the experiments is shown. More pictures are shown in the TRAFCON video available on the ECHORD youtube channel.

5 Simulations and Experiments

The performance of the TRAFCON traffic manager has been evaluated using both simulations and experiments. Simulations allowed to compare the performance of the TRAFCON traffic manager with the ones obtained by Elettric80 using a set of manually optimized rules on a big size warehouse of a customer of Elettric80. Experiments allowed to validate the effectiveness of the TRAFCON traffic manager on a real warehouse, the arena, where unexpected events can take place.

5.1 Simulations

Currently, in Elettric80 warehouses AGVs are not allowed to dynamically change the path they have been assigned and the optimized traffic rules produce a coordination among the AGVs. Thus, in order to evaluate the performance of the coordination strategy introduced in TRAFCON, we disabled the dynamic routing and we compared the results with those that can be obtained by the Elettric80 coordination. The index used for evaluating the efficiency of the coordination is the time necessary for transporting the goods.

The simulation framework emulates the real control framework currently adopted by Elettric80. Three processes are executed on the same PC (an Intel Pentium D 3.20 GHz): the *mission manager emulator* (MME), which generates

Table 1. Computation times the proposed TMS [ms]

	Number of vehicles							
	10		15		20		25	
	Max.	Avg.	Max.	Avg.	Max.	Avg.	Max.	Avg.
Update $\mathcal{F}$	93	33	201	58	336	131	629	95
Seg. Res.	213	16	266	27	547	56	908	67

the transportation requests according to a statistical model of the material flow within the considered plant; the *vehicle manager emulator* (VME) which interfaces the Coordination algorithm with the MME and the AGVs. The state of the AGVs is internally emulated by this process; the Traffic Manager System (TMS), where the coordination algorithm is implemented, which communicates with the VME through a UDP protocol. Our TMS is implemented in MATLAB. For testing the computational requirements of the coordinator algorithm (see Section 3.2), the vehicles are supposed to move on a roadmap used in a real industrial plant which is composed both of curvilinear and straight line segments as shown in Fig. 6. The size of the physical layout is 220m x 150m. The map has 988 pick/drop stations, 4515 trajectory segments. The maximum driving speed of AGVs is 1.5 m/second. The average number of segments constituting a path is 25. Four simulations running 10, 15, 20 and 25 AGVs have been executed. The average and maximum time required to define the forbidden regions $\mathcal{F}$ and to update the lists of reserved segments (time to execute the while loop in Alg. 1) are reported in Tab. 1. The AGVs send their positions and status to the TMS every $\delta = 0.5$ seconds. In some cases the coordinator takes a longer time run. The messages that are received when the TMS is busy executing the coordinator algorithm, are discarded. However, when a vehicle is allowed to advance, the list of reserved segments is long enough to let it to advance for 5 seconds. Thus the loss of messages from the AGVs does not affect the motion of the vehicles. Thus, the TRAFCON coordinator is computationally efficient enough to be executed on a real, big size, plant. In order to evaluate the performances with which the transportation tasks are executed is based on two standard measures denoted as *time to pick* (TTP) and *time to drop* (TTD). The TTP is the time a load has to wait at the source station before it is picked by an AGV. The TTD is the time required by an AGV to transport the load to the destination. These tests are executed running a simulation on a smaller plant (size: 260m x 45m, 179 pick/drop stations and 1191 trajectory segments) with 7 AGVs. The engineers of Elettric80 have decided to test the TMS on this layout since it is more affected by traffic congestions. The ratio of loads generation is 140 per hour. The duration of the simulations is about 70 minutes. The proposed algorithm is compared to the one proposed by Elettric80 considering the results from three simulation runs. The average and maximum values of the TTP and TTD are reported in Tab. 2. Our TMS allows to obtain TTP values that are lower than the ones obtained by Elettric80. Considering the TTD values there is not a clear advantage of one algorithm over the other. However, we should consider that the original requires up to 15 days of engineer work to be fine tuned. Differently, the proposed algorithm is able to coordinate vehicles on different roadmaps without the need of any additional engineering work. More simulations, on simpler plants, can be found in [23].

Table 2. Performances comparison. The high variability of traffic conditions explains the big difference between average values and maximum values.

Proposed TMS				
Sim. No.	TTP [s]		TTD [s]	
	Max.	Avg.	Max.	Avg.
1	251	77	191	86
2	242	81	245	89
3	152	69	200	85
Elettric80 TMS				
Sim. No.	TTP [s]		TTD [s]	
	Max.	Avg.	Max.	Avg.
1	241	90	253	87
2	223	78	159	83
3	256	93	176	85

5.2 Experiments

In order to evaluate the effectiveness of the dynamic routing, the roadmap reported in Fig. 7 has been designed and implemented in the arena where three AGVs are circulating.

A total of 20 missions have been assigned to each vehicle. The total time for completing all the missions has been chosen as an efficiency indicator. The vehicles are coordinated using the strategy proposed in Section 3.2 and the routing strategy proposed in Section 3.3 has been implemented. During the experiment, it has been chosen $\bar{\rho} = 10$, $T_{dev} = 10$ seconds, $T_{block} = 25$ seconds.

Three different scenarios have been considered. First, no unexpected events take place in the plant and all the stops are due to coordination. In the second and third scenarios, three and five unexpected events lasting a random time have been simulated. In these cases some AGVs are blocked for external causes due to external events. The results are shown in Fig. 7 where the time to complete all the missions is on the y-axis. It can be seen that the implementation of the routing strategy allows always to decrease the overall completion time. Furthermore, the advantage due to routing increases when some unexpected events take place.

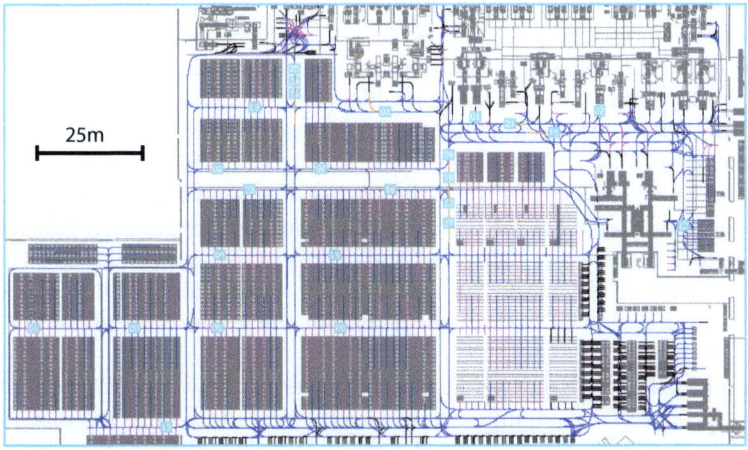

Fig. 6. Layout of the roadmap used for the test with 25 AGVs. This roadmap is used on a real plant.

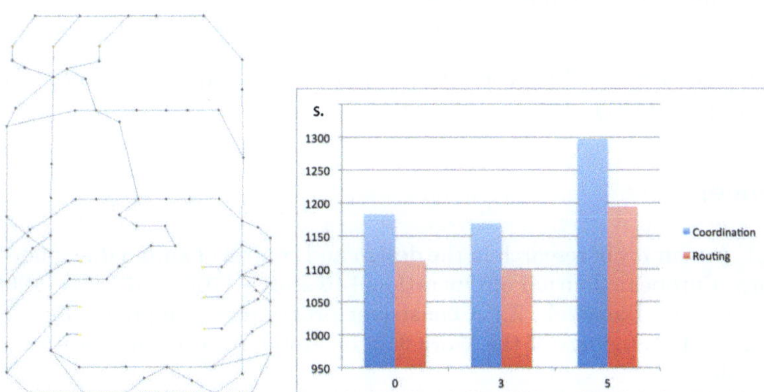

Fig. 7. The roadmap used for testing the routing strategy and the results of the routing simulations. The completion time using only coordination (blue) and using coordination together with dynamic routing (red).

This is due to the fact that, using only coordination, a vehicle blocked by an unexpected event can block other following vehicles as well. Using the routing strategy, a blocked vehicle chooses a deviation after a while and this prevents from creating traffic jams.

6 Conclusions

The TRAFCON experiment has seen a very fruitful cooperation between the ARSControl group and Elettric80. After an initial phase where the Elettric80 personnel and the ARSControl staff have "tuned" their languages and their approaches to research, the problem has been clearly defined. This has been one of the most important phases of the experiment, where the industrial and the research partners defined together the problem to solve. Once the problem has been defined, the ARSControl team could mathematically formalize it and exploit advanced techniques as the coordination diagram and time windows to solve the problem, always keeping an eye to the real problem and its constraints, thanks to the continuous feedback of the industrial partner. In parallel, Elettric80 has taken care of the necessary industrial validation phase and it has built the arena. The TRAFCON traffic manager has been experimentally validated and it has shown good performance both in terms of efficiency and in terms of installation time and cost.

The impact of the TRAFCON traffic manager in the world of automatic warehouses is expected to be significant. In fact it allows to reduce installation costs, to eliminate the error prone manual tuning of traffic rules and to increase the efficiency of the system. The improved quality of service and the cost reduction of AGV based logistics will pave the way to a greater and faster diffusion of automatic warehouses and will leverage more and more humans from tedious and dangerous tasks as pallets handling.

Acknowledgement. This work was supported by the EU Project EC FP7-ICT-231143 ECHORD.

References

1. Vis, I.F.A.: Survey of research in the design and control of automated guided vehicle systems. European Journal of Operational Research 170(3), 677–709 (2006)
2. Borowiecki, T., Banaszak, Z.: A constraint programming approach for AGVs flow control. In: Proceedings of the Workshop on Robot Motion and Control, pp. 153–158 (1999)
3. Peng, J., Akella, S.: Coordinating multiple double integrator robots on a roadmap: Convexity and global optimality. In: Proceedings of the IEEE International Conference on Robotics and Automation, pp. 2751–2758 (April 2005)
4. Herrero-Perez, D., Martinez-Barbera, H.: Decentralized coordination of autonomous AGVs in flexible manufacturing systems. In: Proceedings of the IEEE/RSJ International Conference on Intelligent Robots and Systems, pp. 3674–3679 (2008)
5. Giridhar, A., Kumar, P.R.: Scheduling automated traffic on a network of roads. IEEE Transactions on Vehicular Technology 55(5), 1467–1474 (2006)
6. Wu, N., Zhou, M.: Modeling and deadlock control of automated guided vehicle systems. IEEE/ASME Transactions on Mechatronics 9(1), 50–57 (2004)
7. O'Donnell, P.A., Lozano-Perez, T.: Deadlock-free and collision-free coordination of two robot manipulators. In: Proceedings of the IEEE International Conference on Robotics and Automation, vol. 1, pp. 484–489 (May 1989)
8. LaValle, S.M., Hutchinson, S.A.: Optimal motion planning for multiple robots having independent goals. IEEE Transactions on Robotics and Automation 14(6), 912–925 (1998)

9. Ghrist, R., O'Kane, J.M., LaValle, S.: Computing pareto optimal coordination on roadmaps. Internationa Journal of Robotics Reseearch 24(11), 997–1010 (2005)
10. Simeon, T., Leroy, S., Laumond, J.P.: Path coordination for multiple mobile robots: a resolution-complete algorithm. IEEE Transactions on Robotics and Automation 18(1), 42–49 (2002)
11. Guo, Y., Parker, L.E.: A distributed and optimal motion planning approach for multiple mobile robots. In: Proceedings of the IEEE International Conference on Robotics and Automation, vol. 3, pp. 2612–2619 (May 2002)
12. Fierro, R., Branca, C., Spletzer, J.R.: On-line optimization-based coordination of multiple unmanned vehicles. In: Proceedings of the IEEE Networking, Sensing and Control, pp. 716–721 (2005)
13. van den Berg, J.P., Overmars, M.H.: Prioritized motion planning for multiple robots. In: Proceedings of the IEEE/RSJ International Conference on Intelligent Robots and Systems, pp. 430–435 (2005)
14. Bennewitz, M., Burgard, W., Thrun, S.: Finding and optimizing solvable priority schemes for decoupled path planning for mobile robots. Robotics and Autonomous systems 41(2), 89–99 (2002)
15. Gayle, R., Sud, A., Lin, M., Manoch, D.: Reactive deformation roadmaps: Motion planning of multiple robots in dynamic environments. In: Proceedings of the IEEE/RSJ International Conference on Intelligent Robots and Systems, San Diego, USA, pp. 3777–3783 (October 2007)
16. Whitty, M., Guivant, J.: Efficient global path planning during dense map deformation. In: Proceedings of the IEEE International Conference on Robotics and Automation, pp. 4943–4949 (2011)
17. Schulze, L., Behling, S., Buhrs, S.: Automated guided vehicle systems: a driver for increased businness performance. In: Proceedings of the International Multiconference of Engineers and Computer Scientists, Hong Kong, China, vol. II (May 2008)
18. Olmi, R., Secchi, C., Fantuzzi, C.: Coordination of multiple AGVs in an industrial application. In: IEEE Int. Conf. on Robotics and Automation, pp. 1916–1921 (May 2008)
19. Balaji, S., Swaminathan, V., Kannan, K.: A simple algorithm to optimize maximum independent set. Advanced Modeling and Optimization 12(1), 107–118 (2010)
20. Olmi, R., Secchi, C., Fantuzzi, C.: An efficient control strategy for the traffic coordination of agvs. In: Proceedings of the IEEE/RSJ International Conference on Intelligent Robots and Systems (IROS), San Francisco, USA, pp. 4615–4620 (September 2011)
21. Olmi, R., Secchi, C., Fantuzzi, C.: A coordination technique for automatic guided vehicles in an industrial environment. In: Proceedings of the IFAC Symposium on Robot Control (2009)
22. Olmi, R., Secchi, C., Fantuzzi, C.: Coordination of multiple robots with assigned paths. In: Proceedings of the 7th IFAC Symposium on Intelligent Autonomous Vehicles (2010)
23. Olmi, R., Secchi, C., Fantuzzi, C.: Coordination of industrial AGVs. International Journal of Vehicle Autonomous Systems 9(1/2), 5–25 (2011)

Kinesthetic Teaching Using Assisted Gravity Compensation for Model-Free Trajectory Generation in Confined Spaces

Jochen J. Steil[1], Christian Emmerich[1], Agnes Swadzba[1], Ricarda Grünberg[1,2], Arne Nordmann[1], and Sebastian Wrede[1]

[1] Institute for Cognition and Robotics (CoR-Lab) and Faculty of Technology Bielefeld University, Universitätsstr. 25, 33615 Bielefeld, Germany
{jsteil,cemmeric,aswadzba,anordmann,swrede}@cor-lab.uni-bielefeld.de
[2] Research Group on Gender & Emotion, Faculty of Psychology Bielefeld University, Universitätsstr. 25, 33615 Bielefeld, Germany
ricarda.gruenberg@uni-bielefeld.de

Abstract. The presented work approaches programming of redundant robots such as the KUKA Lightweight Robot IV in a co-worker scenario from a user-centered point of view. It specifically asks, how the user's implicit knowledge about the scene and the task can be transferred effectively to the robot through kinesthetic teaching. It proposes a new method to visualize the implicit scene model conveyed by the user when teaching a respective inverse kinematics and measures generalization by the robot. Based on these insights and empirical results from a previously performed user study, the present study argues that physical guidance of a task in confined spaces with static obstacles is too difficult to achieve in a single interaction. Summarizing earlier results and putting them into context, it is shown how to assist users to remedy this issue. The key is to divide the process in an explicit *configuration phase* for teaching the implicit scene model and a subsequent already assisted *programming phase* to teach the task based on a particular *assisted gravity compensation* mode. Further results from the user study confirm that this renders kinesthetic teaching in confined spaces feasible and enables a flexible and fast reconfiguration of the robot.

Keywords: kinesthetic teaching, redundant robots, implicit scene modeling, assisted gravity compensation, user study, physical human-robot interaction.

1 Motivation

Assistance systems that allow for close human-robot collaboration are becoming increasingly important in industrial manufacturing for two major reasons. On the one hand, ever increasing demands for more flexible production and smaller charges require fast re-configuration and re-programming of automation systems [1]. On the other hand, the demographic challenge requires to enable

F. Röhrbein et al. (eds.), *Gearing Up and Accelerating Cross-Fertilization between Academic and Industrial Robotics Research in Europe*, Springer Tracts in Advanced Robotics 94,
DOI: 10.1007/978-3-319-03838-4_6,

workers to be longer productive and produce added value particularly in European developed countries and Japan. Robotic co-workers are therefore a vision of many researchers in contemporary robotics [2, 3]. They are envisaged to support humans in assembly processes, simplify workflows, or prevent health risks such as joint and back problems. Fortunately, advanced force-torque sensing integrated into compact actuation units [4] with variable stiffness [5] has led to the development of compliant force-controlled robot manipulators [6, 7]. Complementary, there is a tendency to increase the degrees of freedom towards kinematically redundant manipulators [8]. These robots provide a great degree of flexibility for the realization of complex applications, for example in industrial [9] or service robotics [10, 11]. But the gained flexibility comes at increased engineering costs, because there is additional need for more complex control modes and explicit modeling steps to represent for instance static obstacles in the scene, which are required to define criteria for redundancy resolution. One crucial challenge for breakthroughs towards applications is therefore to hide this complexity from the user and to enable intuitive interactive configuration and programming of such robots by non-experts.

In this paper, we summarize the results of our project MoFTaG -*Model Free Trajectory Generation*, which was conducted as a so called "experiment" in the framework of ECHORD towards this goal. MoFTaG approaches the co-worker scenario from a user-centered point of view and specifically asks, how implicit knowledge of the user can be transferred into the robotic system in efficient ways through kinesthetic teaching, i.e. through physical guidance of the robot. Since long this general idea has been the rationale of teach-in procedures for traditional robots and also for programming-by-demonstration approaches [12], where kinesthetic teaching has been considered as an interface for providing demonstrations [13]. However, mostly task-space movements [14–16] or a particular task trajectory including a specific redundancy resolution [10, 17] have been considered. The execution of tasks in these works is typically performed in an unrestricted workspace using an analytic inverse kinematics. It may then also incorporate further explicit constraints e.g. to avoid self-collisions for a humanoid robot [16] or to add online-obstacle avoidance in dynamic environments [19]. Kormuchev et al. [20] even separate kinesthetic teaching for the task explicitly from the advantage of exploiting redundancy for task execution only: "This advantage does not relate to or affect the teaching, but matters only during the reproduction of the task".

MoFTaG promotes a different point of view: The combination of both the exploitation of the redundancy through interactive learning of the inverse kinematics in an explicit *configuration phase* and interactive guidance of the task-space trajectory in a *programming phase* is regarded as essential for using robots in confined spaces. We will show that the full potential of kinesthetic teaching does not unfold unless we combine the technological features of compliance and redundancy with learning methods in a highly integrated system that offers transparent interaction modes to the user. Consider for instance a fabrication scenario, where the work space is populated by fixed obstacles for instance in

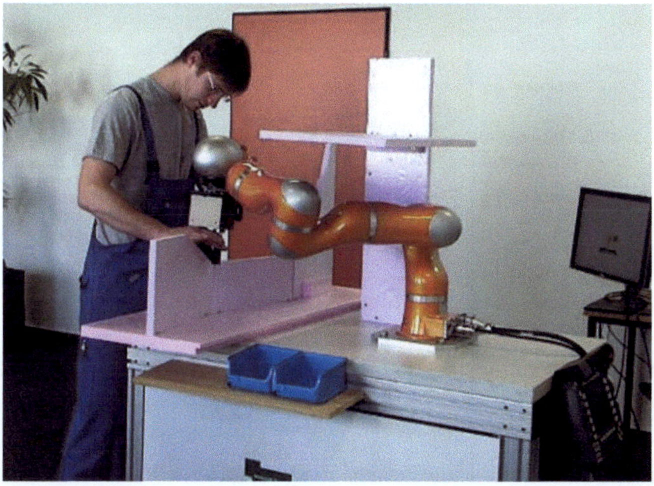

Fig. 1. A worker teaches a 3-D trajectory in a confined workspace while already being assisted by the robot's control system to avoid static obstacles in the scene [18]

form of a wall, a ceiling, a restriction not to reach into forbidden regions etc. Imagine a worker kinesthetically teaching tasks in such a confined scenario, e.g. see Fig. 1 showing a scenario from our user study on kinesthetic teaching with fabrication-floor workers presented in [18][1]. It turned out in this study that this task is not easy. Note that he or she has to deal with the simultaneous teaching of a task space trajectory and a feasible redundancy resolution while respecting the constraints present in the confined space. What knowledge can be transferred in order to avoid explicit modeling of the scene and respective programming effort and how? Furthermore, how to make this feasible for non-expert users?

We systematically investigate and answer these research questions as follows. Section 2 describes our setup using the LWR IV for a reference scenario and introduces the basic interaction modi for recording training data from the system. Section 3 demonstrates what kind of knowledge is transferred to the system when training the redundancy resolution in the confined work space. Furthermore, it investigates the generalization abilities of the learned mapping in terms of the proportion of the workspace that can safely be reached without colliding with the static obstacles. This is the part of the workspace that is eligible for kinesthetic teaching of task space trajectories, which is considered in Section 4 and shown to be difficult for lay-users if not further assisted [18]. This leads us to devise a novel *Assisted Gravity Compensation* mode [21] in Section 5, which

[1] In several parts of this paper we briefly refer to results from this user study with workers from a medium-sized manufacturing company – HARTING KGaA (`http://www.harting.com`) – in Germany. Details, an in depth analysis and results can be found in [18].

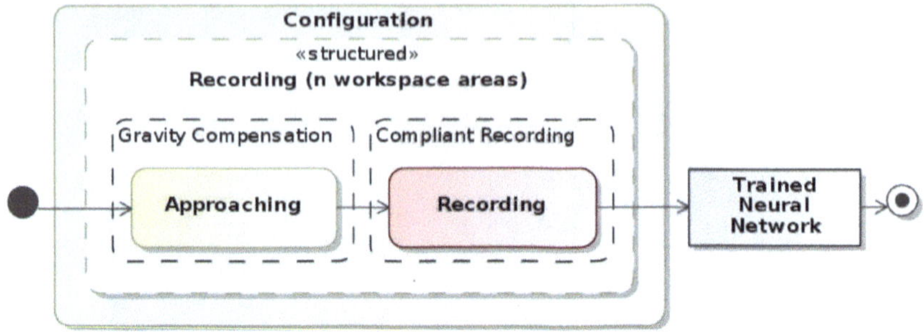

Fig. 2. Recording of training data for redundancy resolution via kinesthetic teaching in the FlexIRob system

is part of a respective two-stage assistive system that we describe and summarize in Section 6. Finally, we discuss some lessons learned in the course of the MoFTaG experiment and perspectives for future work in Section 7.

2 System Setup and Kinesthetic Teaching of Redundancy Resolution

Our robotic system prototype (www.cor-lab.de/flexirob) employs the KUKA lightweight robot IV (LWR IV). The LWR IV is a redundant manipulator with seven degrees of freedom allowing a manifold of configurations in joint-space for a single end-effector position and thus provides high flexibility for complex movements in the workspace. Its impedance-based control scheme [22] enables active compliance of the manipulator which is beneficial for the desired physical human-robot interaction.

In order to resolve the redundancy imposed by the kinematics of the LWR IV, we utilize supervised neural learning based on data recorded in physical human-robot interaction as indicated in Fig. 2. In a first phase of exploration the robot is in its *gravity compensation* mode[2] [22], where forces applied to the robot are not counteracted by the controller. This phase is used by the teacher to position the robot manually in a desired cartesian position $\mathbf{x}_{init}$ with a desired joint angle position $\mathbf{q}_{init}$. Once this configuration is reached, the robot is switched to impedance control and the tutor records joint values $\mathbf{q}_{cur}$ and corresponding end-effector positions $\mathbf{x}_{cur}$ in this training area (see e.g. Fig. 3 center) via kinesthetic teaching as shown in Fig. 2. The teacher repeats these steps for any area of interest, providing different null-space constraints in different areas of the workspace, see Fig. 3 (right). We call this the *configuration phase* for reasons explained in more detail later.

[2] In order to avoid switching control modes on the LWR IV controller during the interaction for technical and safety reasons, we reimplemented the native gravity compensation mode in our software control architecture.

The recorded training data is then passed to a neural network for supervised learning of a respective solution for the inverse kinematics. Such learning has proven to work well on various robot platforms and with different neural network algorithms [23–26]. In this paper, we use an extreme learning machine (neural network) [27], a random-projection based single-hidden layer feed-forward neural network with efficient read-out learning. It implements a static mapping from end-effector positions $\mathbf{x} = (x, y, z)$ in task-space to joint values $\mathbf{q} \in \mathbb{R}^7$. That is, the neural network learns a single solution of the inverse kinematics for the entire workspace from the training data with. Whereas different redundancy resolutions in different parts of the workspace are present, the neural network provides one single redundancy resolution for each particular task-space position. This learning approach is an easy-to-use, fast one-shot learning algorithm which, if combined with appropriate regularization techniques, has very good function approximation and generalization abilities [28]. Learning is completely data-driven and has therefore no explicit model knowledge about the specific robot platform or the environmental scene. Once learning is finished, the trained network is embedded in the hybrid control scheme of the robot as shown in Fig. 4.

Fig. 3. System setup: The LWR IV robot utilized for kinesthetic teaching (left) in a confined workspace with training data recorded in a single training area (center) and with training data recorded in several different areas of the workspace (right).

Any given cartesian end-effector target $\mathbf{x}^*$ is passed to the learned network, which maps it to desired joint values. These values serve as constraints for an hierarchical analytic controller based on ideas by Grupen and Huber [29]. Its primary task is to position the end-effector $\mathbf{x}^*$, the secondary task is to fulfill the null-space constraints given by the network. The redundancy resolution is thereby provided by the network while the precise fulfillment of the task-space task is ensured by the hierarchical controller. In [24], this control architecture was first implemented and evaluated in several obstacle scenarios. Generalization to collision-free movements with a typical position accuracy of approx. 2 mm in untrained areas were reported. We conclude from [24] that in general the learning approach is feasible.

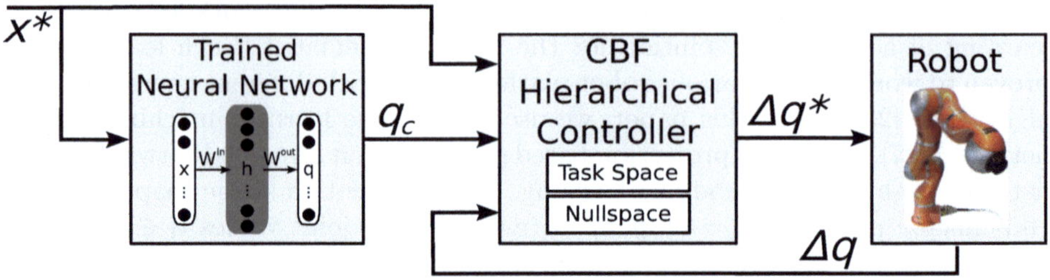

Fig. 4. Hierarchical control scheme of the FlexIRob system

3 Kinesthetic Teaching, Implicit Modeling and Generalization

As far as described in Section 2, the human tutor does not program the robot by teaching a specific task or movement in joint-space. In contrast to conventional kinesthetic teaching procedures, he or she rather *configures* the robot by providing training data in different areas of the robot's workspace for the learning algorithm to infer an appropriate redundancy resolution. We now describe and analyze this *configuration phase* in some depth, because this usage of kinesthetic teaching is rather unconventional and specific to redundant robots.

3.1 Implicit Scene Modeling via Kinesthetic Teaching

In principle, the teacher can transfer knowledge about any area of interest in the work space through kinesthetic teaching. No explicit modeling or programming steps are involved. In the following, this implicit modeling step is illustrated for the scene shown in Fig. 3 where static obstacles confine the workspace to the upper and left side. Once the network is trained according to the environmental obstacles, we visualize those parts of the workspace which are never used by the robot's links, joints or any other parts of its body while maneuvering through the accessible workspace.

For this purpose, first an obstacle setup-independent set of targets T lying on a grid G is defined and those points are subtracted which lay outside the workspace W of the LWR IV as shown in Fig. 5 (left), i.e. $T = G \cap W$. Second, all points from T that are not accessible in the particular obstacle setup are removed to obtain $\hat{T}$. This step excludes all targets located within the obstacles and for the setup in Fig. 3 also targets that are located above the board part of the obstacle, because these are not accessible for the robot due to joint limits. $\hat{T}$ constitutes the accessible workspace of the robot for the particular scene.

During evaluation the robot is commanded to reach for each of the targets in $\hat{T}$ with the learned neural network included in the inverse kinematics control of the robot. For each target the resulting posture, i.e. the positions of all links and joints of the LWR IV, is stored. These positions comprise the *maneuvering space* M, i.e. all points of the space where some part of the robot was for some target position. By 'inverting' this viewpoint, we visualize those parts of the workspace which are never visited by any part of the arm. All points that are in or close to the maneuvering space are subtracted from the obstacle-independent set T to form the *implicit scene model* $S := \{\mathbf{x} \in T : dist(\mathbf{x}, M) > 0.1\}$, which is encoded in the particular learned redundancy resolution. If the system was trained appropriately by the human tutor, these parts should enclose the static obstacles of the environmental scene.

Fig. 5 illustrates this procedure. From left to right the obstacle independent set of targets T and the simulated obstacles are shown. The next figure shows the implicit scene S that was obtained after training the system with only one training area as shown in Fig. 3 (center). Only few parts of the obstacle (the lower part of the box part of the obstacle) are actually covered by the implicit scene model. Large parts of the obstacle remain in the maneuvering space of the trained system. This indicates that the provided training data does not comprise enough knowledge about the environmental scene. The rightmost chart in Fig. 5 shows the implicit scene model obtained after the teacher has trained the system in 12 different training areas, some of which are displayed in Fig. 3 (right). The corresponding implicit scene model S approximates the obstacles very well. It shows that by providing providing meaningful data in several training areas of the workspace the teacher is able to transfer his or her intuitive understanding of the scene to the movement generating system of the robot.

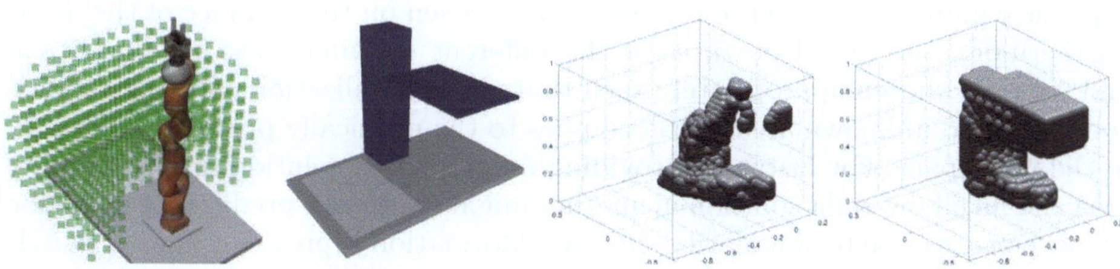

Fig. 5. Visualization of the obstacle-independent workspace T, obstacles in simulated setup and the implicit scene model for one and 12 trained areas (from left to right)

3.2 Generalization and the Number of Training Areas

It is one major goal of the described kinesthetic teaching to enable the robot to move in the predefined, confined workspace. The trained system shall reach at best all targets that are theoretically accessible without colliding with the environment. We now ask the question, how many training areas are actually needed for this purpose?

Fig. 6 shows that the with respect to more training data increasing accuracy of the implicitly scene model (see Section 3.1) also implies that more parts of the workspace are accessible for the robot. This evaluation is again based on the task specific target points $\hat{T}$ defined in Section 3.1. The robot is commanded to reach for each target in $\hat{T}$ with the learned neural network included in the inverse kinematics control of the robot. A target is said to be reached, if a positioning precision of at least 1 cm without colliding with the environmental obstacles is obtained. The reached targets are displayed as green dots from two view points. Fig. 6(a) shows the result after providing data in only one training area as in Fig. 3 (center). Already one training area makes 51.3% of the targets $\hat{T}$ accessible for the robot although the training data itself covers only ca. 5% of that space. Fig. 6(b) visualizes the resulting reached targets for 12 training areas. The robot reached 97% of the defined accessible targets in $\hat{T}$, while the training data itself covered only 36.5% of this space. Note that the entire teaching procedure still requires less than 10 minutes for a skilled person although the number of 12 training areas seems large at first sight. The kinesthetic teaching enables non-expert users to configure a redundant robot to operate in a confined workspace without writing a single line of code.

3.3 Generalization and the Quality of Training Data

In the last sections, the evaluations accessed generalization of the learned mapping if an expert trains the system. That is, the independent variable is the number of training areas and the data within a training area is assumed to be per se suitable for learning. The network generalizes to unseen parts of the work space and the generalization performance is fairly good due to the extrapolation capabilities of the neural network.

Of course, this is only part of the story, as the quality of the training data also depends e.g. on the number and more strongly even on the variance of the data. Furthermore, the initial postures in the different training areas are important and must be chosen appropriately such that the generalization of the null-space constraints to unknown areas also complies to the physically present constraints in the workspace. For instance, providing an elbow-left solution in one training area and an elbow-right solution in another might cause non-predictable behavior or movements close to singularity since no information is provided how to switch between both solutions.

It is therefore crucial that also lay users intuitively understand how to provide such data. From experience we know that guiding the end-effector in a spiral-like motion in a particular area is a good solution and therefore we instructed users to do so in the study described in [18]. Still, Fig. 7 illustrates the diversity of

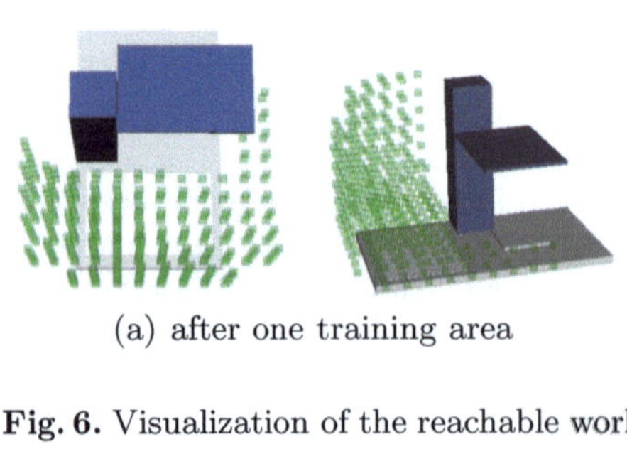

(a) after one training area

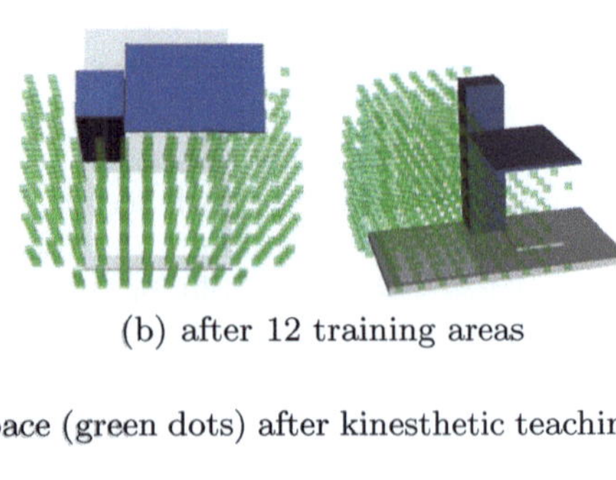

(b) after 12 training areas

Fig. 6. Visualization of the reachable workspace (green dots) after kinesthetic teaching results that can be obtained if non-expert users train the redundancy resolution in the same scenario as before. On the left, the task-space coordinates of the recorded training data are shown as red and green lines. In particular, Fig. 7(a) shows that a too small variance in the training data and badly chosen postures lead to poor generalization performance. During execution of a test trajectory (black line), which should follow a straight line from one training area to the other, no collisions with the environment occurred. But the accuracy of the performed movement is low (Fig. 7(a), left). In contrast, Fig. 7(b) shows the case that users intuitively provide suitable data after some instruction and a quick familiarization phase (for details on the experimental procedure see [18]) such that an accurate and collision-free test movement is performed by the trained robot system. We conclude from this that online feedback about the quality of training data would be an important further ingredient to make the overall procedure even more reliable. This was, however, beyond the scope of the current experiment.

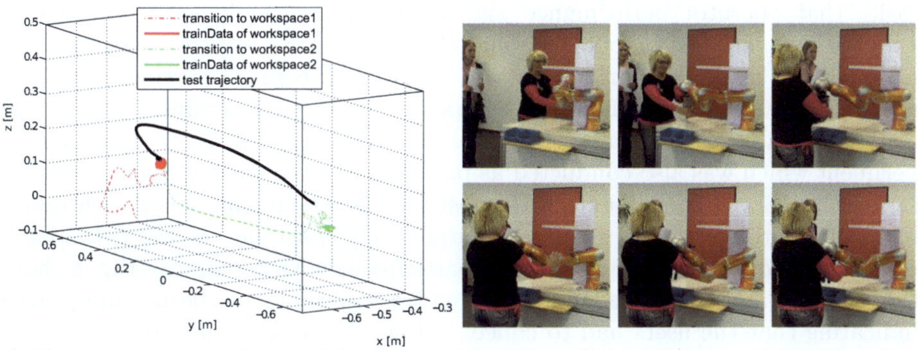

(a) Test person recording training data with only a low variance and inappropriate postures (elbow-left and elbow-right).

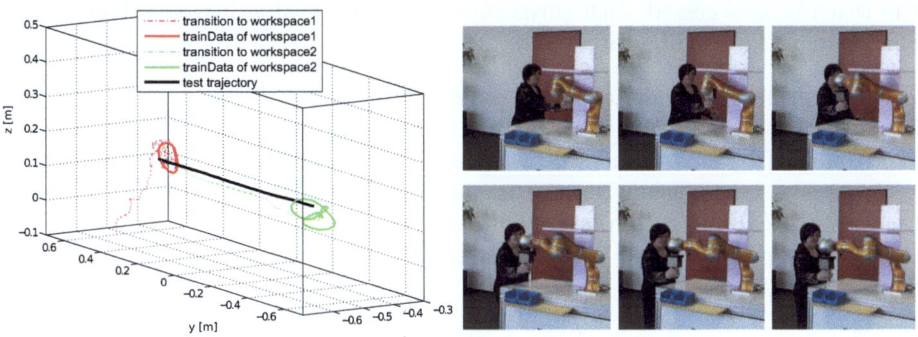

(b) Test person providing good training data.

Fig. 7. Examples of different user behaviors during the configuration phase

4 Kinesthetic Teaching of Configuration and Task Is Difficult: A User Study

Whereas we have considered kinesthetic teaching for redundancy resolution so far, for a concrete application we also need to teach a particular task. However, in current kinesthetic teaching approaches the actual task and the redundancy resolution are not distinguished from each other. The teacher has to deal with both issues simultaneously. For controlled experiments, we have studied this combination of redundancy resolution and teaching a task in an an adapted version of the wire loop game[3]. This is a classical teach-in, which here is performed in a confined workspace. That is, we need to provide simultaneously both an accurate task space trajectory (wire loop game) and a suitable inverse kinematics in order to avoid collisions with the static scene.

We expect that the lack of separation of task-space programming and redundancy resolution causes a high complexity, unsuited for utilizing simple programming-by-demonstration approaches such as kinesthetic teaching. This problem has also been observed before in the literature, but neither explicitly investigated nor solved. Kaber and Riley demonstrated in [30] for a tele-operation task "that operator performance and workload are significantly affected by whether joint or world mode (i.e., end-effector position) control is required [... and that] for example, world mode can reduce task completion times, but may also increase the number of contact errors when working in confined spaces" [31].

We were able to confirm this hypothesis through a kinesthetic teaching experiment which was also conducted as part of the larger user study that was described in [18]. 24 participants with mostly no practical experience with robots had to fulfill the wire loop game task without any assistance for the redundancy resolution and were later compared to participants with assistance. Fig. 8 shows the scenario and some snapshots from video recordings of the teaching phase indicating that the users had to concentrate on both task and redundancy resolution as well as to master the physical effort for controlling end-effector and joint configuration simultaneously. For evaluation, the quantitative success of the teach-in is accessed with three measurements, evaluating effectiveness and efficiency of the teaching procedure [31]:

1. The task-space accuracy is measured by means of the maximum euclidean deviation of the teach-in trajectory from the target.
2. Abidance by environmental constraints is accessed by counting unintended contacts, i.e. collisions, of the manipulator with the obstacle. This is done automatically in a simulation software.
3. Efficiency is measured by the time needed for the teach-in.

The experimental results, of which we briefly report the most important here, reveal a systematic deficit of the participants to successfully teach the robot system the desired trajectory in the confined workspace. Firstly, most of the

[3] The original wire-loop game consists of a metal loop and a serpentine length. The loop has to be guided along the wire without touching it.

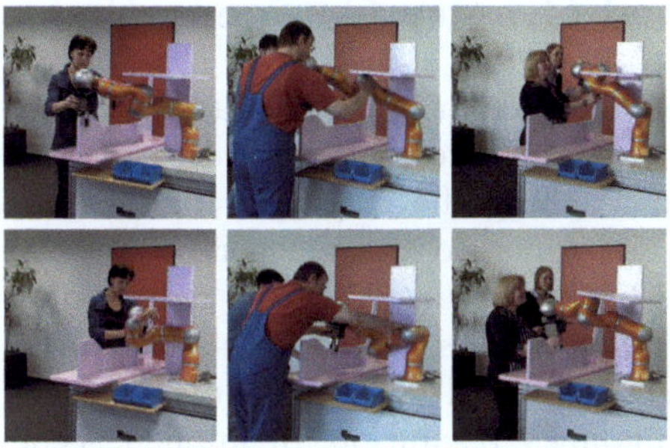

Fig. 8. Non-assisted users during the teach-in task

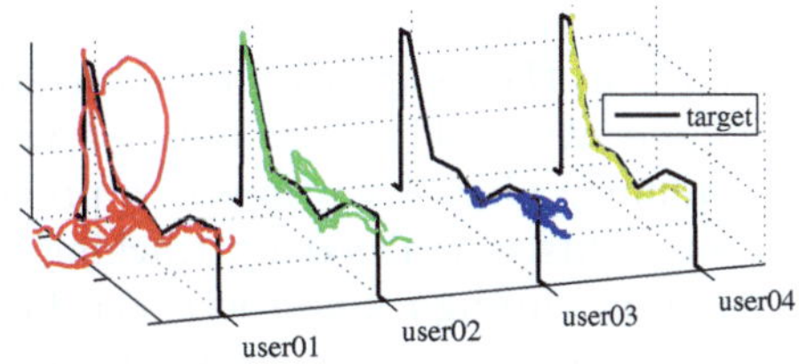

Fig. 9. Exemplary teach-in trajectories of non-assisted users

participants were not able to accurately follow the styrofoam parcours, which is indicated by high deviations from the target trajectory of 0.12 ± 0.11 meters averaged over all participants. Fig. 9 shows exemplary trajectories of four participants indicating very different teach-in experiences and success. Whereas few users achieved a high task-space accuracy, e.g. `user04` with a maximum deviation to the target of approx. 0.05 meters, most of them failed to simultaneously find a valid joint configuration in the confined workspace and move the end-effector accurately along the desired trajectory. As a result, they deviated from the target movement up to 0.52 meters (see `user01` or `user02`). Furthermore, two participants aborted the wire loop game due to the difficulty, e.g. `user03`.

Secondly, there were only two users that did not encounter collisions between the robot arm and the obstacle. That means that regardless of the task-space accuracy 22 of 24 participants failed to kinesthetically teach the robot a target trajectory without colliding with its environment. They fail to transfer their knowledge about the constraints in the environment to the robotic system, compare also Fig. 12.

Finally, we measured an average time of 93.4 ± 44.5 seconds needed for the teach-in. Rather than the pure average value, here the high variance across the participants is interesting. The fastest user managed to complete the task in 27 seconds, but the longest teach-in took about more than 3 minutes. We point out that, although it is possible (and the results showed that few of the non-expert users managed) to solve the described kinesthetic teaching task in joint control mode, most users failed to maneuver the redundant robot in the confined workspace along a target trajectory. In [31] the authors hypothesize that "the operator may have good global situational awareness on the end goal for the manipulator, but may suffer from poor local situational awareness on the position of each manipulator joint", which is illustrated in Fig. 8 (bottom right). We conclude from this experiment that kinesthetic teaching of a redundant robot is indeed difficult.

5 Intuitive Teaching by Assisted Gravity Compensation

To remedy the situation and to enable users to teach in confined spaces, we now propose to separate the learning of the redundancy resolution as described in Section 2 in the *configuration phase* from teaching the trajectory in the *programming phase*. The idea is to provide assistance to the teacher during the programming phase by means of a previously learned or configured redundancy solution. This is realized by tracking *free movements in the task-space* given by the user through kinesthetic teaching, while simultaneously *controlling the joint-space* according to the provided redundancy resolution, or implicit scene model, respectively. We call this mode *assisted gravity compensation* and briefly describe its respective control scheme, which is also displayed in Fig. 10.

5.1 Assisted Gravity Compensation Control Scheme

The assisted gravity compensation mode is designed to assist human co-workers in complex environments, where different inverse kinematic solutions are required in different areas of the robot's workspace. Hence, on the one hand, redundant solutions of the inverse kinematics must exist. On the other hand, in order to allow moving the end-effector in kinesthetic teaching a compliant robot platform is required. Given a user interacting with the robot, we observe a human-intended cartesian displacement $\Delta\mathbf{x}$ from the current end-effector position $\mathbf{x}$. In order to allow the user to freely interact with the robot in task space, this displacement is defined as the new desired cartesian position

$$\mathbf{x}^* = \mathbf{x} + \Delta\mathbf{x}. \tag{1}$$

Next, we need an (estimated) inverse kinematics $IK(\cdot)$ to select a redundancy resolution $\mathbf{q}_c$ for the desired position $\mathbf{x}^*$ according to the (possibly constrained) environment, i.e. confined work space,

$$\mathbf{q}_c = IK(\mathbf{x}^*). \tag{2}$$

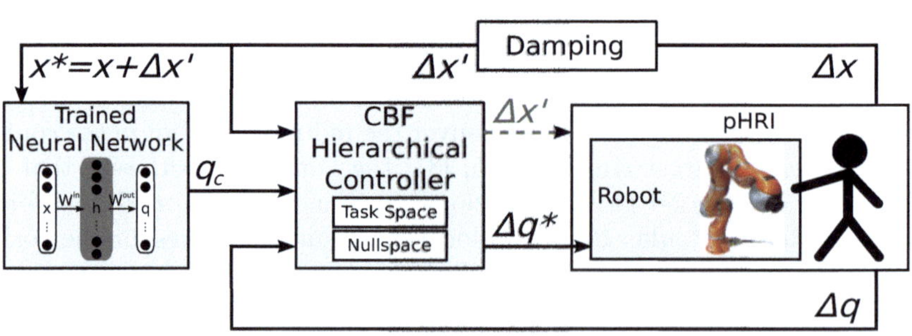

Fig. 10. Block diagram showing the assisted gravity compensation control scheme on the FlexIRob system with its three control loops. The controller output $\Delta\mathbf{q}^*$ sent to the robot corresponds to the user-given cartesian displacement $\Delta\mathbf{x}$.

From a control theoretical point of view this redundancy assistant provides constraints for the robot controller in the null-space of the redundant manipulator. In principle, any redundancy resolution approach can be used for assistance, e.g. approaches based on optimization techniques [32], key-frame based approaches [10], specific closed-form solutions [33] or learned redundancy resolutions [24]. In our approach, the inverse kinematics is the learned mapping from task-space to joint-space according to the *configuration phase* (cf. Section 2).

Finally an analytic controller is required that fuses the cartesian task $\mathbf{x}^*$ from kinesthetic teaching in task-space with the joint configuration $\mathbf{q}_c$ given by the redundancy assistant in a way that it projects the joint constraints $\mathbf{q}_c$ into the null-space of the redundant manipulator, e.g. such as the general gradient projection method

$$\Delta\mathbf{q}^* = J^\dagger(\mathbf{q})\Delta\mathbf{x}^* + (I - J^\dagger J)\Delta\mathbf{q}_c \quad \text{and} \quad \mathbf{q}^* = \mathbf{q} + \Delta\mathbf{q}^*, \tag{3}$$

where $J^\dagger$ constitutes the Moore-Penrose-Pseudoinverse of the task Jacobian. This projection of the environmental constraints into the null-space of the forward kinematic mapping allows the realization of $\mathbf{x}^*$ while respecting the null-space constraints as described before. This is done in the center box in Fig. 10.

This combination of the robot's compliance and the analytic controller incorporating the joint-space constraints provided by the redundancy assistant allows the user to freely interact with the end-effector. Simultaneously, the joints are controlled according to the redundancy resolution of the current end-effector position.

5.2 Wireloop Teach-In Using Assisted Gravity Compensation

In this section, we present some comparative results from participants in the user study, who were assisted during the wire-loop game (group A), vs. the non-assisted group (N). Figure 11 shows the teach-in trajectories that were recorded by the participants during the wire-loop game. It clearly reveals that the trajectories of group A (see Fig. 11(a)) are smooth, very similar to each other, and

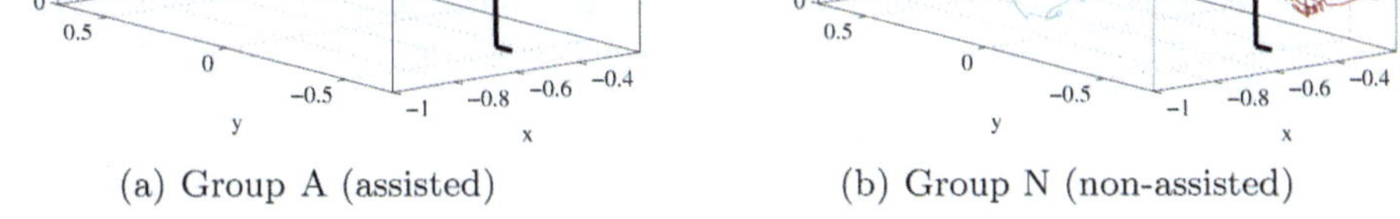

(a) Group A (assisted) (b) Group N (non-assisted)

Fig. 11. Teach-in trajectories of the participants during the wire-loop game close to the target, which was represented by the Styrofoam *parcours* during the study and is plotted in Fig. 11 as black lines. In contrast, participants in group N (see Fig. 11(b)) recorded trajectories that are jerky, deviate a lot from each other, and in some cases exhibit strong error in regard to the target trajectory.

In order to evaluate these differences systematically, we conducted an analysis of system effectiveness, which is assessed by task-space accuracy and avoidance of collisions with the environment, and of system efficiency, which is measured by the time needed by the participants for the teach-in. As for the task-space accuracy, we evaluated two metrics revealing the systematic difference between the two groups. As already mentioned in Section 4, we measured the maximum cartesian deviation of the users' trajectory from the target trajectory. The results in Fig. 12(a) show that assisted participants stay significant closer to the target trajectory than the non-assisted and do not deviate a lot amongst each other, namely 4.9±1.4 compared to 12.2±11.1 centimeters on average. This qualitative difference is also present when comparing the shape of the recorded trajectories by means of a Procrustes analysis [34]: Since we did not explicitly encourage the participants to perform a specific timing during the teach-in, the trajectories are normalized in time, i.e. we remove the velocity profile by re-sampling the data with constant velocity and equal number of points. For comparison, the user trajectory and target trajectory are optimally superimposed by means of translation and rotation (we omit scaling in our analysis). Finally, the procrustes difference is calculated as the average euclidean distance between the points of both trajectories. The results in Fig. 12(b) reveal the significantly higher task-space performance of participants in group A as compared to those in group N. In particular, the average Procrustes distance for group N measures 15.8 ± 1.2 cm, which means that even after optimal translation and rotation the geometrical shapes of teach-in trajectory and

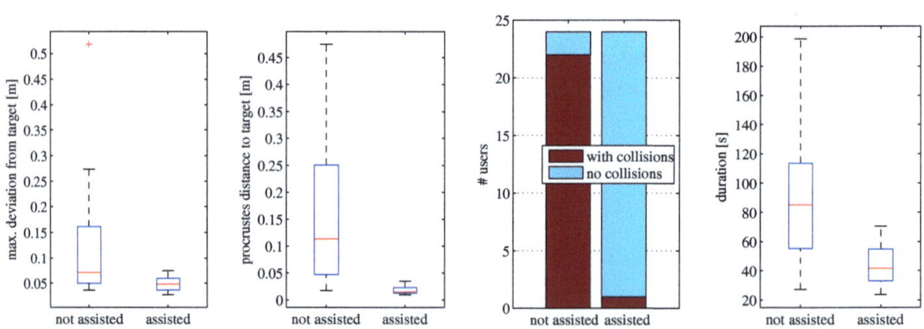

(a) Maximum deviation of the participants' teach-in trajectories to the target trajectory.

(b) Shape comparison of the participants' teach-in trajectories to the target trajectory.

(c) Number of participants with and without collisions.

(d) Time needed by the participants for the teach-in.

Fig. 12. Comparison of the participants' performance during the teach-in of the wire-loop game target trajectory on average differ by 15 cm per point. An ANOVA confirmed that the differences between the groups were significant [18]. Concerning the number of collisions (see Fig. 12(c)) with environmental obstacles, for 22 of 24 participants in group N collisions occurred during the teach-in. In contrast, only one of 24 assisted participants produced contacts between the robot arm and the environment. Also the time needed by the participants for the teach-in reveal a significant advantage of group A (see Fig. 12(d)). On average, participants of group A needed only half the time for teaching the same task.

6 Intuitive and Safe Configuration and Programming of Redundant Robots

Equipped with the *assisted gravity compensation* mode and the results obtained in the previous sections we can now reformulate the process of teaching of redundant manipulators to a two-stage interaction work flow separating into a dedicated *configuration phase* on the one hand and an *assisted programming phase* on the other hand. The idea of this two-stage process is to learn the redundancy resolution first, thereby transferring the user's intuitive understanding of the environmental scene to the robot's movement generating system, and then already use the learned solution for assisting the user in the second phase, i.e. the teach-in of the actual task.

This proposed separation on the one hand allows intuitive teaching of redundant robots. On the other hand, since configuration of the robot and the actual task are treated separately, *re-programming* according to a new task can be done without *re-configuring* the robot. This is a significant advantage in presence of frequently changing tasks and allows for teaching of several tasks in the same environment without repeating the configuration step. Vice versa, also reconfiguration of the robot is possible without overriding or forgetting a taught task.

6.1 Interaction Workflow

We briefly describe and summarize the resulting interaction workflow as depicted in Fig. 13. The first step in the configuration phase is *approaching* the work space area where a desired redundancy resolution will be taught. For this purpose, the robot is set to gravity compensation to be freely movable in its entire joint configuration. Once the desired joint configuration for this workspace area is applied by the user, the robot is switched to joint impedance and commanded to stay in the current joint position. The user can now start *recording* training data by applying force to the end-effector. These two steps can be repeated for an arbitrary number of different workspace areas to kinesthetically teach different redundancy resolutions for different areas of the robot‘s workspace. As described in Section 2, the recorded joint angles $\mathbf{q}$ and corresponding cartesian end-effector positions $\mathbf{x}$ are used for supervised training of the neural network. (In principle, any other trainable function approximator with reasonable inter- and extrapolation capabilities can be used.) The trained neural network then encodes

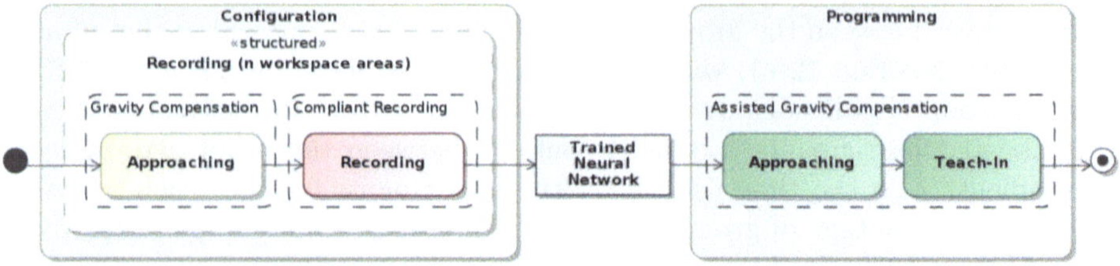

Fig. 13. Interaction work-flow with assisted teach-in using the interaction control modes: gravity compensation (yellow box), compliant recording (red box) and assisted gravity compensation (green boxes). Recording of the training data can be repeated for several workspace areas. The trained neural network from the configuration phase is transferred to the programming phase to be used inside the assisted gravity compensation mode.

an inverse kinematics with the desired redundancy resolution for all workspace areas and is subsequently used in the programming phase, to be included in the assisted gravity compensation or in the hierarchical control scheme as described in Section 2.

The programming phase starts again with *approaching* the teach-in task, this time in the assisted gravity compensation mode. While *approaching* and during the subsequent *teach-in*, the user no longer needs to care for the redundancy resolution, which is now done by the control mode according to the constraints that were implicitly encoded in the neural network during the configuration phase.

6.2 Software Architecture and System Safety

When physically interacting with a robot, safety is a major and primary concern, in particular when lay users are involved. The described interaction system is implemented as a coherent set of learning, interaction, and control components with the Compliant Control Architecture [35]. Lower-level control components run inside a soft real-time execution engine implemented as a Xenomai task in user space, which is interfaced with KUKA's Fast Research Interface (FRI) [36]. Due to the risks in physical contact between the robot and human users, we applied a number of safety measures to ensure safety for humans on the one hand and allow research in system studies on the other hand.

The main risk is collision between a robot and its user [37]. The first means to reduce the instantaneous severity of impacts is the use of the KUKA Lightweight Robot IV, which reduces manipulator link inertia and weight by using lightweight but stiff materials. The interaction control modes introduced above and used throughout the experiments are compliant to external forces which also reduces impact in case of a collision. Low-level control components and the FRI-based communication checks monitor the quality of service of the network communication, providing reliable robot control at all times. The control system immediately stops the robot when communication quality drops. Limits for joint positions, velocities, acceleration and torques were configured and are constantly checked on different layers of the software architecture to prevent injuries from collisions or shearing. The hierarchical controller used in the hierarchical control and assisted gravity compensation modes deals with the uncertainty of insufficient or inadequate training of the system by prioritizing control of the end-effector trajectory and therefore preventing unexpected end-effector movements of the robot. Finally, we implemented a collision detection based on the force-sensing of the LWR IV, which allows for instantaneous control changes to avoid unintended, invasive contact with the robot's environment.

7 Lessons Learned

The strongest lesson learned from the presented work is that physical guidance of a redundant robot in form of kinesthetic teaching is much more than just a "teach-in". This challenges current thinking, which often takes for granted that kinesthetic teaching shall be used to provide an end-effector task trajectory and shall mostly ignore the inverse kinematics. However, kinesthetic teaching can provide both, a task trajectory and a smart and task-appropriate redundancy resolution. It also needs to provide these two aspects in co-worker scenarios with a typical workplace that has a confined environment, because the explicit programming of tasks in such environments is costly and tedious. On the downside, it appears that the increased flexibility of redundant arms, which is necessary to navigate in confined spaces, comes at the cost of making kinesthetic teaching practically difficult. At least for non-expert users, as our user study showed.

The proposed solution is a two-stage interaction procedure that includes as first step an independent teaching of inverse kinematics to implicitly model the environment and as second step an already assisted task learning. This closely integrates different technological, computational and engineering methodologies. Technology-wise it is fundamental to employ redundancy in order to position the robot in the confined space in the first place and to use compliance for both recording of training data and assisted teaching of the task. From the computational point of view, the employed neural network algorithm must be fast and reliable enough for blending smoothly in the interaction workflow. Generalization is used to keep the need for training data low and a fast retraining based on new data is always possible. The engineering finally has to offer a strong integration of hierarchical control and learning, the switching of interaction control modes and also it needs to implement respective safety measures. Only the combination of these three partial perspectives on the system allows for the fluent and apparently simple interaction to perform the teaching in a confined environment with a redundant robot. The second lesson therefore is that the users can do it, if properly assisted through the assisted gravity compensation, which we were able to verify in our study as well.

From a more general perspective, the presented work is an example that systemic integration of technology, learning methodology, and interaction in a user-centered perspective allows for advanced assistance systems. This is important for an SME environment, where flexibility and fast reconfiguration must be possible without consulting engineering experts. The key is to hide the actually increased technological complexity of the system through a combination of intuitive interfaces, a cognitive systems and software engineering that enables smooth interaction, and an iterative user-interface design, which takes into account the needs and preferences of the users. While our system already goes a long way towards this goal, there are still some missing pieces. First of all, the safety measures have to be leveraged to a level where the whole system can be certified. Currently, it is legally feasible to work with users based on the verification of the particular experimental scenario and the fact that still one experimenter can execute an emergency stop at any time although this never happened in one full week of experimentation. But in the longer term this policy needs to be substituted by an autonomous safety system in the future. Further integration with mobile systems for mobile manipulation and vision for visually guided assistance is desirable.

The third lesson, however, that we learn from the user experiments is about the role of feedback while learning. As discussed in the text, the quality of the training data is important. That is a non-intuitive issue for non-experts. While with some training and good instruction it was feasible to train the system for non-experts, clearly both users and experimenters felt the need for a systematic improvement through feedback about the quality of the training data and the learning success. From again a more general perspective, this observation makes a case for increasing user-adaptation of the robotic system. This can be done for instance by giving the right feedback at the right time, by adjusting movements to the body height of the user, or for instance by adjusting compliance to the guiding force. In our future work, we plan to go along these lines to further improve the quality of the interaction in the described system with the goal to bring it to the factory floor in the nearer future, once respective hardware is achievable at reasonable costs also for the SME domain.

Acknowledgements. J.J.S. and S.W. designed research, J.J.S. and C.E. wrote paper, all authors contributed. C.E. and J.J.S. conducted core research in learning, A.S., C.E. and S.W. designed interaction loop, A.N. and S.W. designed software architecture and system integration, A.N., A.S. and C.E. implemented software components for the user study, all authors collaborated for user study, C.E. and R.G. did the data analysis. This work was funded by the European Communitys Seventh Framework Programme FP7/2007-2013 Challenge 2 Cognitive Systems, Interaction, Robotics through the projects ECHORD with the experiment Model-free flexible trajectory generation (MoFTaG) (A.S, C.E) and AMARSi - Adaptive Modular Architectures for Rich Motor Skills under GA No 248311 (A.N). We thank the HARTING Technology Group (www.harting.de) for generous support in conducting the user study, which was performed at HARTING's production site with HARTING factory-floor workers. We also thank OWL-MASCHINENBAU for teaming-up with us to disseminate results of the project, in particular through a workshop in the OWL-MASCHINENBAU

Academy (www.owl-maschinenbau.de) dedicated to reach SME decision makers and production technology experts.

References

1. European Strategic Robotics Platform. Robotic visions to 2020 and beyond – The strategic research agenda (SRA) for robotics in Europe. Technical report, European Robotics Technology Platform, Brussels (2009)
2. Haddadin, S., Suppa, M., Fuchs, S., Bodenmüller, T., Albu-Schäffer, A., Hirzinger, G.: Towards the robotic co-worker. In: Pradalier, C., Siegwart, R., Hirzinger, G. (eds.) Robotics Research. STAR, vol. 70, pp. 261–282. Springer, Heidelberg (2011)
3. Groothuis, S.S., Stramigioli, S., Carloni, R.: Lending a helping hand: Toward novel assistive robotic arms. IEEE Robotics & Automation Magazine 20(1), 20–29 (2013)
4. Albu-Schäffer, A., Haddadin, S., Ott, C., Stemmer, A., Wimböck, T., Hirzinger, G.: The DLR lightweight robot: Design and control concepts for robots in human environments. Industrial Robot 34(5), 376–385 (2007)
5. Tsagarakis, N., Sardellitti, I., Caldwell, D.: A new variable stiffness actuator (CompAct-VSA): Design and modelling. In: Proc. IROS, pp. 378–383 (2011)
6. Bischoff, R., Kurth, J., Schreiber, G., Koeppe, R., Albu-Schäffer, A., Beyer, A., Eiberger, O., Haddadin, S., Stemmer, A., Grunwald, G., Hirzinger, G.: The KUKA-DLR Lightweight Robot arm a new reference platform for robotics research and manufacturing. In: Joint 41st International Symposium on Robotics and 6th German Conference on Robotics, pp. 741–748 (2010)
7. Grzesiak, A., Becker, R., Verl, A.: The bionic handling assistant: A success story of additive manufacturing. Assembly Automation 31(4), 329–333 (2011)
8. Conkur, E.S., Buckingham, R.: Clarifying the definition of redundancy as used in robotics. Robotica 15(5), 583–586 (1997)
9. Daimler, A.G.: Leichtbauroboter im Piloteinsatz im Mercedes-Benz Werk Untertürkheim (Lightweight robots employed in pilot application at Mercedes-Benz site Untertürkheim), Press release. Daimler AG (2009), `http://goo.gl/RSeYl`
10. Akgun, B., Cakmak, M., Yoo, J.W., Thomaz, A.L.: Trajectories and keyframes for kinesthetic teaching: A human-robot interaction perspective. In: Proceedings of the Seventh Annual ACM/IEEE International Conference on Human-Robot Interaction, pp. 391–398 (2012)
11. Breazeal, C., Siegel, M., Berlin, M., Gray, J., Grupen, R., Deegan, P., Weber, J., Narendran, K., McBean, J.: Mobile, dexterous, social robots for mobile manipulation and human-robot interaction. In: Proc. ACM SIGGRAPH (2008)
12. Schaal, S., Ijspeert, A., Billard, A.: Computational approaches to motor learning by imitation. Philosophical Transactions of the Royal Society of London Series B Biological Sciences 358(1431), 537–547 (2003)
13. Billard, A., Calinon, S., Dillmann, R., Schaal, S.: Robot Programming by Demonstration. In: Siciliano, B., Khatib, O. (eds.) Springer Handbook of Robotics, ch. 59, pp. 1371–1394. Springer, Heidelberg (2007)
14. Calinon, S., Billard, A.: Incremental learning of gestures by imitation in a humanoid robot. In: Proceedings of the ACM/IEEE International Conference on Human-Robot Interaction, pp. 255–262. ACM (2007)
15. Mühlig, M., Gienger, M., Hellbach, S., Steil, J.J., Goerick, C.: Task-level imitation learning using variance-based movement optimization. In: IEEE Conf. Robotics and Automation, pp. 1177–1184. IEEE (2009)

16. Mühlig, M., Gienger, M., Steil, J.J.: Interactive imitation learning of object movement skills. Autonomous Robots 32, 97–114 (2012)
17. Vijayakumar, S., D'souza, A., Shibata, T., Conradt, J., Schaal, S.: Statistical learning for humanoid robots. Autonomous Robots 12, 55–69 (2002)
18. Wrede, S., Emmerich, C., Grünberg, R., Nordmann, A., Swadzba, A., Steil, J.J.: A User Study on Kinesthetic Teaching and Learning for Efficient Reconfiguration of Redundant Robots. Journal of Human-Robot Interaction 2, 56–81 (2013)
19. Khansari-Zadeh, S., Billard, A.: A dynamical system approach to realtime obstacle avoidance. Autonomous Robots, 1–22 (2012)
20. Kormushev, P., Calinon, S., Caldwell, D.G.: Imitation learning of positional and force skills demonstrated via kinesthetic teaching and haptic input. Advanced Robotics 25(5), 581–603 (2011)
21. Emmerich, C., Nordmann, A., Swadzba, A., Steil, J.J., Wrede, S.: Assisted gravity compensation to cope with the complexity of kinesthetic teaching on redundant robots. In: International Conference on Robotics and Automation, Karlsruhe, pp. 4307–4313 (2013)
22. Albu-Schäffer, A., Ott, C., Hirzinger, G.: A Unified Passivity Based Control Framework for Position, Torque and Impedance Control of Flexible Joint Robots. In: Thrun, S., Brooks, R., Durrant-Whyte, H. (eds.) Robotics Research. STAR, vol. 28, pp. 5–21. Springer, Heidelberg (2007)
23. Neumann, K., Rolf, M., Steil, J.J.: Learning inverse kinematics for pose-constraint bi-manual movements. In: Proceedings of the 11th International Conference on Simulation of Adaptive Behavior: from Animals to Animats, pp. 478–488 (2010)
24. Nordmann, A., Emmerich, C., Rüther, S., Lemme, A., Wrede, S., Steil, J.J.: Teaching Nullspace Constraints in Physical Human-Robot Interaction using Reservoir Computing. In: IEEE Int. Conf. on Robotics and Automation, pp. 1868–1875 (2012)
25. Phung, A.S., Malzahn, J., Hoffmann, F., Bertram, T.: Data based kinematic model of a multi-flexible-link robot arm for varying payloads. In: IEEE. Conf. Robotics and Biomimetics, pp. 1255–1260 (2011)
26. Rolf, M., Steil, J.J., Gienger, M.: Efficient exploration and learning of whole body kinematics. In: IEEE 8th International Conference on Development and Learning (ICDL 2009), Shanghai, CH, pp. 1–7 (2009)
27. Huang, G.B., Wang, D.H., Lan, Y.: Extreme learning machines: a survey. International Journal of Machine Learning and Cybernetics 2(2), 107–122 (2011)
28. Neumann, K., Emmerich, C., Steil, J.J.: Regularization by intrinsic plasticity and its synergies with recurrence for random projection methods. Journal of Intelligent Learning Systems and Applications 4(3), 230–246 (2012)
29. Grupen, R.A., Huber, M.: A framework for the development of robot behavior. In: AAAI Spring Symposium Series: Developmental Robotics. Stanford University (2005)
30. Kaber, D.B., Riley, J.M.: Effects of visual interface design, and control mode and latency on performance, telepresence and workload in a teleoperation task. In: Proc. XIVth Triennial Congress of the International Ergonomics Association and 44th Annual Meeting of the Human Factors and Ergonomics Society, pp. 503–506 (2000)
31. Steinfeld, A., Fong, T., Kaber, D.: Common metrics for human-robot interaction. In: HRI 2006: Proceedings of the 1st ACM SIGCHI/SIGART Conference on Human-Robot Interaction, pp. 33–40 (2006)
32. Martin, D.P.: Resolution of kinematic redundancy using optimization techniques. IEEE Transactions on Robotics and Automation 5(4), 529–533 (1989)

33. Iossifidis, I., Steinhage, A.: Controlling a redundant robot arm by means of a haptic sensor. In: VDI BERICHTE: ROBOTIK 2002, Leistungsstand - Anwendungen - Visionen, pp. 269–274 (2002)
34. Kendall, D.G.: A survey of the statistical of shape theory. Statistical Science 4(2), 87–99 (1989)
35. Nordmann, A., Rolf, M., Wrede, S.: Software abstractions for simulation and control of a continuum robot. In: Noda, I., Ando, N., Brugali, D., Kuffner, J.J. (eds.) SIMPAR 2012. LNCS, vol. 7628, pp. 113–124. Springer, Heidelberg (2012)
36. Schreiber, G., Stemmer, A., Bischoff, R.: The fast research interface for the KUKA Lightweight Robot. In: IEEE ICRA 2010 Workshop on Innovative Robot Control Architectures, pp. 15–21 (2010)
37. Alami, R., Albu-Schäffer, A., Bicchi, A., Bischoff, R., Chatila, R., Hirzinger, G.: Safe and dependable pHRI in anthropic domains: State of the art and challenges. In: Proceedings of the 4th IARP/IEEE-RAS/EURON Workshop on Technical Challenges for Dependable Robots in Human Environments, vol. (1) (2006)

Part II

Robotic Grasping

Part II, Robotic Grasping, summarizes the results of five experiments funded by ECHORD focused on one of the most challenging research areas in robotics, i.e. grasping and manipulation. Besides the methodological and technological advancements that the consortia (most of them saw the cooperation between academic partners and an industrial-or SME partner) achieved in the course of their projects, the peculiarity of the research work conducted within such experiments is their strict relation to real-world applications, from bin picking in industrial environments to deboning operations in food industry.

The challenge in robotic manipulation is to replicate the human dexterity. Humans can grasp and manipulate an object dexterously through coordination of the hand-arm system, a highly redundant system, which we learn how to control through daily experience. What distinguishes human skills and current industrial robotic grasping systems is the ability of the humans to grasp and manipulate several objects with different shape, size, material, while most of the existing industrial robotic systems need to change grippers for grasping of different objects. The reason for such a big dissimilarity is in the dexterity of human fingers, that allows us to handle a pencil, a keyboard, a cup, a book and a fork with the same hand. A concrete example of what an artificial anthropomorphic hand can achieve can be found in the DEXDEB experiment, where the authors propose to adopt the highly dexterous Shadow hand as a robotic assistant in a deboning operation.

Even though the mechanism used for grasping is certainly one of the most relevant issues, it is not the only one. Grasping an object in a safe and robust way requires specific skills in planning the motion of the arm and the fingers in a coordinated fashion. Such planning highly relies on the availability of a good object model and on the knowledge of the object location with respect to the gripper. Therefore, many efforts need to be spent on object detection and recognition. Experiments ActReMa, LearnBiP and GRASPY propose a number of solutions for reliable object recognition and localization, even when a large number of similar objects are placed close to each other in a disordered pile.

Once the object is recognized and accurately located, depending on its shape, grasp and manipulation planning is a big challenge and most of the existing techniques depend on the specific kinematics of the hand/gripper. Grasp generation using dynamic simulators is proposed in the LearnBiP experiment an A*-based algorithm which follows a reachability map analysis is found in the GRASPY experiment, whereas the problem of grasp planning for a mobile manipulator is tackled in the ActReMa experiment via local multi-resolution height maps.

All the above grasp planning solutions are strictly dependent on the specific gripper or hand used, while the availability of a programming framework common to any hand kinematics would greatly simplify the entire process of robotic grasping. This is the objective of the HANDS.DVI experiment, which proposes a synergistic approach to easily program a paradigmatic hand and an algorithm to map the control action on the available hand is devised.

Following the work of the cited research teams allowed us to have a strict overview of the European robotics research dealing with grasping and manipulation and we have to admit that the dexterity of robots is still far behind that of humans. With this overview, we now provide a brief synopsis of each experiment contributing to Part II.

The experiment ActReMa addressed the problem of grasping individual objects from an unordered pile in a box with a mobile manipulator. The core of the approach is the representation of objects as compounds of simple shape and contour primitives, which allows both robust object perception and efficient grasp planning. The active perception technique of objects ensures robustness against noise, occlusions and missing information. In addition, it is able to offline learn new objects by single scans or CAD models. The local multi-resolution height map provides efficiency of the grasp planning process. Experimenters tested their algorithms with real objects like: cross-clamping pieces, drain pipe connectors, drilled plates.

The experiment LearnBiP aimed at improving performance of automatic bin-picking processes, which have a relevant role in industry; specifically the research results could allow to minimize cycle time and setup costs, thus potentially keeping production in countries with high labour cost. The main contribution of this experiment relies on a fast selection process of the most suitable grasp to take one of the objects out of the bin. The basis of the method is a dynamic grasp simulator used in conjunction with a method for learning grasp priorities of all the good grasps generated by the planner and thus achieve a fast and efficient automatic grasp selection process.

The experiment GRASPY a new functionality in the commercial robot NAO, i.e. an online object manipulation system, which allows the robot to grasp an object out of a human hand and give it back in real time. Object detection is realized via a stereo vision system that was specifically designed for the NAO head and had to take into account the severe constraints of space and computational power of the robot. Object grasping was tackled by resorting to the reachability map approach and it was the first time that the small NAO grasped objects with a single hand rather than using both arms emulating a two-finger gripper. The second part of the functionality, handing over the object to the human, was achieved by combining the speech recognition unit of the robot with its force sensing capabilities.

The experiment HANDS.DVI, had the objective to develop a common framework for programming of robotic hands independently from their kinematics, mechanical construction and sensor equipment. The proposed approach is based on the exploitation of grasp synergies, i.e. a reduced set of parameters that can be used to control all the degrees of freedom of the hand. Theoretical tools have been studied to design a suitable mapping function of the control action (decomposed in its elemental actions) from a "paradigmatic hand" model onto the articulated robotic hand co-domain. Interestingly enough, experimental tests have been conducted by grasping with different robotic hands a sensorized object capable of measuring contact forces.

The last experiment of Part II is DEXDEB, which tackled for the first time an application study of using dexterous robotic hands for deboning operations to establish a human-robot co-working platform for cutting, deboning and muscle extraction operation in meat industry. The intention was to substitute the left hand of the operator with a robotic gripper, hence the manual operation was deeply studied to define requirements that a robotic gripper should meet for assisting a human during the deboning operation. During the experiment, a metamorphic hand was designed and produced and compared with the commercial anthropomorphic hand C6M by Shadow robotics. The results gave useful insight into the benchmarking of utilizing dexterous hands in deboning robotized operation.

Active Recognition and Manipulation for Mobile Robot Bin Picking

Dirk Holz, Matthias Nieuwenhuisen, David Droeschel, Jörg Stückler, Alexander Berner, Jun Li, Reinhard Klein, and Sven Behnke

Computer Science Institute, University of Bonn, Germany
{holz,nieuwenhuisen,droeschel,stueckler,behnke}@ais.uni-bonn.de
{berner,lij,rk}@cs.uni-bonn.de

Abstract. Grasping individual objects from an unordered pile in a box has been investigated in stationary scenarios so far. In this work, we present a complete system including active object perception and grasp planning for bin picking with a mobile robot. At the core of our approach is an efficient representation of objects as compounds of simple shape and contour primitives. This representation is used for both robust object perception and efficient grasp planning. For being able to manipulate previously unknown objects, we learn object models from single scans in an offline phase. During operation, objects are detected in the scene using a particularly robust probabilistic graph matching. To cope with severe occlusions we employ active perception considering not only previously unseen volume but also outcomes of primitive and object detection. The combination of shape and contour primitives makes our object perception approach particularly robust even in the presence of noise, occlusions, and missing information. For grasp planning, we efficiently pre-compute possible grasps directly on the learned object models. During operation, grasps and arm motions are planned in an efficient local multiresolution height map. All components are integrated and evaluated in a bin picking and part delivery task.

Keywords: Mobile bin picking, contour and shape primitives, active object perception, grasp planning.

1 Introduction

Removing individual parts from unordered piles in boxes—*bin picking*—is one of the classical problems of robotics research [1–3]. So far, bin picking robots have been stationary. Typical setups consist of a 3D sensor mounted above the box, an industrial robot arm that is equipped with a gripper, and a data processing unit for detecting objects and planning grasping motions. Sometimes, the sensor is mounted at the manipulator, allowing for close view of selected parts. In order to extend the workspace of the robot and to make bin picking available for environments designed for humans, we implement bin picking using an autonomous mobile manipulation robot. Mobile bin picking is made feasible by the advances

F. Röhrbein et al. (eds.), *Gearing Up and Accelerating Cross-Fertilization between Academic and Industrial Robotics Research in Europe*, Springer Tracts in Advanced Robotics 94,
DOI: 10.1007/978-3-319-03838-4_7,

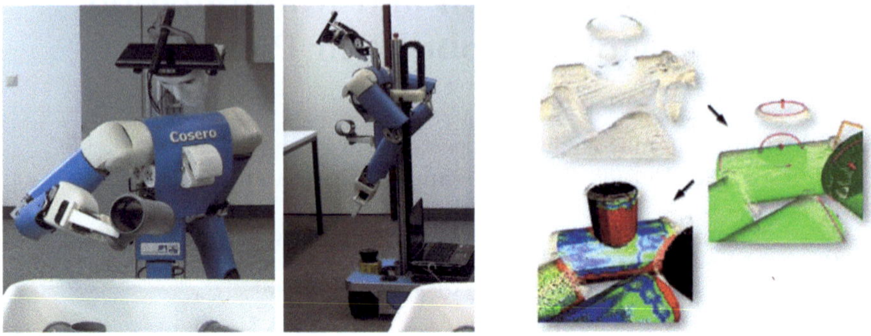

Fig. 1. Mobile bin picking scenario. Objects are grasped from a transport box (left) and placed on a processing station (middle). Both grasp planning as well as object detection and pose estimation are based on contour and shape primitives (right).

in sensing, computing, and actuation technologies, but still poses considerable challenges to object perception and motion planning.

Our scenario is motivated by industrial applications. We consider the task of grasping objects of known geometry from an unordered pile of objects in a transport box as well as transporting it to a processing station and placing it. Solving this mobile manipulation task requires the integration of techniques from mobile robotics like localization and path planning, and manipulation, like object perception and grasp planning. Our robot Cosero has been designed for mobile manipulation and intuitive human-robot interaction tasks, which were tested successfully in RoboCup@Home competitions [4].

While much research investigates the recognition of generic and organic objects in images and point clouds, such methods do not exploit properties of man-made objects that are often encountered in industrial and service robotics applications. Typically, man-made objects are composed of shape primitives such as cylinders, spheres, and planes as well as certain contour primitives (see, for example, the object in Fig. 1). We exploit this property and represent objects as compounds of shape and contour primitives. This representation is used for both robust object perception and efficient grasp planning. For object perception, we follow a popular line of work by matching graph models [5]. Both the object being searched for and the scanned scene are represented as compositions of geometric primitives in graphs. Object hypotheses are generated by identifying parts of a search graph in the graph of a captured scene. Using the established correspondences, one is able to determine the pose of the object in the scene. Here, we extend our previous work [6, 7] by using a combination of contour and shape primitives to increase robustness and reliability.

For being able to manipulate previously unknown objects, we present an approach that allows the robot to learn object models from single scans or CAD models in an offline phase. To cope with severe occlusions, we employ active perception that considers previously unseen volume and interest in particular regions in the form of primitive and object detection results. The combination of shape and contour primitives makes our object perception approach particularly robust even in the presence of noise, occlusions, erroneous measurements and missing information.

For mobility, we extend global navigation techniques by precise local alignment with the transport box and the processing station. For manipulation, we efficiently pre-compute possible grasps directly on the learned object models. During operation, grasps and arm motions are planned in an efficient local multi-resolution height map. We integrate all components to perform the complete task and evaluate the performance of the integrated system.

2 Related Work

Despite its long history, stationary bin picking is still an active research area. One recent implementation of Papazov *et al.* [8] utilizes a Microsoft Kinect sensor mounted above a table to acquire depth images of the scene. Object models are matched to the measured point cloud by means of geometric feature descriptors and RANSAC [9]. Papazov *et al.* consider tabletop scenes where multiple objects are arranged nearby, including the stacking of some objects. They select the object to be grasped based on the center of mass height (high objects are preferred). Each object is associated with a list of predetermined grasps, which are selected according to the orientation of the gripper. The grasping is performed by a compliant lightweight robot arm with parallel gripper. Bley *et al.* [10] propose another approach of grasp selection by fitting learned generic object models to point cloud data. In contrast to our approach they manipulate separated objects. Choi *et al.* [11] proposed a Hough voting-based approach that extends point-pair features [12, 13], which are based on oriented surface points, by boundary points with directions and boundary line segments. Choi *et al.* use a structured-light 3D sensor mounted on an industrial arm to acquire point clouds of small objects in a transport box, and grasp them with a high success rate. Another extension of Drost *et al.* [13] has been proposed by Kim and Medioni [14]. They consider visibility in between the paired surface elements to sort out false matches.

While the above methods for object detection work best with objects containing distinct geometric features, other approaches rely on the decomposition of point clouds into geometric primitives. The method proposed by Schnabel *et al.* [15] is based on RANSAC and efficiently detects planes, spheres, cylinders, cones, and tori in the presence of outliers and noise. Another work in this direction is Li *et al.* [16], who build a graph of primitive relations and constraints. Assuming symmetry and consistent alignments of shapes in man-made objects, the orientations and positions of detected shapes are iteratively refined. The above approaches require dense depth measurements. In contrast, Liu *et al.* [3] developed a multi-flash camera to estimate depth edges. Detected edges are matched with object templates by means of directional Chamfer matching and objects are grasped with a three-pin gripper that is inserted into a hole at a success rate of 94 %.

Manipulation in constrained spaces like boxes and shelves requires manipulators with more than the minimal 6 DoF and leads to difficult high-dimensional motion planning problems. Cohen *et al.* [17] proposed a search-based motion planning algorithm that combines a set of adaptive motion primitives with motions generated by two analytical solvers.

Chitta *et al.* [18] proposed an approach to mobile pick-and-place tasks integrating 3D perception with grasp and motion planning. This approach has been used for applications like tabletop object manipulation and the transportation of objects. In these applications, objects stand well-separated on horizontal surfaces or are ordered in feeders. Klingbeil *et al.* [19] utilized a Willow Garage PR2 robot to grasp unknown objects from a pile on a table and read their bar-codes to demonstrate a cashier checkout application. Because the dense packing of objects in a pile poses considerable challenges for perception and grasping, Chang *et al.* [20] proposed pushing strategies for the interactive singulation of objects. Gupta and Sukhatme [21] estimate how cluttered an area is and employ motion primitives to separate piled Lego bricks.

Other systems for which mobile pick-and-place has been realized include HERB [22], developed at the Intel Research Lab Pittsburgh. HERB navigates around a kitchen, searches for mugs and brings them back to the kitchen sink. Rollin' Justin [23], developed at DLR Oberpfaffenhofen, grasped coffee pads and inserted them into the coffee machine, which involved opening and closing the pad drawer. The Armar robots [24], developed at KIT, demonstrated tasks in a kitchen scenario that require integrated grasp and motion planning. In the health care domain, Jain and Kemp [25] present EL-E, a mobile manipulator that assists motor impaired patients by performing pick and place operations. Beetz *et al.* [26] used a PR2 and the robot Rosie, developed at TU Munich, to cooperatively prepare pancakes, combining mobile manipulation and tool-use.

In most of these mobile manipulation demonstrations, the handled objects are well-separated. To the best of our knowledge, mobile bin picking has not yet been realized.

3 Representing, Detecting, and Learning Object Models

In our approach, objects are represented by compositions of contour and shape primitives. We model such compositions by a graph. Detected primitives, and the model parameters describing them, form the vertices of the graph. Spatial relations between neighboring primitives are encoded in the edges. Given the graph of a query object, we convert an input point cloud into a graph modeling the captured scene and find sub-graphs that match the graph of the query object.

3.1 Scan Acquisition and Preprocessing

Detection of Transport Boxes. In addition to the 3D sensor for part perception, we use three horizontally mounted laser scanners in the robot's base and torso for localization and collision avoidance. One scanner in the robot's torso is mounted slightly above table height and used for detecting transport boxes and aligning the robot to detected boxes (see Fig. 2). To detect a transport

Fig. 2. Two views on a scanned transport box with the extracted line segment (green) corresponding to the box's front side and the estimated transport box model (red)

box, we continuously extract line segments from the 2D laser scan. We check the straightness of line segments by principle component analysis and neglect those exceeding a given curvature. The closest remaining line segments that fit the dimensions of the transport box correspond to its rim.

Acquisition and Alignment of Point Clouds. For perception of the content of the transport box, we use a Microsoft Kinect camera that is mounted on the robot's pan-tilt unit. To compensate for the sensor's limited field-of-view, we acquire three overlapping point clouds from different views (left, middle, right). For obtaining a consistent scan of the transport box, we register the three point clouds with the Iterative Closest Point (ICP) algorithm [27]. For efficiency, we remove duplicate points from overlapping areas as well as outliers.

Preprocessing. For the detection of geometric primitives we rapidly calculate local surface normals for the residual points. For being able to detect contour primitives we extract all points belonging to sharp edges and occlusion boundaries. We employ the algorithm by Bendels *et al.* [28] that computes, for every point, a contour probability using various criteria: 1) the angle between the query point's normal and the normals of the neighbors, 2) the relative positions of the neighbors to the query points and 3) the shape of the underlying surface at the query point (encoded in the eigenvalues of the covariance matrix).

3.2 Primitive Detection

Shape Primitive Detection. In order to detect simple geometric shape primitives such as planes, cylinders, and spheres, we employ the algorithm by Schnabel *et al.* [29] based on random sampling. It decomposes a point cloud $P = p_1, \ldots, p_N$ into a set of geometric shape primitives ϕ_i with support points S_{ϕ_i} and a set of remaining points R:

$$P = S_{\phi_1} \cup \cdots \cup S_{\phi_A} \cup R. \qquad (1)$$

Each support set S_{ϕ_i} is a connected component of points that are 1) close to the primitive (distance $< \epsilon$) and 2) compatible w.r.t. the angle (angle $< \alpha$) between

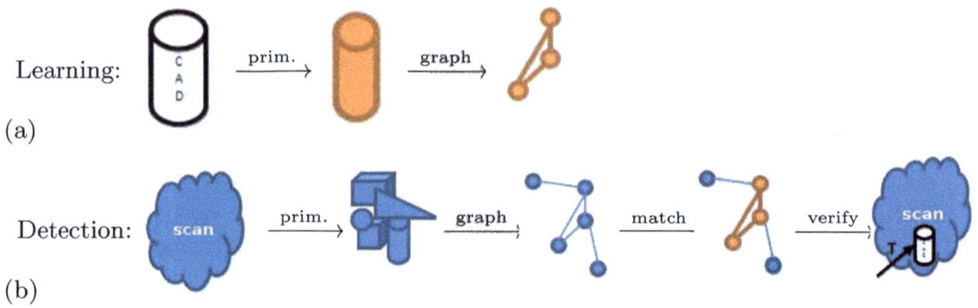

Fig. 3. Object learning and detection: (a) We detect contour and shape primitives and construct a graph encoding their composition. (b) Object hypotheses are obtained from constrained sub-graph matching and verified in the original point cloud.

surface normals at the point n_s and on the primitive at the closest point on the primitive $n(\phi_i, s)$:

$$s \in S_{\phi_i} \Rightarrow \|s, \phi_i\| < \epsilon \wedge \angle\left(n_s, n(\phi_i, s)\right) < \alpha. \quad (2)$$

As an important extension to [29], we distinguish two different phases of detecting (shape) primitives: offline learning of new objects and online detection of known objects (see Fig. 3). For learning of new objects given CAD models, we first sample the model uniformly to obtain a 3D point cloud and aim to find an optimal decomposition into arbitrary shape primitives (open model parameters). In the online phase, we save computations by constraining the primitive detection to only find primitives existing in the query object (fixed model parameters).

Contour Primitive Detection. Given the set of points lying on sharp edges and occlusion boundaries in a point cloud, we aim to find contour primitives that help in detecting objects and resolving ambiguities. To this end, we only consider circular contours since our objects of interest either contain cylinders or cylindric holes that result in circular contours. Additionally, as circles have only one shape parameter excluding the position and orientation—the radius—we are able to detect them very robustly in noisy and occluded data. As for the shape primitive detection, we use a RANSAC-based approach and distinguish between online and offline phases. In the offline phase, we fit circle hypotheses by sampling three points, estimating the common plane, and determining the center of the circle through the points. In the online phase, the points are chosen according to the searched radii and efficiently tested using a fast octree implementation that is also used in the shape primitive detection. We only accept hypotheses with a sufficient number of supporting inliers.

3.3 Object Detection

Graph Construction. We efficiently create the annotated shape graph of the scanned point cloud and apply our sub-graph matching approach following ideas of [15]. For each detected shape primitive ϕ_i a vertex is added to the topology graph $G(\Phi, E)$, i.e., $\Phi = \phi_1, \ldots \phi_a$. An edge $e_{ij} = (\phi_i, \phi_j)$ is added if the support sets of the primitives ϕ_i and ϕ_j are neighboring, i.e.:

$$\exists p \in S_{\phi_i}, q \in S_{\phi_j} \ : \ \|p - q\| < t \tag{3}$$

with t denoting a distance threshold.

Three types of constraints are encoded in the graph: *node constraints* for the similarity of primitives (model parameters such as type and size), *edge constraints* for the similarity of spatial relations (e.g., the angle between two planes), and *graph constraints* given only implicitly by the topology in the graph (e.g., parallelism of disconnected planes). Detected contour primitives are added to the graph $G(\Phi, E)$ already containing detected shape primitives. False positives in the online contour detection will be pruned in our graph matching approach, as their relative pose is inconsistent with the object model graph.

Graph Matching. The input to our graph matching method is a graph of contour and shape primitives of the query object to be searched. We start with a random edge in the query graph and find similar edges in the scene graph. We compare edges by their relative pose and nodes by their shape properties, and compute a score for each match. This score is zero if the types of primitives do not match and increases with the similarity in relative pose and shape parameters. For each matching edge, we expand the match to adjacent edges (and nodes) in the query and scene graph if they also match in relative pose and shape properties. Expansion is stopped if either the whole query graph matches or no further corresponding edges can be found in the scene graph. The process is repeated in order to find multiple objects in the scene.

Pose Estimation and Verification. We determine object poses from partial matches between the model and scene graph. Depending on the type of the shape primitive, each correspondence determines some of the six degrees of freedom of the pose. A circle-to-circle correspondence, for example, completely determines the translation between the circle centers and two rotational degrees of freedom by the alignment of the circle planes. It does not, however, determine the rotation about the axis perpendicular to the circle plane through its center. Hence, we require several correspondences between primitives until the pose of the object is fully retrieved. For symmetric objects, we can take any transformation around the self-symmetry axis and compute a valid transformation. It is possible to detect all self-symmetries when learning a new object model (cf., [30]).

Computed object poses are refined by registration of the model to the measured points using ICP [27]. As the previous steps only consider the matching between contour and shape primitive but not their consistency with the overall scan, false-positives may be generated. In a verification step, we check the overlap of object hypotheses with the actual point cloud and remove those with insufficient overlap (15% in our experiments). In case several hypotheses overlap, we remove the hypotheses with lower overlap and only keep the best.

3.4 Learning New Object Models

In industrial applications, CAD models of the objects are often available beforehand and can be used for creating new object models and for computing feasible grasps. For applications where no CAD models are available, we propose a simple approach for learning new object models. Multiple object exemplars are placed on a designated learning board in different, roughly known orientations, such that a single scan is sufficient to obtain all information needed.

In a preprocessing step, we compute surface normals and remove outliers. Referring to Fig. 4, we first use the shape primitive detection to find the dominant plane and to register the scanned scene to our model of the board. We then segment all points above the dominant plane and separate them into individual point clouds. The result is equivalent to the same number of scans taken from various viewpoints, with the difference that we know a rough transformation for each scan. In order to obtain a complete 3D model of the exemplar, we register segments with similar orientations in a pair-wise fashion following a coarse-to-fine approach. For an initial alignment, we compute point-pair features [31] on both point clouds to be registered, match the resulting feature descriptors and apply RANSAC to find a consistent set of matches and the corresponding transformation aligning the two clouds. We only allow transformations that are consistent with the ones given by the learning board. The found transformation is refined using ICP [27]. The segmented point clouds are then merged using the estimated inter-pair transformations. For being able to visually inspect the result, we reconstruct the surface of the scanned object using Poisson surface reconstruction [32] and present it to the user.

Regardless of whether we construct a 3D point model of the object as described above or sample the point cloud from the surface of an available CAD model, the final object model is constructed by first detecting shape and contour primitives, and then creating the graph modeling the spatial relationships between the detected primitives. Fig. 4 shows examples of learned object models.

4 Active Object Perception

With increasing geometric part complexity or depth in the pile, large surface parts are likely to be occluded and cannot be acquired from a single viewpoint. A single scan is thus prone to being incomplete and may provide only fragments of the actual part surfaces leading to considerable uncertainty in object detection and pose estimation. In order to achieve robustness of object perception with respect to occlusions, we have developed an active object perception method for

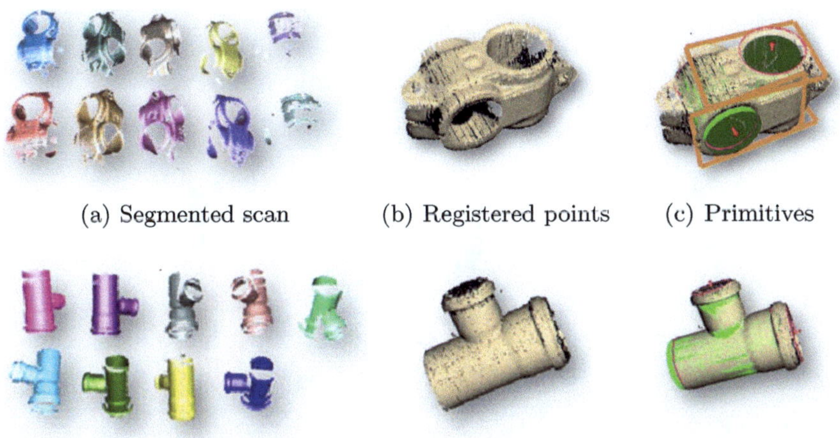

(a) Segmented scan (b) Registered points (c) Primitives

(d) Segmented scan (e) Registered points (f) Primitives

Fig. 4. For learning new object models in the offline phase, we first segment the scan (a+d), then register the residual points (b+e), and detect primitives (c+f)

actively moving the sensor to various view poses. Referring to Fig. 5, we apply the following procedure to explore the transport box and actively detect objects: we register newly acquired scans and update the so far built model of the transport box content. From the model, we can deduce previously occluded volumes. In addition, we represent the outcomes of primitive and object detection in the model as a measure of interest in the corresponding region. Both measures are considered when planning the next view from which a scan is to be acquired. For planning the next best view, we apply a sampling-based approach. We consider the box to be explored and stop the exploration if no more unknown or unseen volume exists within the transport box.

Preprocessing, Registration, and Transport Box Modeling. For fusing and aligning acquired scans, we incrementally build a volumetric model of the transport box (see Fig. 6) and register newly acquired scans. For registration, we have extended an approach to incremental registration from our previous work [33]. The aligned scan points are added to the model while avoiding to add duplicate points. For efficiency, we restrict both the processing of scans and the volume being modeled to the extents of the transport box. From the laser-based box detection and the acquired 3D scan, we deduce an oriented bounding box for the transport box and discard all points lying outside. Moreover, the laser-based measurements of the rim allow us to predict the so far unseen transport box volume. The model is represented as a multiresolution voxel grid map based on [34]. It is organized as an octree with leaves that model multiple attributes of the underlying volume, e.g., the object detection's interest in that region or the volume's occupancy. In addition to constraining the modeled volume to the

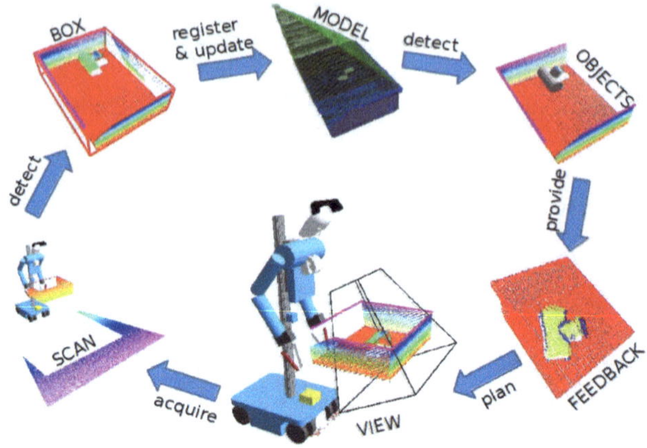

Fig. 5. Active perception pipeline: In a newly acquired scan (bottom), we first detect the transport box, then register the scan and update the so far built model. Detected objects provide feedback for focusing view planning to interesting regions. We then approach the planned view and acquire a new scan.

transport box, we further increase efficiency by performing a lazy evaluation by first building the tree, and then recursively updating the content of the changed cells. We employ efficient ray-casting within the bounding box boundaries that allows for fast integration of new measurements and adapting the model to scanned changes in the scene.

Sample Generation and Travel Cost. We apply a sampling-based approach for determining the next best view. We first generate a set of sample view poses $\mathbf{V}$ and then estimate, for every sample view $\mathbf{v} \in \mathbf{V}$, the involved traveling cost $L(\mathbf{v}, \mathbf{r})$ from the robot's current pose $\mathbf{r}$ and the expected information gain $I(\mathbf{v})$. The view with highest utility, i.e., with a high information gain and low traveling cost, is selected as the next best view $\mathbf{v}^\star$:

$$\mathbf{v}^\star = \operatorname*{arg\,max}_{\mathbf{v} \in \mathbf{V}} I(\mathbf{v})\ e^{-\lambda L(\mathbf{v},\mathbf{r})} \tag{4}$$

where λ is a parameter to trade off traveling cost and information gain. As approaching a new view pose close to the box only involves local navigation with the robot's omnidirectional base, we approximate the involved traveling cost $L(\mathbf{v}, \mathbf{r})$ by the Euclidean distance between the robot's current pose and the base poses of the sampled views. Typical camera view samples and a summary of the applied scheme for encoding the interest in certain regions are visualized in Fig. 7.

Identifying Regions of Interest. Our approach for detecting contour and shape primitives as well as objects composed of such primitives gives us detailed

(a) Raw input scans (b) Aligned scans (c) Updated model

Fig. 6. Raw input scans (a) are incrementally registered (b) to update a consistent model of the transport box and its content (c). The model distinguishes seen free space (green) and previously unseen volume, and encodes—for each cell—the probability of being occupied and the object detection's interest in this region.

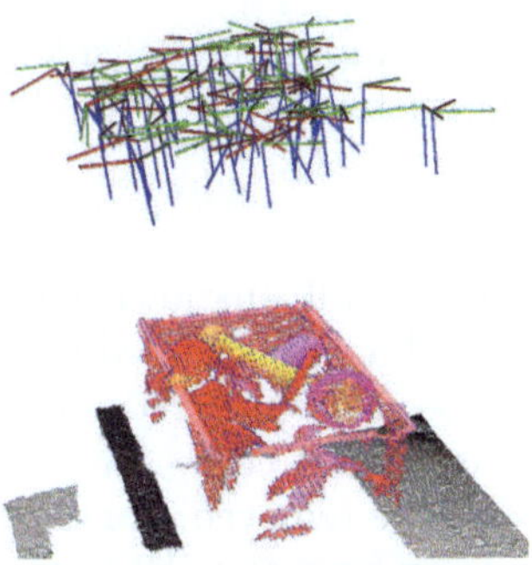

(a) Sampled view poses

Class	Primitive detection	Object detection	Interest
1	None	None	High (0.75)
2	Positive	None	High (1.0)
3	Positive	Positive	Low (0.25)
4	Negative	–	None (0.0)

"Positive": point lies on searched primitive/object
"Negative": point does not lie on searched primitive

(b) Coding *region interest*

Fig. 7. View planning: (a) examples of sample camera view poses; (b) and scheme for encoding the object detection's interest in certain regions

feedback on all regions in an input point cloud to guide the further acquisition of scans to regions having no or only little confidence in detected objects. Referring to the coding scheme in Fig. 7(b), when a point is detected to lie on a primitive that is not belonging to the object's shape primitive compound we are searching for (class 4), the region around the point is less interesting than regions where object detections are still possible (classes 1 and 2). We model interest as an attribute for each cell in the model in addition to the cell's occupancy.

Information Gain Estimation. Classic approaches to view planning consider previously unseen volume and previously seen surface. Given a model m, the expected information gain $I_{\text{model}}(\mathbf{v})$ from adding measurement z at view pose $\mathbf{v}$ is then approximated by the change in entropy H before and after adding z:

$$I_{\text{model}}(\mathbf{v}) = H(m) - H(m|z). \tag{5}$$

Several simplifying assumptions can be made [34] that finally allow for approximating (5) using ray-casting from the sampled view pose $\mathbf{v}$ towards the modeled transport box and computing the information gain $I_{\text{model}}(\mathbf{v})$ in closed form by considering the number of unknown cells along the ray.

For actively guiding object detection, we focus view planning on sensing regions of interest in addition to considering previously unseen volume. For this purpose, we integrate primitive and object detection results in the model. Referring to the applied coding scheme, regions that possibly contain an object but where no object has been detected yet are more interesting and have a higher interest score, respectively. In order to draw the robot's attention to interesting regions, we compute an interest score related measure $I_{\text{interest}}(\mathbf{v})$ as the sum of interest values over all cells visible from the sensor pose, and combine both measures for the final information gain $I(\mathbf{v})$:

$$I(\mathbf{v}) = \alpha \frac{I_{\text{model}}(\mathbf{v})}{I_{\text{model}}(\mathbf{v})^{\star}} + \beta \frac{I_{\text{interest}}(\mathbf{v})}{I_{\text{interest}}(\mathbf{v})^{\star}} \tag{6}$$

where $I_{\text{model}}(\mathbf{v})^{\star}$ and $I_{\text{interest}}(\mathbf{v})^{\star}$ are, respectively, the maximum model information gain and interest score over all samples. The parameters α and β can be used to weight the two measures (in our experiments $\alpha = \beta = 1$).

Implementation Details. For efficiency, we first extract all unknown cells (previously unseen volume) from the transport box, as well as all occupied cells (previously seen surface). If regions of modeled free space are considered interesting, they are handled extra and in addition to the extracted cells. All cells outside the sensor's view frustum (at the sampled view $\mathbf{v}$) are removed. For each of the the remaining cells, we conduct a reverse ray-casting from the cell to the view pose to determine the cell's visibility and update $I_{\text{model}}(\mathbf{v})$ and $I_{\text{interest}}(\mathbf{v})$ accordingly. To further speed up this step, we limit resolution and depth in the tree. Furthermore, we only consider those segments of the ray that are contained in the oriented bounding box of the transport box model.

Overall, our approach allows focusing the acquisition of new range scans in regions where we expect to find objects. For planning these views, we can compute and evaluate 100 samples per second.

5 Mobile Manipulation

For manipulating objects in our mobile bin picking scenario, we employ efficient grasp planning, motion planning for reaching objects on collision-free paths, and a global-to-local navigation strategy for moving between the transport box and the processing station.

Grasp Planning. We plan grasps in an efficient multistage process that successively prunes infeasible grasps using tests with increasing complexity. In the first stages, we find collision-free grasps on the object, irrespective of the pose of the object and not considering its scene context (see Fig. 8(a)). These poses can be pre-calculated efficiently in an offline planning phase. We sample grasp poses on the shape primitives. From these poses, we extract grasps that are collision-free from pre-grasp pose to grasp pose according to fast collision check heuristics.

During online planning, we examine the remaining grasp poses in the actual poses of the objects to find grasps where a collision-free solution of the inverse kinematics in the current situation exists. We filter grasps before evaluation against our height map and finally search collision-free inverse kinematics solutions for the remaining ones. We allow collisions of the fingers with other parts in the transport box in the final stage of the grasp, i.e., in the direct vicinity of the object to grasp. The shape of the fingers allows for pushing them into narrow gaps between objects. Our grasp planning module finds feasible, collision-free grasps at the object. The grasps are ranked according to a score that incorporates efficiency and stability criteria.

Motion Planning. We distinguish two types of motions that need to be planned: the motion for grasping the object and the motion for removing it from the transport box. For planning the grasping motion, we identify the best-ranked grasp that is reachable from the current posture of the robot arm. We solve this by successively planning reaching motions for the found grasps. We test the grasps in descending order of their score. For motion planning, we employ LBKPIECE [35]. In order to further increase performance, the grasping motion is again split into multiple segments. This allows for a quick evaluation if a valid reaching motion can be found by planning in descending order of probability that planning for a segment will fail.

After the execution of the reaching motion, we check if the grasp was successful. If the object is within the gripper, a removal motion is planned with the object model attached to the end-effector using the detected object pose. We allow minor collisions of the object and the end-effector with the collision map in a cylindrical volume above the grasp pose.

To reach the processing station, global navigation and local alignment are used in the same way as for approaching the box. Finally, the work piece is deposited at the processing station.

Multiresolution Height Map. We employ a multiresolution height map that extends our prior work on multiresolution path planning [36] for efficiently evaluating collision-free grasp postures and planning trajectories. The height map is represented by multiple grids that have different resolutions. Each grid has $M \times M$ cells containing the maximum height value observed in the covered area (Fig. 8(b)). Recursively, grids with quarter the cell area of their parent are embedded into each other, until the minimal cell size is reached. With this approach, we can cover the same area as a uniform $N \times N$ grid of the minimal cell size with only $\log_2((N/M) + 1)M^2$ cells.

Planning in the vicinity of the object needs a more exact environment representation than planning farther away from it. This is accomplished by centering the collision map at the object. This approach also leads to implicitly larger safety margins with increasing distance to the object.

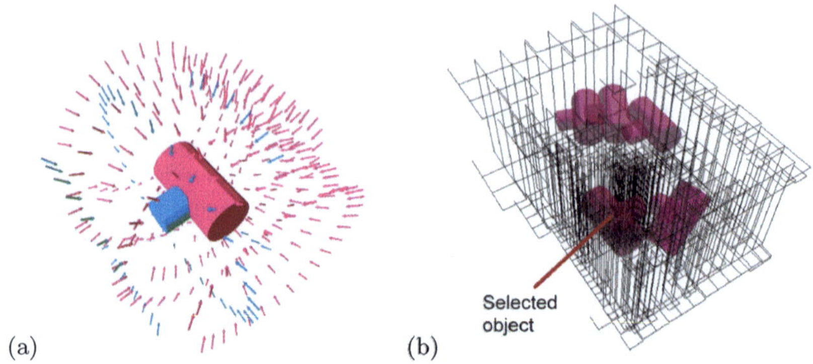

Fig. 8. Grasp planning: (a) We sample possible grasps and pre-grasp poses (visualized as arrows by color of primitive) for each shape primitive according to its parametric description, and discard those that are in collision within the object. (b) For grasp selection and motion planning we use a multiresolution height map.

Navigation. We use a global-to-local strategy for approaching the location where transport boxes are delivered to and the location where work pieces are processed. This removes the need of knowing the exact locations. Instead we assume an environment model in the form of a 2D map and a rough estimate of the locations. We navigate globally to the approximate locations, and accurately align the robot locally with the transport box and the processing station, respectively.

For global navigation, we employ state-of-the-art methods for localization and mapping in 2D representations of the environment. Adaptive Monte Carlo Localization [37] is used to estimate the robot's pose in a given occupancy grid map using a laser-range finder. We use A* search [38] to find the shortest obstacle-free path from the estimated pose in the map to the target location.

In order to maximize the workspace of the robot and allow for active perception, an accurate alignment between the robot and, respectively, the transport box and the processing station is necessary. We achieve this alignment by locally navigating to a predefined pose relative to the deduced box model (or model of the processing station) or to poses planned by the active perception component. This reactive alignment makes our approach robust against variations in the poses of transport boxes and processing station.

6 Experiments and Results

We tested the integrated system with our cognitive service robot Cosero [4]. For the detection experiments, we have chosen three types of objects (see Fig. 9). The objects are similar to parts found in real-world construction applications—a rectangular wooden plate containing two drilled holes (manually created CAD model), a cross clamping piece (CCP), a typical construction part to connect poles (CAD models are freely available), and a drain pipe connector (scanned from different perspectives and registered to obtain a 3D model).

Model Learning. For every object we have first computed a query graph using the offline variant of our approach. All reconstructions could be performed within 12 to 15 seconds. We compared the parameters of the detected primitives in the reconstruction to ground truth parameters and observed deviations of about 2% to 3%, using a precise 3D scanner. To judge the quality of our detected primitives for object detection, we compared the models learned from the scans with handcrafted models based on ground truth data. The observed recognition results were similar.

Object Detection and Pose Estimation. To assess the perception of our object detection approach, we acquired—for each object model—scans in five differently piled heaps of objects. We compared our shape and contour approach to our previous approach using shape primitives only [6, 7], and to an implementation of the state-of-the-art approach by Papazov and Burschka [31] based on point-pair features (PPF) with parameters as recommended by the authors. For qualitative analysis and visual inspection we show example scans and object detections for all approaches in Fig. 9(a). Quantitative results are summarized in the table in Fig. 9(b). As the PPF-based approach is randomized, it is run for ten times on every input scan to determine the average detection rate. For the class of objects in our problem setting our method clearly outperforms the other two methods.

Compared to the shape-only approach [6, 7], adding contour primitives improves detection results by resolving object pose ambiguities. In case of the CCP, we often find only a single cylinder that does not suffice for determining the object pose as it leaves two degrees of freedom undetermined. Adding circular contour primitives yields unique pose estimates and successful detections. The worst case for the shape-only approach is the scan of the wooden plates lying nearly flat on the table. The objects cannot be detected as the only found primitives are indefinitely extending planes with three open degrees of freedom. By combining the planes with the contour primitives, all objects with estimable pose can be detected.

Compared to the PPF-based approach (showing outstanding performance for objects of generic shape) we also achieve better detection rates. Man-made objects as addressed in our work are usually formed by compositions of few simple geometric primitives. These are easy to detect for our approach but also have the property of less varying local surface normals. This is disadvantageous for geometric features such as PPFs since feature descriptors computed in these areas do not provide unique transformations and can cause false positives. Our approach does not produce any false positives, because it requires only a minimal number of shape and contour matches. A restriction of our solution is that it is only suitable for objects that can be described by contour and shape primitives. It cannot be applied to arbitrary organic objects.

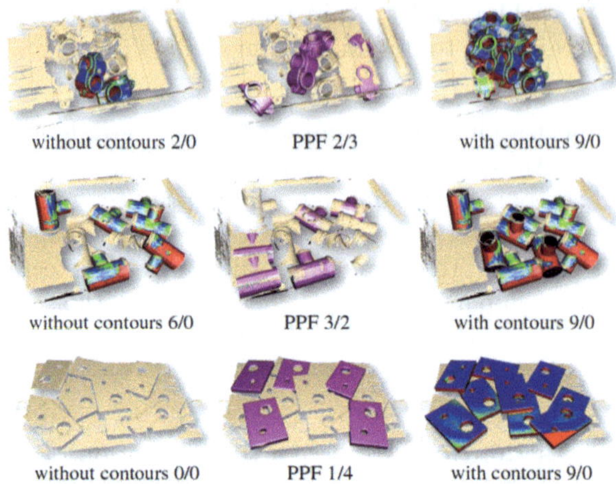

(a) Examples (true/false positives)

	Ours	Only 3D	PPF
CCP	**0.81**	0.23	0.29
Pipe	**0.89**	0.47	0.27
Wood	**0.82**	0.21	0.08
Overall	**0.84**	0.30	0.22

$$\text{acc.} = \frac{\text{true positives}}{\text{t. pos.} + \text{f. neg.} + \text{f. pos.}}$$

(b) Detection accuracy

Fig. 9. Results for object detection: (a) typical detections for the different objects used in our experiments, (b) overall average detection accuracies of the evaluated methods

Active Perception. For a proof-of-concept and a qualitative analysis of the active component of our perception pipeline, we conducted a series of roughly 50 experiments where the robot had to explore different arrangements of pipe connectors in transport boxes (ranging from one to ten objects per box). After successful verification object hypotheses were accumulated and tracked using a multi-hypotheses tracker. Regions where no object was detected, but which could have contained objects drew the robot's attention when planning the next view. In all experiments, active perception resulted in a fully explored transport box and at least one graspable object. Fig. 10 shows typical examples where the active perception component leads to new object detections and a complete model of the transport box. In cases of severe occlusions, the robot has been able to find occluded objects after acquiring a second or third view. No more than three views were needed to find an object. In most cases an object was already detected in the first view.

Mobile Bin Picking. For assessing the performance of the overall system, we have recorded 32 runs, in which the robot picks up one pipe connector object from the transport box (filled with 10 pipe connectors) and delivers it to the processing station. In 28 runs, the robot could successfully grasp and deliver the object. In nine of these successful runs, the robot first failed to grasp an object, detected its failure, and performed another grasp. This was the case, for instance, when the object slipped out of the gripper after grasping. In four runs, the object was not successfully delivered to the processing station. In three out of

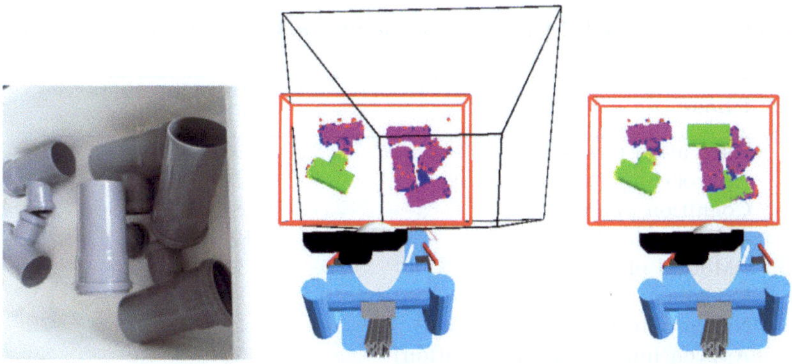

(a) Scene (b) Detections + NBV (c) Second view

Fig. 10. Active perception: occlusions in the scene (a) hinder the robot in finding all objects. Considering the primitive (magenta) and object (green) detections when planning the next best view (b), allows for finding more objects in the second view (c).

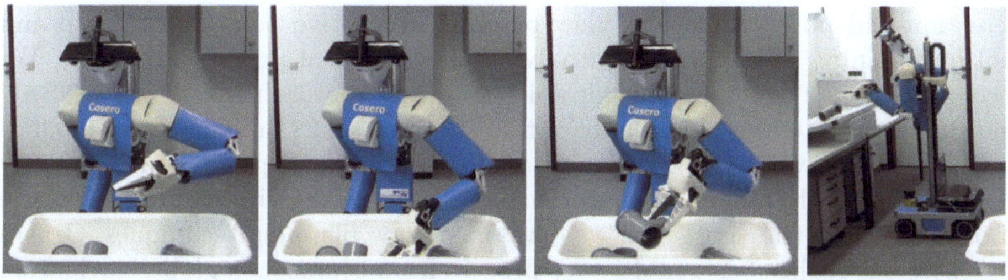

Fig. 11. Example of a mobile bin picking and delivery run. From left to right: the robot grasps an object out of the transport box and puts it on the processing station.

the four failed runs the last object could not be detected. In one run, the object slipped out of the gripper after lifting. This is caused by the collisions between the gripper and other objects that we have to allow during the grasp. These minor collisions can cause changes in the object's pose that can make the chosen grasp impossible or unstable. Some images from one of the runs are shown in Fig. 11. Table 1 shows the mean and standard deviation of the measured phase durations for the 32 individual runs. Please note, that the timings for the grasp selection and motion planning within the cognition phase are averaged over the ten runs it took to clear one completely filled box. One can see that the longest phase is the cognition phase where objects are detected and the grasping motion is planned. This phase also includes the transmission of the sensor data to the object recognition module on a physically distinct computer on the robot.

Table 1. Time needed for phases of the mobile bin picking demonstration

Phase	**Duration** (in sec.) Mean	Std. dev.
Navigation (transport box)	20	8
Approaching (transport box)	16	11
Cognition phase	83	41
- Grasp selection	19.9	14.4
- Motion planning	3.8	2.6
Grasping	36	7
Navigation (processing station)	26	9
Approaching (processing station)	22	9
Putting the object on the processing station	18	2

7 Conclusion

In this work, we presented an integrated system for a mobile bin picking application. It combines manipulation and navigation skills, including: 3D object detection and pose estimation, view planning for active perception, grasp and motion planning as well as global navigation and local precise alignment. We recognize objects using an efficient noise-resistant approach based on RANSAC and subgraph matching. In order to obtain the necessary shape and contour composition graphs, we derive models automatically from 3D point clouds or CAD files. To cope with imperfect measurements and occlusions in the unordered pile of objects in the transport box, we developed view planning techniques. Grasping objects is realized as a multistage process from coarse, i.e., global navigation in the environment, to fine, i.e., planning a collision-free end-effector trajectory within a multiresolution collision map. Intermediate steps align our robot to the transport box and the processing station using local sensing and navigation, and evaluate the graspability of objects using fast heuristics.

We showed the applicability of our approaches in a mobile bin picking and part delivery task in our lab. A video summarizing our work is available on our website[1]. Among other skills, we demonstrated mobile bin picking in the @Home final of RoboCup 2012 in Mexico, where our robots convinced the high-profile jury and won the competition.

Acknowledgments. This work was supported by the EU Project EC FP7-ICT-231143 ECHORD.

References

1. Ikeuchi, K., Horn, B.K.P., Nagata, S., Callahan, T., Feirigold, O.: Picking up an object from a pile of objects. In: Robotics Research: The First International Symposium, pp. 139–162. MIT Press (1984)

[1] `www.ais.uni-bonn.de/ActReMa`

2. Rahardja, K., Kosaka, A.: Vision-based bin-picking: Recognition and localization of multiple complex objects using simple visual cues. In: Proc. IEEE Int. Conf. on Intelligent Robots and Systems (1996)
3. Liu, M.-Y., Tuzel, O., Veeraraghavan, A., Taguchi, Y., Marks, T.K., Chellappa, R.: Fast object localization and pose estimation in heavy clutter for robotic bin picking. Int. J. of Robotics Research 31(8), 951–973 (2012)
4. Stückler, J., Holz, D., Behnke, S.: RoboCup@Home: Demonstrating everyday manipulation skills in RoboCup@Home. IEEE Robotics & Automation Magazine 19(2), 34–42 (2012)
5. Felzenszwalb, P.F., Huttenlocher, D.P.: Pictorial structures for object recognition. Int. J. Comput. Vision 61(1), 55–79 (2005)
6. Nieuwenhuisen, M., Stückler, J., Berner, A., Klein, R., Behnke, S.: Shape-primitive based object recognition and grasping. In: Proc. 7th German Conference on Robotics (2012)
7. Nieuwenhuisen, M., Droeschel, D., Holz, D., Stückler, J., Berner, A., Li, J., Klein, R., Behnke, S.: Mobile bin picking with an anthropomorphic service robot. In: Proc. IEEE Int. Conf. Robotics and Automation, pp. 2319–2326 (2013)
8. Papazov, C., Haddadin, S., Parusel, S., Krieger, K., Burschka, D.: Rigid 3D geometry matching for grasping of known objects in cluttered scenes. Int. J. of Robotics Research 31(4), 538–553 (2012)
9. Fischler, M.A., Bolles, R.C.: Random sample consensus: a paradigm for model fitting with applications to image analysis and automated cartography. Commun. ACM 24(6), 381–395 (1981)
10. Bley, F., Schmirgel, V., Kraiss, K.-F.: Mobile manipulation based on generic object knowledge. In: Proc. IEEE Int. Symp. on Robot and Human Interactive Communication (2006)
11. Choi, C., Taguchi, Y., Tuzel, O., Liu, M.-Y., Ramalingam, S.: Voting-based pose estimation for robotic assembly using a 3D sensor. In: Proc. IEEE Int. Conf. Robotics and Automation (2012)
12. Wahl, E., Hillenbrand, U., Hirzinger, G.: Surflet-pair-relation histograms: a statistical 3D-shape representation for rapid classification. In: Proc. Int. Conf. on 3-D Digital Imaging and Modeling (2003)
13. Drost, B., Ulrich, M., Navab, N., Ilic, S.: Model globally, match locally: Efficient and robust 3D object recognition. In: Proc. IEEE Conf. on Computer Vision and Pattern Recognition (2010)
14. Kim, E., Medioni, G.: 3D object recognition in range images using visibility context. In: Proc. IEEE Int. Conf. on Intelligent Robots and Systems (2011)
15. Schnabel, R., Wessel, R., Wahl, R., Klein, R.: Shape recognition in 3D point-clouds. In: Proc. Int. Conf. on Computer Graphics, Visualization and Computer Vision (2008)
16. Li, Y., Wu, X., Chrysathou, Y., Sharf, A., Cohen-Or, D., Mitra, N.J.: Globfit: Consistently fitting primitives by discovering global relations. ACM Trans. on Graphics 30, 52:1–52:12 (2011)
17. Cohen, B.J., Subramanian, G., Chitta, S., Likhachev, M.: Planning for manipulation with adaptive motion primitives. In: Proc. IEEE Int. Conf. Robotics and Automation (2011)

18. Chitta, S., Jones, E.G., Ciocarlie, M., Hsiao, K.: Perception, planning, and execution for mobile manipulation in unstructured environments. IEEE Robotics & Automation Magazine 19(2), 58–71 (2012)
19. Klingbeil, E., Rao, D., Carpenter, B., Ganapathi, V., Ng, A.Y., Khatib, O.: Grasping with application to an autonomous checkout robot. In: Proc. IEEE Int. Conf. Robotics and Automation (2011)
20. Chang, L., Smith, J.R., Fox, D.: Interactive singulation of objects from a pile. In: Proc. IEEE Int. Conf. Robotics and Automation (2012)
21. Gupta, M., Sukhatme, G.S.: Using manipulation primitives for brick sorting in clutter. In: Proc. IEEE Int. Conf. Robotics and Automation (2012)
22. Srinivasa, S.S., Ferguson, D., Helfrich, C.J., Berenson, D., Collet, A., Diankov, R., Gallagher, G., Hollinger, G., Kuffner, J., Van de Weghe, M.: HERB: a home exploring robotic butler. Autonomous Robots 28(1), 5–20 (2010)
23. Bäuml, B., Schmidt, F., Wimböck, T., Birbach, O., Dietrich, A., Fuchs, M., Friedl, W., Frese, U., Borst, C., Grebenstein, M., Eiberger, O., Hirzinger, G.: Catching flying balls and preparing coffee: Humanoid Rollin'Justin performs dynamic and sensitive tasks. In: Proc. IEEE Int. Conf. Robotics and Automation (2011)
24. Vahrenkamp, N., Asfour, T., Dillmann, R.: Simultaneous grasp and motion planning: Humanoid robot ARMAR-III. IEEE Robotics & Automation Magazine 19(2), 43–57 (2012)
25. Jain, A., Kemp, C.C.: EL-E: an assistive mobile manipulator that autonomously fetches objects from flat surfaces. Autonomous Robots 28(1), 45–64 (2010)
26. Beetz, M., Klank, U., Kresse, I., Maldonado, A., Mösenlechner, L., Pangercic, D., Rühr, T., Tenorth, M.: Robotic roommates making pancakes. In: Proc. Int. Conf. on Humanoid Robots (2011)
27. Mitra, N.J., Gelfand, N., Pottmann, H., Guibas, L.: Registration of point cloud data from a geometric optimization perspective. In: Symp. Geometry Processing (2004)
28. Bendels, G.H., Schnabel, R., Klein, R.: Detecting holes in point set surfaces. Journal of WSCG 14(1-3) (February 2006)
29. Schnabel, R., Wahl, R., Klein, R.: Efficient RANSAC for point-cloud shape detection. Computer Graphics Forum 26(2), 214–226 (2007)
30. Berner, A., Bokeloh, M., Wand, M., Schilling, A., Seidel, H.-P.: A graph-based approach to symmetry detection. In: Proc. IEEE/EG Int. Symp. on Volume and Point-Based Graphics (2008)
31. Papazov, C., Burschka, D.: An efficient RANSAC for 3D object recognition in noisy and occluded scenes. In: Proc. Asian Conf. on Computer Vision (2011)
32. Kazhdan, M., Bolitho, M., Hoppe, H.: Poisson surface reconstruction. In: Proceedings of the Fourth Eurographics Symposium on Geometry Processing, SGP 2006, pp. 61–70 (2006)
33. Holz, D., Behnke, S.: Sancta simplicitas – on the efficiency and achievable results of SLAM using ICP-based incremental registration. In: Proc. IEEE Int. Conf. Robotics and Automation, pp. 1380–1387 (2010)

34. Hornung, A., Wurm, K.M., Bennewitz, M., Stachniss, C., Burgard, W.: OctoMap: An efficient probabilistic 3D mapping framework based on octrees. Autonomous Robots (2013)
35. Şucan, I.A., Kavraki, L.E.: Kinodynamic motion planning by interior-exterior cell exploration. In: Chirikjian, G.S., Choset, H., Morales, M., Murphey, T. (eds.) Algorithmic Foundation of Robotics VIII. STAR, vol. 57, pp. 449–464. Springer, Heidelberg (2009)
36. Behnke, S.: Local multiresolution path planning. In: Polani, D., Browning, B., Bonarini, A., Yoshida, K. (eds.) RoboCup 2003. LNCS (LNAI), vol. 3020, pp. 332–343. Springer, Heidelberg (2004)
37. Fox, D.: Adapting the sample size in particle filters through KLD-sampling. I. J. Robotic Res. 22(12), 985–1004 (2003)
38. Hart, P.E., Nilsson, N.J., Raphael, B.: A formal basis for the heuristic determination of minimum cost paths. IEEE Trans. on Systems Science and Cybernetics 4(2), 100–107 (1968)

Automatic Grasp Generation and Improvement for Industrial Bin-Picking

Dirk Kraft[1], Lars-Peter Ellekilde[2], and Jimmy Alison Jørgensen[1]

[1] Cognitive and Applied Robotics Group, The Mærsk Mc-Kinney Møller Institute, University of Southern Denmark, Campusvej 55, 5230 Odense, Denmark
{kraft,jimali}@mmmi.sdu.dk
[2] Scape Technologies A/S, Kochsgade 31C, 5000 Odense C, Denmark
lpe@scapetechnologies.com

Abstract. This paper presents work on automatic grasp generation and grasp learning for reducing the manual setup time and increase grasp success rates within bin-picking applications. We propose an approach that is able to generate good grasps automatically using a dynamic grasp simulator, a newly developed robust grasp quality measure and post-processing methods. In addition we present an offline learning approach that is able to adjust grasp priorities based on prior performance. We show, on two real world platforms, that one can replace manual grasp selection by our automatic grasp selection process and achieve comparable results and that our learning approach can improve system performance significantly. Automatic bin-picking is an important industrial process that can lead to significant savings and potentially keep production in countries with high labour cost rather than outsourcing it. The presented work allows to minimize cycle time as well as setup cost, which are essential factors in automatic bin-picking. It therefore leads to a wider applicability of bin-picking in industry.

Keywords: bin picking, industrial robotics, grasping, dynamic simulation, robust grasp quality measure, grasp learning.

1 Introduction

Automation has been an important topic in industrial production since at least the late 1940s. Within that time-frame many individual production steps have been automated in the form of individual machines, which are then connected to form a complete production work flow. In large scale production this is usually achieved by deploying conveyor systems, which is not feasible for small scale or spatially distributed facilities. A common approach taken in this situation is to place the in-between products in containers or on pallets. The individual parts then need to be placed into feeding stations to be reintroduced into the automated production process. Today this is still a process with high human labor involvement. These tasks of transportation and placing of objects into feeding stations are actually a significant part of the tasks performed in production facilities [1]. An alternative to human labor for these tasks is to use a bin-picking

F. Röhrbein et al. (eds.), *Gearing Up and Accelerating Cross-Fertilization between Academic and Industrial Robotics Research in Europe*, Springer Tracts in Advanced Robotics 94,
DOI: 10.1007/978-3-319-03838-4_8,

(a) Platform 1 (b) Platform 2

Fig. 1. Experiment platforms. See Section 4.1 for a detailed description.

system, that can automatically locate objects in the containers, grasp them and transport them to the feeding station. Two examples of such systems are shown in Fig. 1.

The bin-picking problem can be broken down into the following sub parts[1]:

S1 Use a sensor system (typically based on a camera, range scanner or a combination of these) to detect an object in the bin and its pose (position and orientation).
S2 Select an appropriate way to grasp the object.
S3 Execute the appropriate robot motion and grasp the object.
S4 If the grasp was successful, move the object to a desired location.

In this paper, we address the process of grasp selection and generation (S2). We assume therefore that the object to be picked has been selected and its pose $p \in SE(3)$ has been estimated.

The two major factors end-users take into account when evaluating the use of a bin picking system are cost and cycle time. Cycle time can be broken down into two aspects: (1) the time it takes to run one cycle (S1-S4) in case no errors occur and (2) the penalty times introduced by recovering from a failure situation (e.g. grasping of the object failed). The second aspect is heavily influenced by the failure rates of the individual systems. Systems that influence the cycle time are for example: pose estimation (e.g. [2–5]), path planning (e.g. [6]) but also grasp selection. Grasp failures occur due to several reasons including incorrect or imprecise pose estimates, hindering placements of neighboring objects and the chosen grasp strategy. These reasons cannot be modeled directly, hence appear as unmeasurable confounders in the system.

[1] For simplicity we assume from here on out that all objects are of the same known (e.g. through a 3D CAD model) type.

Our core interest in this work is to address the process of grasp selection (S2) to reduce setup time (initial cost) and to optimize failure rates (and thereby average cycle time). This problem can be broken down into two issues:

I1 Choose a set of "good grasps" $G = \{g_1, \ldots, g_n\}$ that covers the object in $SE(3)$ as well as possible.
I2 For each grasp g_α, a priority π_α should be defined based on an estimate of the success probability of that grasp.

An arbitrary grasp parameterization can be used for g. For the two devices considered in this work (parallel gripper and suction cup) it is sufficient to define g as a pose $g \in SE(3)$ to which the device needs to be moved before being activated (closing of the gripper or applying of suction). In case multiple grasps are applicable in a given situation, the priority π_α can be used to select a specific grasp (normally the one with the highest priority). Grasp quality, simulated or real or both, is relevant for both (I1) and (I2).

In this paper, we will present:

P1 A simulation framework for automatically generating a set of robust grasps that cover the object. (I1)
P2 An approach to calculate the grasp quality needed to define feasible priorities for these generated grasps. (I2)
P3 A process for refining priorities through learning from prior real world experiences. (I2)
P4 An experimental evaluation of these processes using two different bin picking systems.

This paper presents an extension of the work shown in [7]. The novelty in this paper lies within the automatic grasp generation process using a dynamic simulator (P1, P2) and an experimental evaluation and comparison of that process (P4). Our learning approach (P3) has previously been presented in [7], but is reiterated in Section 3.4 and additional learning results are presented.

The paper is organized in the following way: Section 2 gives an overview of the state of the art in grasp generation (I1) and grasp learning (I2). Our grasp generation methods are discussed in Section 3, which also reintroduces the grasp learning approach, Section 4 presents the experimental results achieved using two different robotics platforms and finally Section 5 presents a conclusion and discusses potential future work.

2 State of the Art

The problem of choosing a set of good grasps (I1) for a specific object is well known, and has been studied intensely over the last decade. Dominating the early literature are the defining of metrics and efficient computation of grasp qualities, with high quality grasps being referred to as "good" grasps. A popular notion of a good grasp is typically defined with an approximation of the form or force closure of a grasp, meaning how well a grasp can resist externally applied wrenches (forces, torques and combinations thereof).

A popular formalization of grasp quality originates from the work in [8], where the quality of a grasp was defined by the ratio between the maximum wrench that a grasp can resist (over all directions) and the contact forces that a gripper can apply. This metric has been used directly or with minor variations in several state of the art data-driven grasp planners [9–13], i.e. planners that select a feasible grasp from a previously offline generated set of grasps (a grasp database). However, often the force closure metric is combined with task specific metrics such as in [14], where a quality metric was used to describe how well tactile sensor data can be used to correct for pose uncertainties. General surveys on grasp quality metrics can be found in [15, 16].

Grasp databases are usually generated offline, but they can be combined with other grasp quality metrics in an online selection stage where information regarding the scene context can be taken into account. This is demonstrated in [10], where a grasping score function is introduced for prioritizing grasps that have a high clearance to the environment, thereby minimizing the risk of collisions.

It is crucial for any data-driven grasp planner that the quality of a grasp reflects the success probability when executing the grasp. With the previously mentioned approaches for calculating grasp quality, the uncertainty of the grasping system is not taken into account. Hence, the success probability of a grasp may be poorly represented with force closure based metrics. This has been investigated in [17, 18] where both kinematic and dynamic (physically based) simulation was used to calculate grasp qualities that, to some degree, reflect how well grasps handle uncertainties. We adopt a similar approach for grasp quality computation using dynamic simulation to calculate qualities reflecting uncertainty. However, we use a force closure based metric to filter low quality grasps from the database and we optimize the set of grasps to equally cover the space of successful grasps on an object.

Unlike the method presented in this paper, none of the above data-driven grasp planning approaches modifies the quality of grasps in the grasp database once it has been created. Hence, high quality grasps that fail frequently during execution can severely limit the overall efficiency of applications that often repeat specific grasps (e.g. bin-picking).

In [19] an approach to improve grasp strategies based on failed executions is presented. The approach uses Dynamic Movement Primitives (DMPs) to control the grasping of an object, and for each failed grasp the DMP is changed to improve the grasp strategy. The DMP approach targets issue (I2) by refining the actual grasps. It does not enable the system to choose between the potentially feasible grasps; instead it uses a standard grasp planner to plan a single grasp which is then optimized. This approach may yield better grasps than those originally found by the grasp planner. However, it is a local optimization which does not influence the actual selection of the grasp and can therefore not prevent conceptually bad grasps to be repeatedly used.

Work on so called grasp densities has been introduced in [20]. These express grasp affordances associated to objects probabilistically and each experimental batch on a specific object updates the grasp affordances. Grasp densities can express a complete set of grasping options associated to an object with associated success probabilities. They are learned from successful grasping attempts, and can therefore be used to target both (I1) and (I2). Based on these grasp densities, efficient and flexible grasp strategies can be developed.

In contrast to the grasp density approach this paper presents a simple learning method, which successfully incorporates the knowledge of successful and failed grasp attempts into the offline generated grasp database. The method is specifically designed to be applicable in industrial contexts, which often do not allow for online learning strategies. We document its performance in two industrial setups and demonstrate its applicability in industrial bin-picking.

In Section 3.4 we present a method to refine the grasp database by readjusting quality labels using success probability estimates from real world executed grasps. This effectively targets (I2) and we show that this refinement increases the overall system performance.

3 System Description

The main focus of this work is to address the grasp generation (P1), the grasp priority determination (P2) and the priority update problem (P3). These problems (P1-P3) need to be seen within the context of the entire bin-picking system (see Fig. 2). The three main components here are *automatic grasp generation* (P1, P2), *manual grasp generation* and *in-production learning* (P3). The purpose of both manual- and automatic grasp generation is to create a grasp database based on the knowledge about the object and the gripper system available for grasping. This database is then used by an industrial bin-picker setup. Once objects are detected in the bin, a grasp is selected based on its priorities (quality)

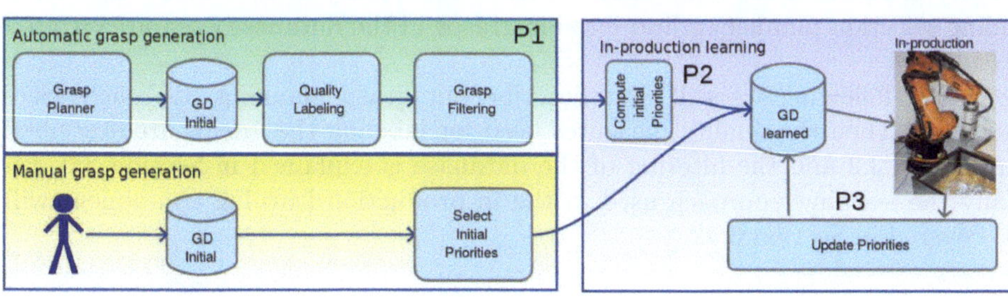

Fig. 2. Flow of the construction of a high quality grasp database. The database used in the online learning component can be generated by either a fully automatic approach or by a manual selection of intuitively good grasps. The automatic approach first computes an initial grasp database using a grasp planner, then it computes grasp qualities and finally it filters to preserve high quality grasps and coverage of the object.

and the selected grasp is executed. If successful the object can then be manipulated as required (e.g. placed in a feeding station). The main purpose of the in-production learning component is to improve the existing database based on the outcomes of previously tried grasps. The outcome of each executed grasp is therefore stored in the execution phase and when sufficiently many outcomes have been registered, a learning algorithm updates the priorities of the grasps in the database.

In our previous work [7] the generation of grasps, including the selection of the grasp qualities/priorities, was performed by an expert user, which is today a typical approach in industrial applications. In this work we compute performance-wise comparable grasp databases automatically by using a heuristic for grasp sampling, a dynamic simulator, several quality metrics and a grasp filtering approach which ensures that the object is well covered with grasps.

The latter is essential to bin-picking applications, because the possible approach directions when grasping objects in a bin are severely limited due to the robot workspace and possible collisions with the surrounding objects or the walls of the bin itself. Most of the previous work done within grasp generation does not take this coverage property of a grasp database into account. One example is the Columbia grasp database [21], which is a publicly available grasp database for several thousand objects using different grippers. For each object-gripper pair roughly 15 grasping poses are defined. Even though these 15 grasps are optimal according to some given force-closure metric, they are useless if none of them can be applied in a specific setting due to collisions with the environment. This is also true for the KIT grasp database [22], that has between 20–100 grasps per gripper-object pair.

Our approach to grasping differs from previous data driven approaches by firstly being completely autonomous and at the same time guaranteeing coverage; and secondly by updating the priorities (qualities) of the grasps in the grasp database even after the setup is placed in production. By emphasizing coverage our approach to build a grasp database distinguish itself from that of the Columbia grasp database and the KIT grasp database by allowing lower quality grasps in areas were there are now high quality grasps. This enables the online selection planner to find feasible grasps in the database even in cluttered scenes.

In the following we will first describe our grasp planning approach (Section 3.1). Then the quality measures used for labeling the grasps are described in Section 3.2 and the filtering of the database is explained in Section 3.3. Finally the learning approach used in the in-production learning component will be covered in Section 3.4.

3.1 Grasp Planning and Simulation Setup

The grasp planner used in this work is composed of two components namely a grasp sampler and a dynamic simulation validator. The grasp sampler is used to generate promising grasps automatically, while the dynamic simulation validator is used to filter away infeasible grasps and to generate so-called aligned grasp targets.

The grasp sampler randomly samples grasp targets using a heuristic, which is specific for the gripper type that is used. A generic grasp target is defined as the 3-tuple

$$g = \left\{T_{obj}^{base}, q_{open}, q_{close}\right\}, \tag{1}$$

where T_{obj}^{base} is the pose of the gripper base relative to the object, q_{open} is the initial configuration of the gripper jaws/fingers and q_{close} is the configuration of the jaws that the gripper controller will move towards when grasping.

In this work both a parallel gripper and a suction cup gripper is used. In case of the suction cup gripper, only T_{obj}^{base} is relevant since the suction cup itself does not have any controllable degrees of freedom.

The dynamic simulation validator performs a full dynamic simulation of each grasp target generated by the sampler. The simulations are performed in an obstacle and gravity free environment and serve both to remove infeasible grasps (present due to artifacts of the sampler) and to generate aligned grasp. For each successful simulation the grasping will have made the object move relative to the gripper until a stable grasp is maintained. This grasp configuration is used to generate a new target which corresponds to what is called an aligned grasp target.

Our definition of an aligned grasp target is illustrated in Fig. 3, where the left grasp is an automatically generated grasp g_α. After executing g_α in a dynamic grasp simulation the right configuration is achieved, illustrating how the object has been aligned by the gripper.

It is important to stress the motivation behind aligned grasp targets. A perfect execution of an aligned grasp will avoid any movement of the object relative to the gripper during the grasping process. This is essential since object movements

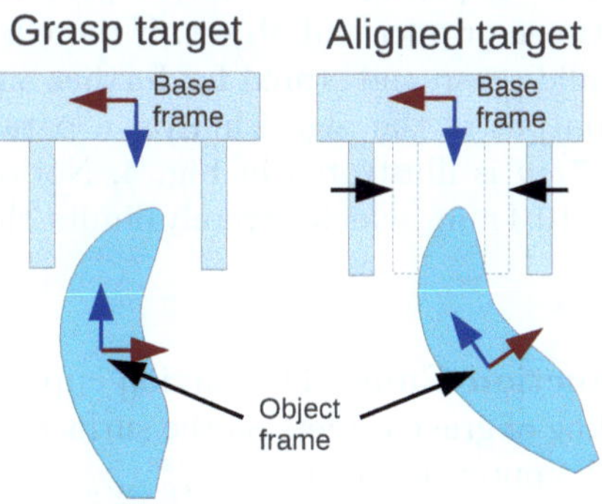

Fig. 3. Illustration of an aligned grasp target. To the left an automatically generated target is illustrated. Using this target as starting point, a dynamic simulation produces the configuration to the right. This new configuration can be used as a grasping target which we denote the aligned target.

might be hindered or influenced by the environment, which is unknown. Hence, aligned grasps are inherently robust toward different environments.

Aligned grasps tend to be more robust toward uncertainties, but they also tend to provide less coverage, which is important in data-driven grasping. Both aligned and successfully simulated target grasps are therefore stored as potential grasp candidates.

In the next three sections the grasp samplers for parallel and suction cup grippers as well as the simulation environment used in the validator will be described.

Grasp Sampler for Parallel Grippers. The parallel gripper grasp sampler is based on random sampling of features on the surface of the object geometries. As input the sampler requires the object and gripper geometry, the minimum and maximum distance between the gripper jaws (d_{closed}, d_{open}) , the tool center point (TCP) of the parallel gripper relative to the base (the z-axis should point in the direction of the approach and the x-axis in the direction of jaws movement), and the open and close configurations of the gripper (q_{open}, q_{close}).

The sampled features are pairs of surface points where the associated surface normals are close to opposite, defined as when the angle θ between them is in the interval $[130°; 230°]$, and where the distance between the surface points are in the interval $[d_{closed}, d_{open}]$. A surface point is defined as a 2-tuple of position and normal and the feature is therefore a 2x2-tuple $\{\{p_1, n_1\}, \{p_2, n_2\}\}$.

For each feature a grasp target pose, T^{obj}_{target}, is calculated relative to the object where the position is

$$P^{obj}_{target} = p_1 + (p_2 - p_1)/2, \tag{2}$$

while the orientation of T^{obj}_{target} is determined by first defining the closing direction as $(p_2 - p_1)/||p_2 - p_1||$ and align the x-axis of T^{obj}_{target} with it. The last part of the rotation is then generated as a random rotation around the x-axis.

Finally, the gripper TCP is placed in the target pose T^{obj}_{target} and it is tested if there is a collision between gripper and object. If there is no collision then the target is considered a valid grasp and stored for further analysis.

An example of the output of the algorithm on a parallel jaw gripper with $d_{close} = 6\,cm$, $d_{open} = 7\,cm$ is illustrated in Fig. 4. Notice that the maximum stroke of the gripper is 10.4 mm, which severely limits the number of feasible grasps on the object.

Grasp Sampler for Suction Cups. The suction cup based grasp sampler is based on random sampling of grasp targets on the surface of the object. As input the sampler requires the approach direction $P^{base}_{approach}$ in which the suction cup will move during grasping, the tool center point (TCP) of the suction cup relative to the gripper base P^{base}_{tcp}, the distance of the approach movement $l_{approach}$, the max diameter of the suction cup d and the CAD models of both suction cup and object. Based on these parameters, grasps for the suction cup are created in the following way:

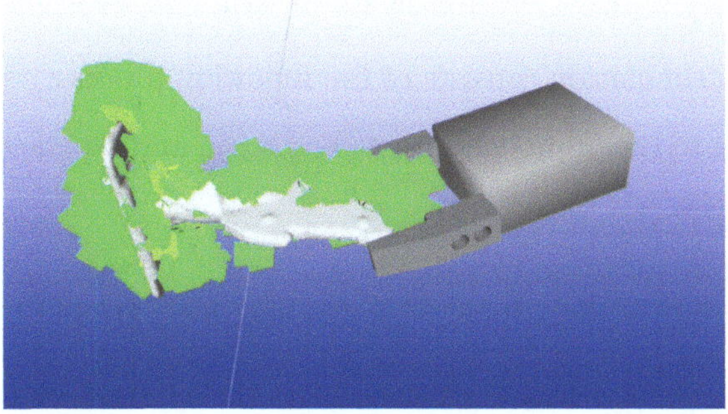

Fig. 4. 500 grasps generated for the parallel gripper (The green markers indicates the grasp targets)

1. Uniformly sample a random point p_{surf} on the object surface.
2. Create a uniformly sampled random direction vector v_{dir}.
3. Calculate the approach target point $P^{obj}_{approach} = p_{surf} + v_{dir} * (ran(-d, d) + l_{approach})$
4. Calculate the grasp target point $P^{obj}_{target} = p_{surf} + v_{dir} * ran(-d, d)$
5. The gripper orientation R^{obj}_{target} can then be calculated such that $P^{base}_{approach}$ points in direction of $-v_{dir}$.

Each generated grasp is verified by placing the suction cup gripper TCP P^{base}_{tcp} in $P^{obj}_{approach}$ with orientation R^{obj}_{target}. If the suction cup and object do not collide the grasp is stored for later analysis, otherwise it is rejected.

Dynamic Simulation Based Validator. Grasps generated using the above heuristics typically suffer from artifacts that are not easy to detect. An example could be the already discussed artifact of the difficulty of sampling aligned poses. However, kinematic heuristics are fast and may generate thousands of grasp targets per second.

Dynamic simulation of a grasp provides a good validation filter and also provides additional aligned targets. In this work we use the dynamic grasp simulator RobWorkSim [23]. Dynamic grasp simulation is computationally expensive compared to kinematic simulation, typically two to eight grasps can be simulated per second on a single core machine. It therefore makes sense to "intelligently" sample grasps before evaluating them in dynamic grasp simulation as described in the previous sections.

The environment that is used for the simulation is an obstacle and gravity free environment. That is, the object is not affected by any forces other than those applied by the gripper. This is clearly unrealistic; hence grasps that are successfully performed in such an environment may not necessarily be successful in a bin-picking environment. In this work we handle this miss alignment between simulation and real world in the later learning step that identifies the truly good and bad grasps.

A single grasp simulation consists of the following process:

1. Move gripper to the approach target, $T_{approach}$, and set the configuration to q_{open}.
2. Move gripper to target pose, T_{target}.
3. Set gripper configuration to q_{close}.
4. When the fingers of the gripper stop moving, move gripper to retract target, $T_{retract}$.
5. If the fingers touches the object, then label the grasp as successful otherwise discard it.
6. For successful grasps store both the original grasp target T_{target} and the found aligned grasp target.

When a suction cup gripper is used, the setting of q_{open} and q_{close} in step 1 and 3 has no effect, since it has no active degree of freedom.

The final set of stored grasps (called $G_{initial}$) is the output of the grasp planner component.

3.2 Quality Measures

The output of the grasp planner, $G_{initial}$, consist of a large amount of grasps without any estimate of their individual quality. Ideally, the quality of a grasp should reflect how likely it is in succeeding when executed for some task and environment. Naturally, this is dependent on the task, the environment and the uncertainties of the system in which it is executed. In practice the grasp quality is usually only a rough approximation calculated without considering the influence of environment and uncertainties. Typically the grasp quality is based on an estimate of how well the grasp of an object can resist externally applied wrenches. These are intuitively good grasps, but do not take into account the uncertainties of the system.

In this work we use a combination of a traditional wrench based measure Q_{gws}, and a robustness measure Q_{robust}. Q_{robust} estimates the quality of a grasp, when considering a Gaussian uncertainty distribution of the pose of the gripper relative to the object. The basic method to compute Q_{robust} is to first define a threshold on Q_{gws} and cut away grasps, that are not robust to even small external wrenches and secondly to label the remaining grasps with the robustness quality, Q_{robust}. The next two subsections will present the calculation of the two mentioned quality measures.

Wrench Space Measure. The grasp wrench space (GWS) measure is a commonly used measure first introduced in [24]. The definition of the GWS, which resembles that of [25] is repeated here for the reader's convenience.

The GWS, Q_{gws}, is determined by the friction cones of the contact points that compose a grasp, where the friction cone of contact point i is approximated by a set of m contact boundary wrenches, $\{w_{i,j}|_{j=\{1...m\}}\}$. These are defined such that the torque is scaled by the radius r of the object:

$$w_{i,j} = \begin{pmatrix} f_{i,j} \\ \frac{1}{r} \cdot d_i \times f_{i,j} \end{pmatrix} \tag{3}$$

where $f_{i,j}$ is one of the force vectors in contact i, and d_i is the vector from the torque origin to the i-th point of contact. The cross product $d_i \times f_{i,j}$ is the torque $\tau_{i,j}$. The GWS is represented as the convex hull over the union of each set of contact boundary wrenches. Finally, the grasp quality measure, Q_{gws}, is determined by the radius of the inscribed 6-sphere of the GWS, which reflects the maximum perturbing wrench that the grasp can counterbalance, given the maximum forces of the fingers.

Robustness Measure. Our robustness measure resembles the approach taken in [17] where they for a specific grasp g use kinematic simulation to calculate success outcomes of thousands of grasps with poses perturbed by a small distance relative to g. Their robustness quality of g is then based on the success percentage of the simulated outcomes, where a successful outcome is defined when Q_{GWS} is larger than a certain threshold.

In our work we use a dynamic simulator, which has the advantage of being more realistic than the kinematic simulators used in [17]. The outcome is in our work determined by evaluating if the object is able to be lifted (move opposite the gripper approach direction) by the grasp. The robustness measure of a grasp is then computed by performing N simulations of grasping an object and for each simulated experiment perturbing the target pose in SE(3) by adding noise following a Gaussian distribution. Based on all N experiments the quality of the grasp Q_{robust} can then be computed as the average outcome of the experiments:

$$Q_{robust} = \frac{1}{N}\sum_{i=0}^{N} o_i \tag{4}$$

where o_i is the outcome of the i-th experiment with a value of 1 if the experiment was successful and 0 otherwise. The grasp simulation is executed with the same process and in the same gravity and obstacle free simulation environment as described in Section 3.1.

3.3 Filtering

The outcome of the grasp generation process is a large number of grasps with varying quality and spatial location. A grasp filter is used to reduce the number of grasps in the database, keeping the grasps of highest quality, without compromising the coverage of the database. To achieve this we use a greedy algorithm which recursively picks the highest quality grasp, $g_{highest}$, from the initial set $G_{initial}$, adds it to the result set G and then removes all grasps in $G_{initial}$ that are within a specified distance (using the metric in (6)) of $g_{highest}$. When $G_{initial}$

is empty the algorithm terminates and G is the set of grasps that is to be used in the online execution.

The filter computes the distance between two grasp poses by using a weighted infinity metric. Each grasp pose is defined by a 7-tuple where the first three values correspond to the position in meters and the last four correspond to the equivalent angle and axis (EAA)[2] representation of the orientation.

$$g_1 = \{x_1, y_1, z_1, eaa_{x1}, eaa_{y1}, eaa_{z1}, eaa_\theta\}^T \tag{5}$$

As input a distance threshold d_{thres} and an angle threshold θ_{thres} is used in the computation of the distance:

$$D(g_1, g_2) = ||W \cdot (g_2 - g_1)||_\infty \tag{6}$$

where $W = \{d^{-1}, d^{-1}, d^{-1}, \theta^{-1}, \theta^{-1}, \theta^{-1}, \theta^{-1}\}^T$. If the distance $D(g_1, g_2)$ is lower than 1 then the grasps are too close and one of them is to be removed. In this work the parameters (d, θ) were selected to control the number of grasps in the database and they were not based on any specific analysis or modeling of the system. Fig. 5 gives an impression of the results we achieved.

(a) (b)

Fig. 5. Result of applying filtering to a database of 1400 grasps. The markers indicate the pose of grasps and the color shows the quality (green, yellow, orange, red having decreasing quality). (a) Before filtering (1400 grasps). (b) After filtering (240 grasps).

3.4 Learning Approach

Based on the previous successes and failures in grasping an object with specific grasps, we are able to compute the success probability of the individual grasps. This should allow us to in the future select grasps that proved to be the most successful in the past. For this we have previously presented a learning approach [7], which is reiterated here.

When running experiments a grasp, g_α, will only be chosen if the object pose x is inside Ω_α. Ω_α is defined as the subspace of object poses for which g_α is

[2] EAA represents orientation as a unit vector (the axis around which the rotation takes place) and the angle (the magnitude of the orientation around the axis).

applicable. The grasps success probability can be calculated as an average of the grasp outcomes:

$$\widehat{P}_{\alpha}(\Omega_{\alpha}) = \frac{1}{N_{\alpha}} \sum_{k=1}^{N_{\alpha}} o(g_{\alpha}, x_k) \tag{7}$$

where $o(g_{\alpha}, x_k)$ is the outcome (success=1, failure=0) of experiments applying g_{α} to picking the object at location x_k. The value of $\widehat{P}_{\alpha}$ is generally referred to as the empirical success probability of g_{α}, which goes towards the true success probability, P_{α}, as N_{α} goes towards infinity.

When running bin picking experiments one cannot ensure that all grasps are tried an equal number of times. Even though objects are thrown randomly into the bin, they have a tendency to lie and pack in certain ways making some grasps applicable more often than others. The learning approach thus needs to take into account how many times a grasp has been tried and take proper precautions. In this work this is done by defining a minimum number of trials, N_{min}, that a grasp has to be performed before being considered a candidate for updating.

Once grasps are generated we do not wish to change their object relative location as this would jeopardize the coverage of the object. The learning should therefore focus on updating how grasps are selected. To facilitate this each grasp has associated a discrete priority, π_{α}. When multiple grasps are applicable for the same object, the grasp with the highest priority is chosen. The learning method therefore focuses on updating these priorities such that they reflect the true performance of the grasps as well as possible.

Even though several hundred grasps are available in the database, G, the set of grasps applicable for a given object pose is usually quite small. This work has thus only applied three different levels of priority, but extension to an arbitrary number is straightforward. The new priorities are assigned by first calculating the overall empirical success probability of all grasps by

$$\hat{P} = \frac{1}{N} \sum_{k=1}^{N} o(g_k, x_k) \tag{8}$$

where N is the total number grasp attempts performed and g_k is the specific grasp tried to grasp the object at pose x_k. The grasp success range is then divided into three by defining an upper and lower bound

$$P_{upper} = \hat{P} + \frac{1}{3}(1 - \hat{P}) \tag{9}$$

$$P_{lower} = \hat{P} - \frac{1}{3}(\hat{P} - 1) \tag{10}$$

The priorities of grasps for which $N_{\alpha} \geq N_{min}$ are updated to priority 1 if $\widehat{P_{\alpha}} > P_{upper}$, priority 2 if $P_{upper} \geq \widehat{P_{\alpha}} \geq P_{lower}$ and finally priority 3 when $P_{lower} > \widehat{P_{\alpha}}$.

4 Experiments

Experiments with the automatically generated grasps have been performed on two platforms (see Section 4.1 and Fig. 1). Section 4.2 presents details of how the grasp databases for the two setups have been defined. The actual experiments with platform 1 and platform 2 are described in Sections 4.3 and 4.4, which also discuss the results. The experiments with automatically generated grasps are compared to experiments with hand defined grasps, which are based on our previous work published in [7].

4.1 Platform Descriptions

Experiments have been performed on two Scape Technologies bin picking systems.

Platform 1: The first platform is shown in Fig. 1a and is comprised of a Kuka Kr5 equipped with a SCAPE tool unit with two pneumatic parallel grippers. The stroke of both grippers is 10.4 mm which significantly limits the variety of grasps each of them can perform. The jaws on the left gripper are created such that it has a narrow grab making it feasible for grasping parallel to the planar surfaces of the test object, whereas the jaws on right side are created to enable wide grasps across the object. The grippers are mounted on pneumatic sliders, which are used to deploy the gripper as needed for a given grasp. The test object used is an approximately 11 cm tall T-shaped metal object, which can be seen in Fig. 5 and in the lower right part of Fig. 1a.

Platform 2: The second platform (shown in Fig. 1b) uses a Kuka Kr30HA. The tool unit is equipped with a standard round 32 mm, $2\frac{1}{2}$ bellow suction cup, also mounted on a pneumatic slider. The test object consists of aluminium sheet metal that has been punched and bent into shape. Its size is approximately 28 cm times 10 cm. The object can be seen in Fig. 8.

The sensor used for object detection and pose estimation is a SCAPE 3D grid scanner. The sensor is characterized by a high accuracy in depth, but with a fairly low resolution (74×74 3D data points), giving rise to a non uniform uncertainty of the 3D data points. A detailed evaluation of this sensor can be found in [26]. The system is furthermore influenced by a number of unmeasurable confounders, which affect the pose estimates as well as the grasping process. This includes, but is not limited to, undesired reflections and occlusion caused by surrounding objects.

4.2 Grasp Databases

The grasp databases used for the experiments are generated in two ways. Firstly grasps are hand-defined, where the user selects the desired poses and specifies priorities based on how well he expects the individual grasps to perform. Secondly grasps are automatically generated and a quality is found based on the methods described in Sections 3.1–3.3. Table 1 summarizes the number of grasps in the databases broken down by type and platform. The number of grasps are selected to ensure a fairly uniform coverage of the object, which in particular for the parallel gripper with automatically generated grasps required more instances than in the hand defined case.

Table 1. Number of grasps in the databases used for experiments

	Hand defined	Automatically generated
Platform 1	163	371
Platform 2	263	252

The parallel grippers have a 180° symmetry around the z-axis and the round suction cup is fully symmetric around the approach direction. These symmetries give the grasps an invariance, which is utilized in the grasping process, enabling picking objects which would otherwise have been outside the reach of the robot or led to a collision between robot and environment. In this analysis, grasps, which only differs in the rotational symmetry, are considered to be the same grasp.

The Scape system does not support a continuous value as grasp quality measure, hence the success probabilities computed in Section 3.2 need to be converted into discrete priorities. The conversion is done by dividing the range of success probabilities into three equally sized intervals and then labeling the grasps as priority 1, 2 or 3 depending on which interval they belong to. Examples of this priority discretization can be seen in Figs. 6, 7 and 8 where grasps with attached priorities are visualized.

4.3 Experiments with Platform 1—Parallel Gripper

Initially 1029 grasp attempts were performed with the hand defined grasp database and new priorities were learned according to the learning scheme presented in Section 3.4. Figs. 6b and 6c show the grasps with priorities before and after the update.

To show the effect of the learning it is important to limit the effect of unmeasurable confounders, such as changes to illumination and how the objects are stacked in the bin. The verification is thus performed by loading both databases and then randomly switching between choosing grasps from one or the other, thereby removing the effect of systematic differences. The results of these experiments can be found in Table 2.

The same approach was then performed for the automatically generated grasp database. Since the database contains approximately twice as many grasps, we used twice as many experiments to perform the learning. The grasps with priorities before and after the update can be seen in Figs. 7b and 7c. The results are presented in Table 3.

Table 2. Results of experiments with test platform 1 using hand defined grasps. These results were already presented in [7]

	Total	Success	Failure	$\widehat{P}$	Confidence Interval (95%)
Baseline priorities	449	234	215	0.521	[0.478; 0.567]
Updated priorities	469	298	171	0.635	[0.592; 0.679]

Analysis of Results. To evaluate and compare the performance of two databases, we need to show, that the differences are statistically significant, hence that the results are not due to chance. The experiments performed have the characteristics of a Bernoulli process, as each trial only has two outcomes which are mutually exclusive [27]. The sampled mean, $\hat{P}$, of such a bi-modal distribution approximately follows a normal distribution around the true mean with a variance given by $\hat{P}(1-\hat{P})/N$, with N being the number of trials. Using this we can calculate confidence intervals as $\hat{P} \pm z\sqrt{\hat{P}(1-\hat{P})/N}$, with z defining the confidence level based on the standard normal distribution. The values for the 95% confidence intervals are shown in Tables 2 and 3. These results can also be illustrated graphically by plotting the expected distributions around the mean of the samples, as shown in Figs. 6a and 7a.

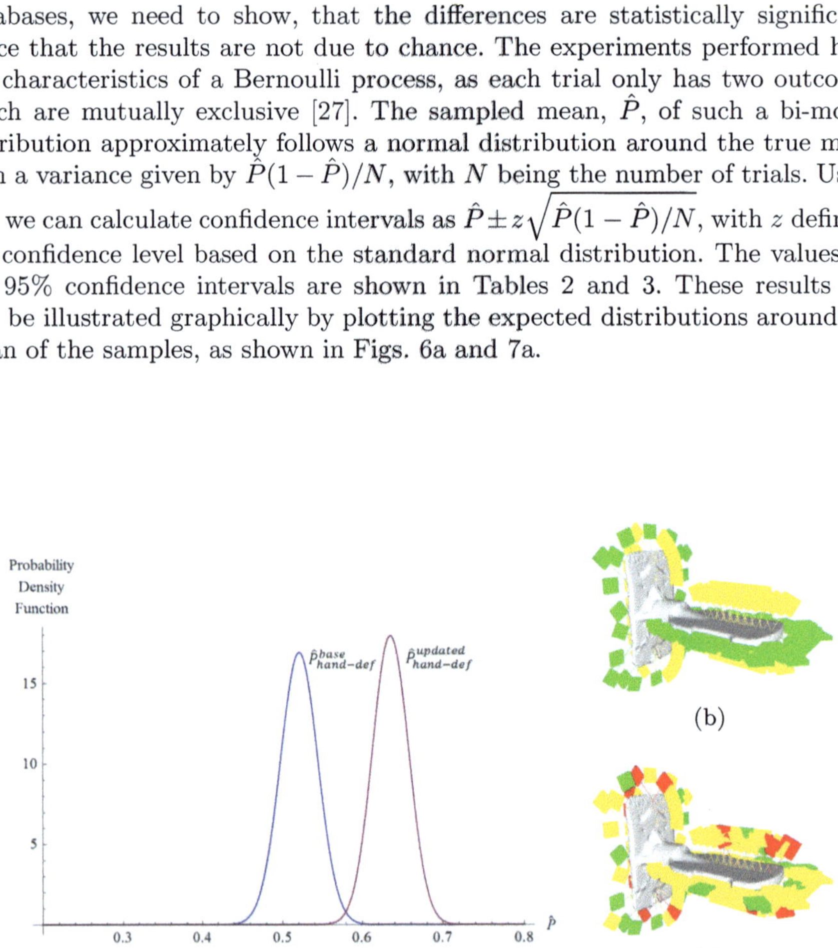

(a) (b) (c)

Fig. 6. Results for the hand defined grasps on platform 1. (a) Normal distributions representing the expected values of $\hat{P}^{base}_{hand-def}$ (blue) and $\hat{P}^{updated}_{hand-def}$ (purple). (b) Grasps with priorities selected by hand. (c) Grasps with priorities after updating (Green, yellow and red corresponds to 1st, 2nd and 3rd priority grasps, respectively). Figs. (b) and (c) originate from [7].

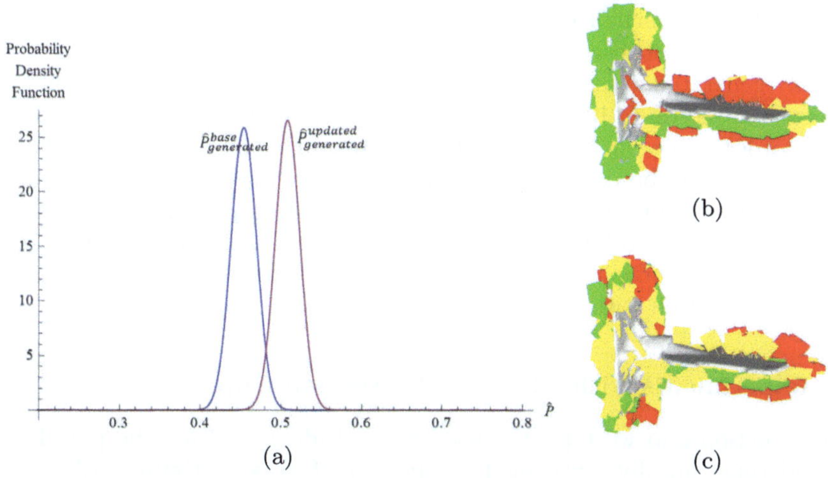

Fig. 7. Results for the automatically generated grasps on platform 1. (a) Normal distributions representing the expected values of $\hat{P}^{base}_{generated}$ (blue) and $\hat{P}^{updated}_{generated}$ (purple). (b) Grasps with initial priorities. (c) Grasps with updated priorities (Green, yellow and red corresponds to 1st, 2nd and 3rd priority grasps, respectively).

Table 3. Results of experiments with test platform 1 using automatically generated grasps

	Total	Success	Failure	$\widehat{P}$	Confidence Interval (95%)
Baseline priorities	1043	474	569	0.454	[0.424; 0.485]
Updated priorities	1108	563	545	0.508	[0.479; 0.538]

A more in depth statistical evaluation can be obtained through hypothesis testing for proportions [28]. First the null-hypothesis that $P^{updated} \leq P^{base}$ is stated. The appropriate test statistics for this case is

$$z = \frac{\widehat{P}^{base} - \widehat{P}^{updated}}{\sqrt{\frac{P_c(1-P_c)}{N^{base}} + \frac{P_c(1-P_c)}{N^{updated}}}} \tag{11}$$

in which P_c is the combined success probability and N^{base} and $N^{updated}$ are the number of trials from the baseline and the updated databases, respectively.

For the hand defined grasps we get $z = -3.50498$. Computing the area under the standard normal distribution to the left of this value gives 0.00023, which is the probability of making an error in rejecting the null-hypothesis. We can thus be very confident in rejecting the null-hypothesis, thereby concluding that $P^{updated} > P^{base}$ for the hand defined grasps. Repeating the calculating for the automatically generated grasps gives $z = -2.32635$ and a probability of rejecting a true null-hypothesis of 0.00640. The null-hypothesis is therefore again rejected with less than a 1% chance of error.

A comparison between the experiments with the hand defined and the automatically generated grasps shows that the hand defined perform slightly better. However, one needs to be careful making this comparison as this data has not been recorded while randomly switching between the databases. The differences could thus in part be due to changes in the environment (e.g. illumination) as well as differences in sensor calibration. An interesting property of the results is that learning seems to have a much greater impact on the hand defined grasps. This could be an indication that the grasp quality measure derived from simulation correlate well with the real-world performance.

4.4 Experiments with Platform 2—Suction Cup

For the suction cup we repeated the approach described for the parallel gripper, and thus initially performed a large set of grasps attempts ($N = 3063$) based on the hand defined database. Based on these the priorities were updated and experiments switching between the baseline and the updated database were performed. The grasps with priorities before and after the update are shown in Figs. 8b and 8c. The general results of these experiments are shown in Table 4.

Table 4. Results of experiments with hand defined grasps on test platform 2. These results were already presented in [7].

	Total	Success	Failure	$\widehat{P}$	Confidence Interval (95%)
Baseline priorities	1137	896	241	0.788	[0.764; 0.812]
Updated priorities	1119	937	182	0.837	[0.816; 0.859]

The initial success rate with the suction cup was significantly higher than with the parallel gripper. However, applying the learning still had a positive impact on the system, managing to improve the probability of grasp success by 4.9% from 78.8% to 83.7%.

For the automatically generated grasps a total of 3045 grasp attempts were performed based on which, the priorities were updated (see Figs. 8d and 8e). Rather than only switching between the baseline and updated priorities to test for improvements, we switched between all four databases, which were the *hand defined baseline*, *hand defined updated*, *generated baseline* and *generated updated*. The results of this 4-way switching are shown in Table 5. The variance for each set has been calculated and the associated normal distributions are plotted in Fig. 8a.

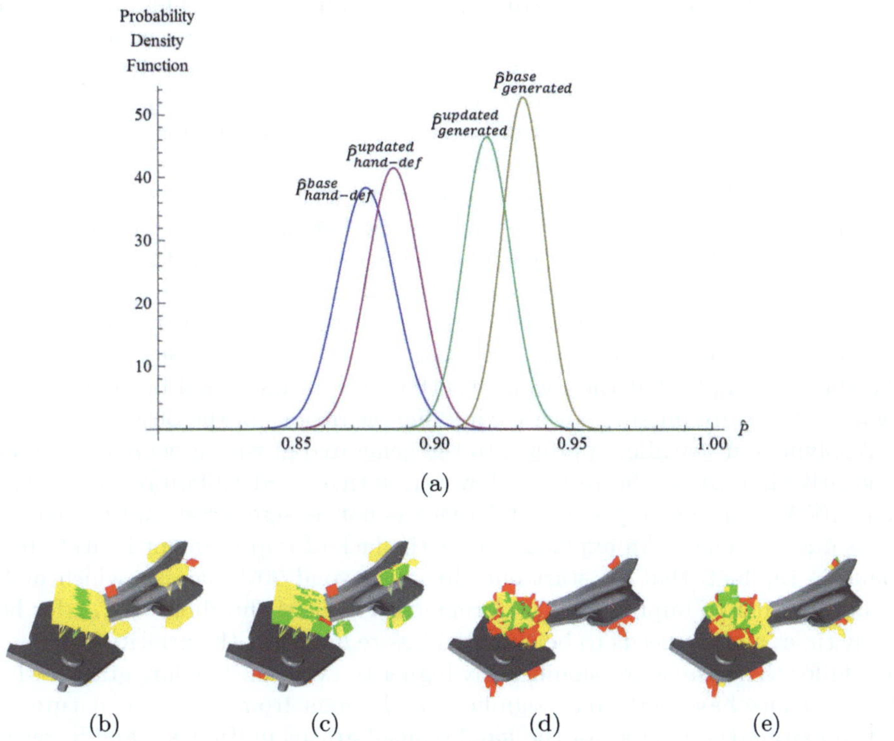

Fig. 8. Platform 2 results. (a) Normal distributions representing the expected values of $\hat{P}^{base}_{hand-def}$ (blue), $\hat{P}^{updated}_{hand-def}$ (purple), $\hat{P}^{base}_{generated}$ (yellow) and $\hat{P}^{updated}_{generated}$ (green) for experiments with the parallel grippers in platform 2, when switching between all 4 strategies. (b) Hand defined grasps with initial priorities. (c) Hand defined grasps with updated priorities. (d) Automatically generated grasps with initial priorities. (e) Automatically generated grasps with updated priorities. Green, yellow and red corresponds to 1st, 2nd and 3rd priority grasps, respectively. Figs. (b) and (c) originate from [7].

Table 5. Results of experiments with test platform 2, when switching between the 4 different grasp databases

	Total	Success	Failure	$\widehat{P}$	Confidence Interval (95%)
Hand defined baseline	1013	886	127	0.875	[0.854; 0.895]
Hand defined updated	1104	977	127	0.885	[0.866; 0.904]
Generated baseline	1110	1035	75	0.932	[0.918; 0.947]
Generated updated	1012	930	82	0.919	[0.902; 0.936]

Discussion of Results. A comparison of the hand defined and the generated grasps shows that the automatically generated grasps outperforms the hand defined. A closer look at why this is the case reveals that the hand defined database contains a few very poorly performing grasps, which were included to enable picking objects in certain difficult configurations. The automatically generated grasp database does not contain the same poor grasps, which results in a small, yet still statistically significant, improvement of the overall performance.

The lack of certain grasps in the database may result in the system failing to be able to grasp an object in certain situations. The switching between databases is implemented such that once an object is found, one of the four databases is chosen randomly and a feasible grasp is selected. If the generated grasps provided a more sparse coverage, then the average number of executed grasp attempts would be lower. Looking at the joined set of performed grasps attempts from the two hand defined and the two generated databases, we see that grasps selected from them have been executed 2117 and 2122 times, respectively. From this we can conclude that no significant difference in the total coverage exist. However, this does not imply that the coverage is the same, as one could imagine that the two databases are missing coverage in different regions of the object.

Applying our learning approach to the generated grasps appears to have had a negative impact on the results. However, with 75 and 82 failures out of more than 1000 grasps attempts, the difference is not as significant and probably in parts due to chance. An explanation for the lack of improvement is likely to be found in the fact, that we start out already beyond 90% success, which makes it much harder to improve the performance. The used threshold, N_{min}, for how many times a grasp needs to be executed before updating the priority is five, and should for this case be set significantly higher to avoid mislabeling grasps, which due to chance have performed significantly different from their true nature.

Comparing the number of the hand defined grasps in Tables 4 and 5, reveals that the general grasp success has improved significantly, whereas the effect of the learning appears to have disappeared. Between these batches of experiments the sensor had to be un- and remounted, requiring it to be re-calibrated. This second calibration appears to have been more accurate, leading to better pose estimates thereby a higher success probability. The learning process is optimizing grasps according to the concrete uncertainties in the system (based on the unmeasurable confounders). By changing the calibration we might also change the characteristics of the uncertainties for which the grasp selection was optimized. To counter this future work should investigate the applicability of an online learning procedure.

5 Conclusion

To increase the applicability of bin picking systems in industry it is important to decrease the cost and cycle time. We have in this paper presented a grasp generation approach that can replace the previously manual selection of grasps, and which thereby provides an easier and more automatic set up procedure. Furthermore we have presented a grasp learning strategy that allows us to decrease grasp failures and thereby decrease cycle time.

The grasp generation mechanism is designed to create grasps that are robust to pose uncertainties and to generate databases, which provide good coverage of the object while minimizing the number of grasps. The results show that we are able to automatically generate grasp databases, that have success probabilities comparable to human generated grasp databases.

In addition a learning approach that allows to adjust grasp priorities based on empirical success probabilities was presented. The experimental evaluation showed that significant success increases were possible in three out of four situations. Two important aspects that could be explored in the future are a repeated application of the described learning procedure and an online learning approach that allows for continuous upgrades.

Acknowledgments. This work was supported by the LearnBiP project (an experiment funded in the context of the ECHORD project (FP7-ICT-231143)) and The Danish National Advantage Technology Foundation through the project Bin Picker.

References

1. Bøgh, S., Hvilshøj, M., Kristiansen, M., Madsen, O.: Identifying and evaluating suitable tasks for autonomous industrial mobile manipulators (aimm). The International Journal of Advanced Manufacturing Technology 61(5-8), 713–726 (2011)
2. Al Hujazi, E., Sood, A.: Range image segmentation with applications to robot bin-picking using vacuum gripper. IEEE Transactions on Systems, Man, and Cybernetics 20, 1313–1325 (1990)
3. Ghita, O., Whelan, P.F.: A bin picking system based on depth from defocus. Mach. Vision Appl. 13(4), 234–244 (2003)
4. Kirkegaard, J., Moeslund, T.B.: Bin-picking based on harmonic shape contexts and graph-based matching. In: Proceedings of the 18th International Conference on Pattern Recognition, vol. 2, pp. 581–584 (2006)
5. Zuo, A., Zhang, J.Z., Stanley, K., Wu, Q.M.J.: A hybrid stereo feature matching algorithm for stereo vision-based bin picking. International Journal of Pattern Recognition and Artificial Intelligence 18(8), 1407–1422 (2004)
6. Ellekilde, L.P., Petersen, H.G.: Motion planning efficient trajectories for industrial bin picking. International Journal of Robotics Research (IJRR) Special Issue on 'Motion Planning for Physical Robots Interaction' (accepted)
7. Ellekilde, L.P., Jørgensen, J.A., Kraft, D., Krüger, N., Piater, J., Petersen, H.G.: Applying a learning framework for improving success rates in industrial bin picking. In: IEEE/RSJ International Conference on Intelligent Robots and Systems (IROS), pp. 1–8 (2012)
8. Ferrari, C., Canny, J.: Planning optimal grasps. In: IEEE International Conference on Robotics and Automation (ICRA), pp. 2290–2295 (1992)
9. Morales, A., Asfour, T., Azad, P., Knoop, S., Dillmann, R.: Integrated grasp planning and visual object localization for a humanoid robot with five-fingered hands. In: IEEE/RSJ International Conference on Intelligent Robots and Systems (IROS), pp. 5663–5668 (2006)
10. Berenson, D., Diankov, R., Nishiwaki, K., Kagami, S., Kuffner, J.: Grasp planning in complex scenes. In: 2007 7th IEEE-RAS International Conference on Humanoid Robots, pp. 42–48 (2007)

11. Goldfeder, C., Ciocarlie, M., Peretzman, J., Dang, H., Allen, P.: Data-driven grasping with partial sensor data. In: IEEE/RSJ International Conference on Intelligent Robots and Systems (IROS), pp. 1278–1283 (2009)
12. Goldfeder, C., Ciocarlie, M.T., Dang, H., Allen, P.K.: The columbia grasp database. In: IEEE International Conference on Robotics and Automation (ICRA), pp. 1710–1716 (2009)
13. Goldfeder, C., Allen, P.: Data-driven grasping. Autonomous Robots, 1–20 (2011)
14. Jørgensen, J.A., Petersen, H.G.: Grasp synthesis for dextrous hands optimised for tactile manipulation. In: Proceedings for the Joint Conference of ISR 2010 (41st Internationel Symposium on Robotics) und ROBOTIK 2010 (6th German Conference on Robotics), pp. 1–6 (2010)
15. Zhang, Y., Gao, F., Gruver, W.: Evaluating the quality of grasp configurations for dextrous hands. In: IEEE International Conference on Robotics and Automation (ICRA), vol. 1, pp. 100–105 (1997)
16. Suárez, R., Roa, M., Cornella, J.: Grasp quality measures. Technical report, Technical University of Catalonia (2006)
17. Weisz, J., Allen, P.K.: Pose error robust grasping from contact wrench space metrics. In: IEEE International Conference on Robotics and Automation (ICRA), pp. 557–562 (2012)
18. Kim, J., Iwamoto, K., Kuffner, J.J., Ota, Y., Pollard, N.S.: Physically-based grasp quality evaluation under uncertainty. In: IEEE International Conference on Robotics and Automation (ICRA), pp. 3258–3263 (2012)
19. Stulp, F., Theodorou, E., Buchli, J., Schaal, S.: Learning to grasp under uncertainty. In: IEEE International Conference on Robotics and Automation (ICRA), pp. 5703–5708 (2011)
20. Detry, R., Kraft, D., Kroemer, O., Bodenhagen, L., Peters, J., Krüger, N., Piater, J.: Learning grasp affordance densities. Paladyn. Journal of Behavioral Robotics 2(1), 1–17 (2011)
21. Goldfeder, C., Ciocarlie, M., Dang, H., Allen, P.K.: The columbia grasp database. In: IEEE International Conference on Robotics and Automation (ICRA), pp. 1710–1716 (2009)
22. Kasper, A., Xue, Z., Dillmann, R.: The kit object models database: An object model database for object recognition, localization and manipulation in service robotics. The International Journal of Robotics Research 31(8), 927–934 (2012)
23. Jørgensen, J., Ellekilde, L., Petersen, H.: Robworksim - an open simulator for sensor based grasping. In: Proceedings of Joint 41st International Symposium on Robotics (ISR) and the 6th German Conference on Robotics, pp. 1–8 (2010)
24. Ferrari, C., Canny, J.: Planning optimal grasps. In: IEEE International Conference on Robotics and Automation (ICRA), pp. 2290–2295 (1992)
25. Miller, A.T., Allen, P.K.: Graspit!: A versatile simulator for robotic grasping. IEEE Robotics and Automation Magazine 11, 110–122 (2004)
26. Møller, B., Balslev, I., Krüger, N.: An automatic evaluation procedure for 3D scanners in robotics applications. IEEE Sensors Journal 13(2), 870–878
27. Petruccelli, J., Nandram, B., Chen, M.: Applied Statistics for Engineers and Scientists. Prentice Hall (1999)
28. Lind, D., Marchal, W., Wathen, S.: Basic Statistics for Business and Economics. McGraw-Hill (2011)

GRASPY – Object Manipulation with NAO

Judith Müller[1], Udo Frese[2], Thomas Röfer[2],
Rodolphe Gelin[3], and Alexandre Mazel[3]

[1] Transregional Collaborative Research Center for Spatial Cognition, SFB/TR8,
University of Bremen, Bremen, Germany
judith.mueller@uni-bremen.de

[2] Deutsches Forschungszentrum für Künstliche Intelligenz, Cyber Physical Systems,
Bremen, Germany
{udo.frese,thomas.roefer}@dfki.de

[3] Aldebaran Robotics, Paris, France
{rgelin,amazel}@aldebaran-robotics.com

Abstract. In this paper we introduce an online object manipulation system for the NAO robot that is able to detect and grasp an object out of a human hand and then give it back in real-time. Known objects are rendered from 3D models and detected stereo contour-based by using a new stereo vision head for NAO. In order to grasp objects, motion trajectories are generated by an A* planner while avoiding obstacles. In order to safely release objects back into a human hand, a combination of tactile and force sensors of the carrying arm is used to detect whether someone touched the grasped object. We performed quantitative experiments in order to evaluate the quality of the detector, the time to grasp an object, as well as the number of successful grasps. We demonstrated the whole system on the real robot.

Keywords: stereo vision object detection, online grasp motion planning.

1 Introduction

The aim of the GRASPY project is to make a small move towards the integration of humanoid robots in our everyday life. In a visionary scenario a small humanoid robot could be a personal assistant that not only is able to organize contacts and emails but also could support a person physically by getting objects for him. This would be especially interesting if a person is not able to get the object for himself. In the scope of this project we wanted to investigate what is necessary to allow the humanoid robot NAO [1] to have this functionality.

In this paper we present an online object manipulation system that is an extension of our previous work [2]. It not only combines a new stereo vision head for NAO with an object detector and an updated version of our grasping function, but also includes the releasing of the object. The objects to grasp are ones that can be completely clasped by the robot's fingers and palm. We focus

F. Röhrbein et al. (eds.), *Gearing Up and Accelerating Cross-Fertilization between Academic and Industrial Robotics Research in Europe*, Springer Tracts in Advanced Robotics 94,
DOI: 10.1007/978-3-319-03838-4_9,

on small objects or objects with a handle, for instance a light standard-sized coffee cup or a pencil.

The first step of the general procedure is to detect the object. Thereby edge detection is performed on the left and right stereo images by computing a contrast-normalized Sobel (cns) image instead of using color segmentation as in the work of Azad *et al.* [3]. Afterwards the contrast images are used to recognize the object by evaluating a range of possible object poses. The detected object is given to the grasp planner as a 6D pose.

The next step of the procedure is to decide whether an object can be grasped. By doing so, possible grasp hand poses as well as possible body positions are evaluated by using a pre-calculated workspace. Once a possible hand grasp pose is found, a motion path from the current hand position is planned using an A*-based algorithm. In contrast to the work of Cotugno and Mellmann [4], who use both arms to emulate a big two-finger gripper, but do not use NAO's real fingers, the grasping function proposed in this paper plans and executes actual single-handed grasps. At last, the resulting path is approximated by a Bezier curve and then executed by a trajectory based motion engine.

Because our system uses speech recognition, which is a part of NAO's standard software package, the robot is able to react to spoken commands. Therefore the robot is able to wait and hold a grasped object until someone asks him to release it. The releasing itself is the last part of our procedure. Thereby the robot uses a combination of tactile and force sensors of the carrying arm and detects whether someone touched the grasped object in order to start a safe object handover.

2 Related Work

A vast number of online manipulation systems can be found in the RoboCup @Home League [5] where robots – among other assignments – have to manage different grasp and detection tasks. One participating team is the German b-it-bots with their robot Jenny [6] from the Bonn-Rheine-Sieg University of Applied Sciences. Jenny is equipped with a 7 DOF arm and a three finger hand with at least one motor per finger. The robot categorizes objects with a so-called *Bag of Features* [7] approach that relies on the extraction of locally invariant features. In 2012, the robot Jenny demonstrated its ability to clean a room. In particular the robot was able to pick up bottles in order to insert them into a stash as well as it was able to wipe tables.

Another online manipulation system can be found in the work of Stückler *et al.* [8], where the robot Cosero from the University of Bonn uses a RGB-D camera to recognize and track objects on a table in real-time. Thereby the main horizontal support plane, i.e. the table, is distinguished from object candidates by the RANSAC [9] algorithm that is applied to a 3D point cloud similar to the work of Rusu *et al.* [10]. Feasible collision-free grasps are derived based on the work of Hsiao *et al.* [11] from the point cloud. The robot Cosero is equipped with two two-finger hands (2 DOF), two 7 DOF arms and an Intel I7 processor, which provides a much higher performance than NAO's Intel Atom processor. At the

RoboCup 2011 the robot demonstrated its ability to carry a table together with a human [12] as well as its ability to cook an omelet in a real pan.

In the work of Kuffner *et al.* [13], a full body motion planner plans a dynamically stable motion in the configuration space using rapidly-exploring random trees. It enables a robot to collision-free pick up a bottle that is placed under a table. Based on that approach Burget *et al.* [14] are using an external computer to plan full body grasp motions that enables NAO to open a drawer door. Since this solution requires a long calculation time before a motion is executed, it does not appear to be suitable for grasping objects out of a human hand.

In general most motion planners are operating either in Cartesian or in configuration space. While motion planners in configuration space as in the works of Kavraki *et al.* [15] and Harada *et al.* [16] are able to guarantee a solution given there is one, planners in Cartesian space as in the approach of Vahrenkamp *et al.* [17] are incomplete and difficult due to redundant kinematics. However, the integration of obstacle avoidance into a path planner operating in Cartesian space is simpler than in configuration space.

Furthermore, motion path planning in Cartesian space can be a very expensive process particularly when the grasping hand is attached to a humanoid robot, which can move in order to reach certain objects. Thus, the reachability needs to be checked by inverse kinematics for many points in order to select a suitable grasp and to validate the reachability along the path. This process can be speed up best by a pre-calculated table as the capability map of Zacharias *et al.* [18,19]. In our work we use the predefined workspace to solve redundant kinematics as well as the reachability along the motion path, which enables our motion path planner to quickly operate in Cartesian space.

Stereo vision based object detection as well as online grasping with NAO constitute as particular problems due to the limited processing power and the under-actuated hand design [20]. While more sophisticated robots have fingers that are controlled by at least one motor per finger, NAO's three flexible fingers per hand are controlled together by a single motor (1DOF). Additionally only if the hand is completely closed the fingers are really stiff. Because of that the hand can realistically only be in the states open or closed. Hence, experiments showed that solid objects such as coffee cups are only graspable if the grasp is form-closure [21]. Furthermore, it seems that performing force-closure single-handed grasps are not possible with NAO, since it is not able to move its fingers individually [22].

3 Stereo Head

The original head of NAO is equipped with two cameras: one in its forehead and one in its chin. This configuration results from the requirement of having a camera oriented upwards to detect people around NAO and another one oriented downwards to detect objects such as a soccer ball on the ground. Because the fields of view of the cameras in the original head design do not overlap each other, stereo vision is impossible with that approach. For the needs of the GRASPY

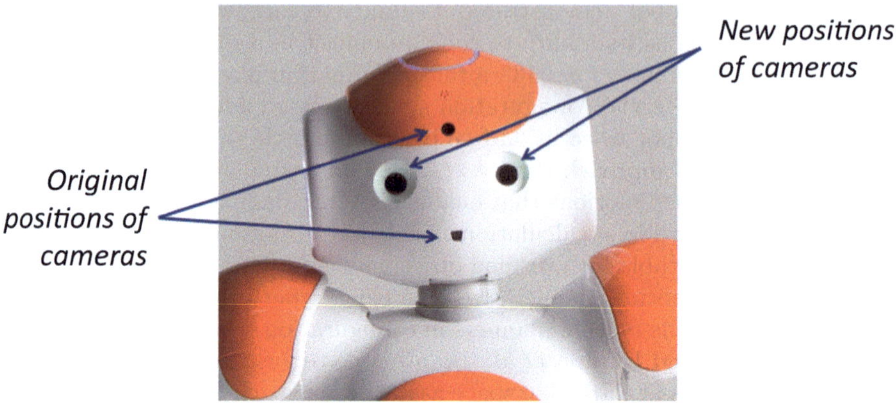

Fig. 1. New positions of the cameras for the stereo head

project, Aldebaran developed a new head with a more natural configuration: the two cameras are positioned in the eyes (Fig. 1). Thereby we selected a different camera model (Aptina MT9M114 [23]) with a higher resolution (1,3 MPixels), a wider field of view (72°) and a better sensitivity (2,24 Lux/(V.sec)) than the previous sensor used in NAO. This new positioning of the cameras in the eyes of NAO required us to suppress the colored LEDS in the eyes. We have to see how we can regain this important feature for man robot interaction in future versions of the stereo head.

The wide overlap (Fig. 2) of the two cameras offers the ability to perform stereo vision in front of the robot. The other major improvement required for stereo vision was the possibility to allow the synchronous acquisition of the two camera images. With the original head design, the application has to select between using the picture from the upper or from the lower camera. This hardware switch is necessary because there is only one video input on the embedded GEODE CPU. With two video inputs, switching between them would require more than 500 ms. As soon as the robot or the environment moves, this delay is generally too big to have a comparison of the two images usable for stereo vision computation.

In the new architecture of NAO's head, we have implemented an FPGA component that makes the acquisition of the two video streams and sends them via an I^2C bus to the ATOM CPU. The delay between two acquired images is less than 33 ms. We are currently working on a hardware synchronization of the two cameras to reduce this delay below 1 ms, but for the GRASPY project we had a delay of up to 33ms. This was reasonable considering the speed of the object to grasp.

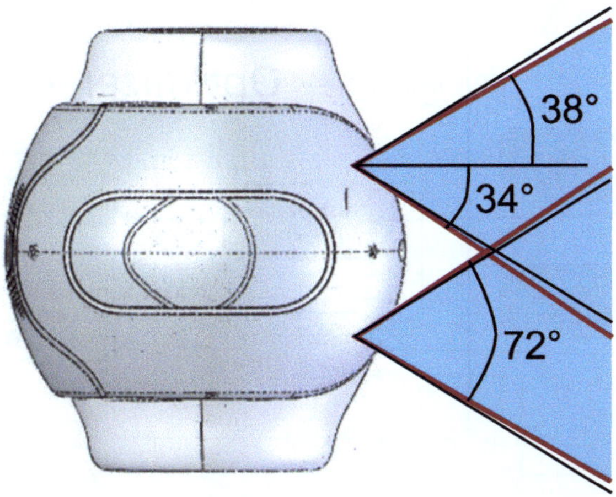

Fig. 2. Overlap of the fields of view with the stereo head

4 Stereo Contour-Based Object Detection

In computer vision there is the general insight that taking hard decisions early impairs robustness. Examples for that are pixel-wise color segmentation or Canny edge detection followed by line segment extraction followed by object detection. Instead, one should take a decision only after considering all relevant input data, in our case the whole stereo image, assessing which interpretation is overall most supported by the data. Compared to mono, stereo gives a better depth perception, and following the above paradigm we do a combined search in both images, not separately.

The first step of our detection is rasterization, i.e. rendering the object in a given hypothetical pose from the perspectives of the left and right cameras. The result is a 2D contour, i.e. a function $[0 \dots 1] \rightarrow \mathbb{R}^2$. The second step is contour evaluation, i.e. computing a response how much the contour is supported by the image. Its definition has already been described in detail in our previous work [2]. Based on this goal function, an optimizer searches through the space of possible poses, finding the cup pose with the largest response.

4.1 3D Object Search Process

Rasterization. The rasterization (Fig. 3) takes a triangle mesh as 3D object model, a camera calibration, and a hypothetical object pose as input and renders the contour of the object at the given pose as viewed from the camera. The first step is to determine which edges of the model form the contour. At the moment, we go through all edges and select those where one adjacent face is viewed from the front and one from the back. This does not consider global occlusion, an extension that could be implemented in the future. As a special rule, faces can

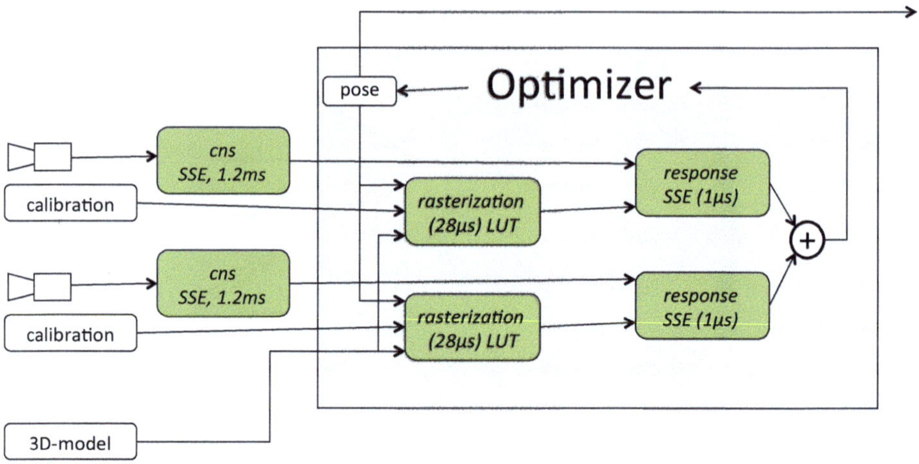

Fig. 3. Dataflow overview of our stereo contour-based object detector

be marked by a color label and edges between visible faces of different labels are also added.

Next, the vertices involved are perspectively projected into the image in an SSE implementation. The projection ignores distortion, which is ≈ 1 pixel only for NAO. The precomputed edge list is sorted such that projected vertices can be used twice.

Global Search Heuristic. The textbook solution for global object search would be to find the maximum response of all poses within the grasping space (6DOF). However, this is computationally beyond scope. Instead, we use an application-specific heuristic. We search only for a single cup orientation by assuming it is roughly vertically aligned and by removing the handle, making it rotationally symmetric. This orientation is obtained from the robot's forward kinematic.

For the position, we go through the image in patches of 64×48 pixels and rasterize the cup at several positions along the center pixel's ray. For each contour, 64×48 responses are computed and the largest overall response is refined.

Then, the cup rotation is determined by evaluating the response of several rotated cups with handle. Finally, the full model is refined. If the response exceeds a threshold (0.65), we switch to tracking mode. If the response in tracking mode falls below 0.5 for 15 frames, we switch back to the global search.

The global search takes ≈ 320ms, so we spread it over several frames, evaluating only between one and two 64×48 blocks in each frame (13–26ms).

Local Search (Refinement and Tracking). During local search, the optimizer (Fig. 3) changes the pose towards growing responses. This procedure is used for tracking as well as to refine a coarse initial pose obtained by our global search heuristic. We use the simple approach to optimize DOFs round-robin one at a time, although there are of course more sophisticated optimization algorithms. However, we exploit that the response computation provides an array of 8×8 responses for shifted contours (2D translation).

So, to refine one DOF, we compute 8×8 responses around the original pose, around a pose changed on step in the considered DOF, and around the inversely changed pose. The subpixel-refined maximum of these $3 \times 8 \times 8$ responses defines the new pose. Therefore the image translation must be converted into a change of pose. This is approximated by a rotation of the object around the camera which moves the object's center in the image according to the obtained image translation.

As image translation is already covered, the 4 remaining DOFs are translation in viewing direction and object rotation around X, Y, and Z (skipped in case of symmetry). The step size is roughly determined to create 3 pixel changes in the image based on object size and distance.

We noticed that the convergence is fairly robust (Sec. 6.3). This motivates to use only one orientation in the global search.

5 Object Manipulation

On the motion side there are two tasks to solve: grasp an object from a human hand and give it back if the human asks for it. In order to grasp an object, we need to calculate a valid motion path from the hand position to the target and avoid obstacles such as the object itself or body parts that may be in the direct path. By doing so we were using the grasp planner of our previous work [2]. Because we had problems with the overlap of detection and grasp space, we modified our old planner to calculate possible body positions online so that the grasp space can be increased without increasing the calculation time. We also added a least-square Bezier fit algorithm in order to smooth the planned way points and increase the execution speed.

The releasing task is to transfer an object to the human hand once a human asks for the object. Thereby, we are using a sensor feedback solution in order to detect whether it is safe to release the object.

5.1 Object Grasping

As described in more detail by Müller *et al.* [2] our grasp motion planning approach is based on a predefined reachability map. This reachability map is a discretization of the workspace with a cube that is divided into equally sized smaller cubes. Each sub cube serves as a region in the workspace. Each region stores a set of reachable lower arm directions for that position (1 DOF).

Thereby we use 4 of the 5 DOF of NAO's arms to define a certain hand pose, and handle the wrist-DOF later. This leaves only one DOF of four for the lower arm direction while a fixed hand position is commanded, less than the two possible DOF. The fifth DOF represents the wrist angle and has only minor

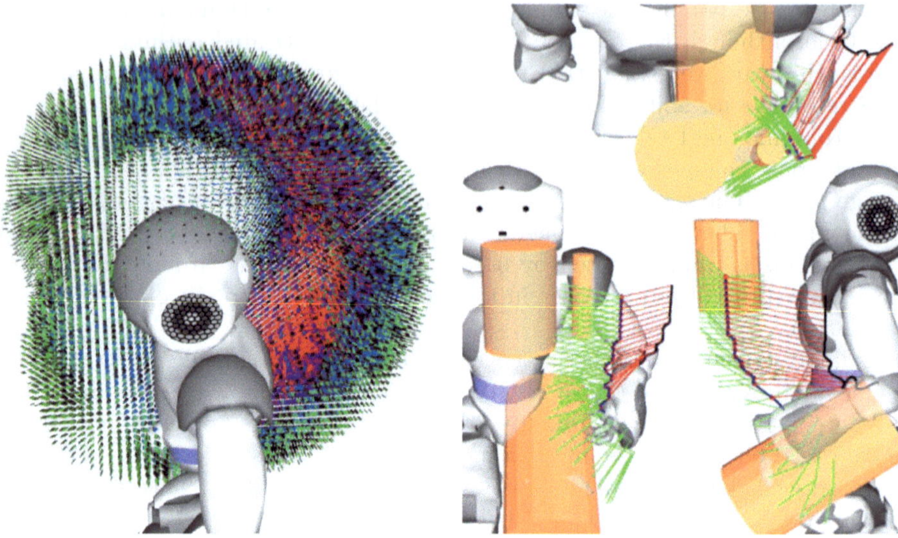

(a) Reachability Map (b) Grasp Path

Fig. 4. (a) The best reachable regions are marked in red, less well-reachable regions are marked in blue, and badly reachable regions are marked in green (b) The linear interpolated motion path of the hand and the elbow with obstacles present

influence on the planning. Because the wrist rotation can be calculated later from the lower arm direction and the joint limits, we only need to store a set of possible lower arm directions per region instead of a set of full hand orientations. Figure 4(a) pictures the reachability map used, where only reachable directions per region are marked.

According to that, the reachability of NAO's hand is clearly very limited and the lower arm direction depends on the hand position. For that reason it is necessary to check for each grasp pose and each point on a motion path whether it can be reached. This leads to the problem that a large number of reachability checks are necessary for motion planning. This can be sped up best by predefining the workspace in a reachability map.

The origin of the reachability map is located in the shoulder of the robot. Thus, it is possible to test with different shoulder positions whether a certain hand position is reachable without the use of inverse kinematics.

Grasp Motion Planning. The first step of the grasp planning is to evaluate a range of grasp poses by using the map. Once a reachable grasp pose is found, the grasp planner plans a path through the grid cells of the map. The reachability map provides the planner with 6D information on the possible hand positions and lower arm directions. Since planning in 6D is very expensive, our A*-based planning algorithm initially only uses the 3D area grid and considers the lower

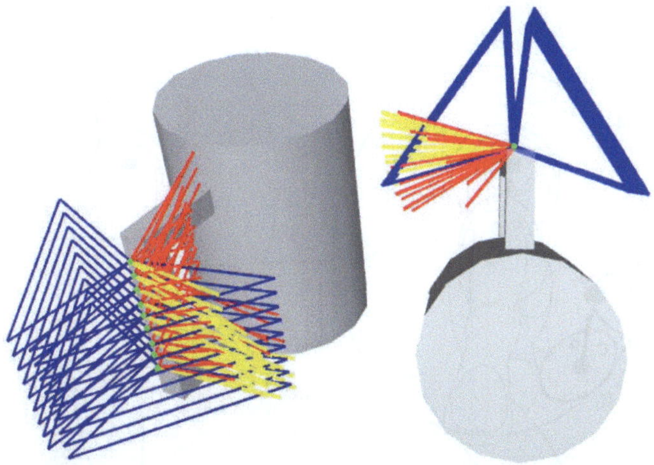

Fig. 5. Each object has its own grasp map, which is generated from a set of predefined grasp rules. Each rule connects a range of lower arm directions to a grasp position (blue). The reachability map (red) is matched to the object's grasp map. Matches are marked yellow.

arm direction only via the cost and heuristic function. Thereby, to be evaluated, nodes are checked for reachability and obstacle collision in order to calculate the heuristics only for verified nodes. In this process, nodes with more suitable lower arm directions are rated better than nodes with greater deviations from the lower arm goal direction. Also the distance between the node evaluated and the goal node in 3D are taken into account.

The output from the planning algorithm is a list of waypoints through the reachability map, which are represented as red dots in Fig 4(b). Since there is a dependence between the hand positions and the directions of the lower arm, a waypoint also includes a direction. Each direction defines the elbow position corresponding to the waypoint and is marked by red lines in Fig 4(b). The final hand orientation is defined by the grasp selection rules as described in the next section and is calculated from the lower arm direction, forward kinematics, and the object pose.

Grasp Selection. In our previous work we selected grasps by testing with a certain amount of predefined body poses, i.e. shoulder positions, whether the resulting reachability map matches with a set of predefined grasp rules. Those grasp rules indicate how an object can be grasped in order to find a suitable grasp point on the object. Each rule is defined by a grasping point and a range of lower arm directions and final hand rotations relative to the object. In Fig. 5, grasp rules are marked with blue triangles. The green dots constitute the position where to grasp and the triangle defines a range of lower arm directions.

In order to select a grasp, the grasp rules are matched with the reachability map. In this process, areas that include a grasping point are examined further

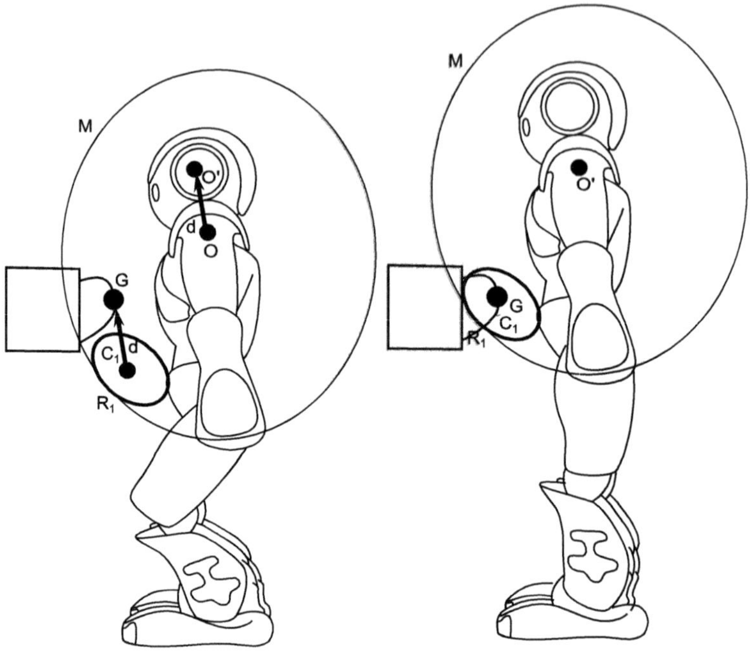

Fig. 6. Schematic depiction of the body pose calculation, M represents the reachability map, O the map origin, R_1 the region of similar cells of the map, C_1 the summed center of a region and d is the translational offset between current and target position G

in order to check whether the corresponding possible lower arm directions are qualified for the grasp. In this process, the possible lower arm directions of the grasp areas are compared to the angle ranges from the grasp rules. The best match is selected.

In our previous approach the search space increases with the amount of predefined body poses and can slow down the planning process. For that reason we investigated – similar to Zacharias *et al.* [19] – each region in the reachability map in detail.

Since the DOF of NAO's arms are very limited, we discovered that we have certain ranges of regions that are very similar to one another, mostly cone-like. So in our new approach, we calculated a fixed amount of center points of similar regions. These center points are used to calculate the body offset between the grasp points defined by the grasp rules and the current body position as it is demonstrated in Fig. 6. The body poses are calculated online by inverse kinematics for a fixed amount of positions. In doing so, we were able to keep the calculation time constant but could increase the grasp space heavily – especially for the highly reachable regions.

Since the robot has two arms, each body offset is calculated for each arm respectively. Thereby body positions, which lead to shorter distances between hand and grasp pose, are rated inferior.

Motion Path Execution. In order to smoothly execute a motion path, it is necessary to minimize the velocity discontinuities at the way points. This can be done by approximating the way points with cubic Bezier curves instead of lines.

In our previous work, we were converting a found grasp plan into a Bezier spline by using the method by DeRose *et al.* [24] in order to generate a trajectory for our motion engine based on the work of Müller *et al.* [25]. In that approach we connected each way point to the next with a cubic Bezier curve using a fixed duration for each sub curve. Although the smoothed path was free of velocity discontinuities, we discovered the problem that sub curves that were short in distance led to a slower movement than sub curves that were longer in distance. As a result the hand accelerates and decelerates unnecessarily.

In order to overcome this problem, we added a least square Bezier fit method as described by Itoh *et al.* [26] and Herold [27] in order to initially approximate all waypoints with a single cubic Bezier spline. This method is using the percentage of the length between each adjacent point of the path with equation

$$t_i = \frac{|d_i - d_{i-1}|}{\sum_{j=1}^{n} |d_j - d_{j-1}|} \tag{1}$$

to synchronize the points with the cubic Bezier curve function

$$B(t, C) = c_0(1-t)^3 + 3c_1t(1-t)^2 + 3c_2t^2(1-t) + c_3t^3 \tag{2}$$

with $C = [c_0, c_1, c_2, c_3]$. The distance d_i to the i-th point is defined by equation

$$d_i = \sum_{j=1}^{i} |P_j - P_{j-1}| \tag{3}$$

with $d_0 = 0$.

We used the residual sum of squares to calculate the error fit. In doing so, in equation

$$E(C) = \sum_{i=1}^{n} (p_i - B(t_i, C))^2 \tag{4}$$

we are summing the squared distance of each waypoint p_i to its Bezier curve approximation defined by the control points $c_0 \dots c_3$.

By setting the derivative of Equation (4) equal to zero, the control points $C = [c_0, c_1, c_2, c_3]$ with minimum error [28] can be found. With Equation (2) the best fitting Bezier curve is defined.

Since the plan is recalculated in each frame, we need to consider that the path changes even if parts of the old plan already had been executed. It is not possible to change a Bezier curve $B(t, C)$ after a certain t^* without changing the whole curve. For that reason we need to split the path at the next point that is to be executed before a replanning is done. Hence we are splitting the path at t^* by using De Casteljau's algorithm in two curves: already executed ($ae(t)$ in Fig. 7) and to be executed ($be(t)$ in Fig. 7).

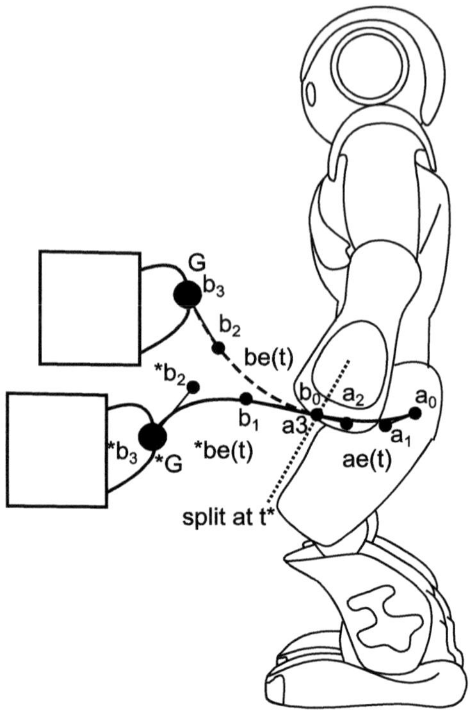

Fig. 7. Schematic depiction of the curve splitting at t^* during the path recalculation; the initial curve from a_0 to G is splitted at the next position to be executed. The new plan is approximated by $^*be(t)$ and connected with the already executed curve $ae(t)$.

The last control point of $ae(t)$ is used as new start point to replan the path to the current goal point. The resulting way points are converted to a new Bezier curve ($^*be(t)$ in Fig. 7) with the condition that the first two control points are fixed in order to keep continuous velocities in the connection point.

Since a plan can be longer or shorter after the replanning, a duration update is also necessary. In doing so, we calculate in each frame i with equation

$$o_i = o_{i-1} \frac{|d_{^*be(1)}|}{|d_{b(1)}|} \tag{5}$$

the change of the path length. Thereby we multiply the previous remaining duration o_{i-1} with the ratio between the path length of the previous sub curve $be(t)$ (with path length $d_{be(1)}$) and the updated curve $^*be(t)$ (with path length $d_{^*be(1)}$).

5.2 Object Releasing

The aim of the release function is to transmit the grasped object to the human in a safe way. The safety mainly concerns the manipulated object that can break if it falls down during the transmission. But it can also concern the safety of the human user: an elderly or immobile person, whom the balance can be unstable or destabilized if the object he should get from the robot falls down during the transmission. We wanted to make sure that the robot opens its hand when the object is correctly caught by the human partner.

The main idea is to detect that the object is pulled from the hand of the robot before releasing it. Because NAOs fingers are not equipped with force sensors, it is not possible to detect the traction force applied by the user pulling the object directly with the fingers. But the traction force is transmitted to the arm, through the rigidity of the wrist. By reducing the stiffness of the arm joints, it is possible to detect the traction force by an unexpected motion of the arm joints. When the robot wants to give its object back, it monitors the position of the arm carrying the object. As soon as this arm moves the robot detects that a traction force is applied on the object and opens its hand.

Of course reacting to an unexpected force with letting go of the object is a dangerous behavior. So we require the user to say *"give it to me"* to put NAO into object return mode and also NAO needs to detect the users hand. This detection is made by one of the two modalities: tactile or vision. The robot expects to see the hand of the user close to its own hand or detects with the tactile sensor on the back of its hand that the user touches its hand. If one of these conditions is fulfilled, the robot can safely open its hand to release its object. To safely release an object in an intuitive way it appears that the robot needs to combine four modalities of perception: audio (*give it to me*), kinesthetic (unexpected motion of the arm), vision (detection of the hand) and tactile (contact of the back of NAO's hand).

When the robot is walking with the object in its hand to bring it to its user, it may happen that the object slips from the fingers and falls. NAO is able to detect this event as well thanks to its proprioceptive sensors. To grasp an object, NAO tries to close its hand to its maximum closed position. Because the object does not allow the complete closing the goal position of the finger is not reached. The difference between the expected finger position and the actual finger position indicates if the object has been grasped or not. If the robot, supposed to be carrying an object, detects that its finger are in the maximum required closed position, it means that the object is no more in its hand.

6 Experiments

The GRASPY experiment concerns the exchange of objects between NAO and a human. The objects to grasp are cylinder-like objects such as a pen or a cup with handles. The places where the robot should grasp them from should be from a human hand or from a table.

The man machine interface is a simple dialog, in which the human tells NAO to grasp a certain object. In case the object is reachable by the robot, NAO confirms this and grasps it. If the robot has the object in its hand, the human may tell it to give it back. Only if the robot has an object in one of its hands

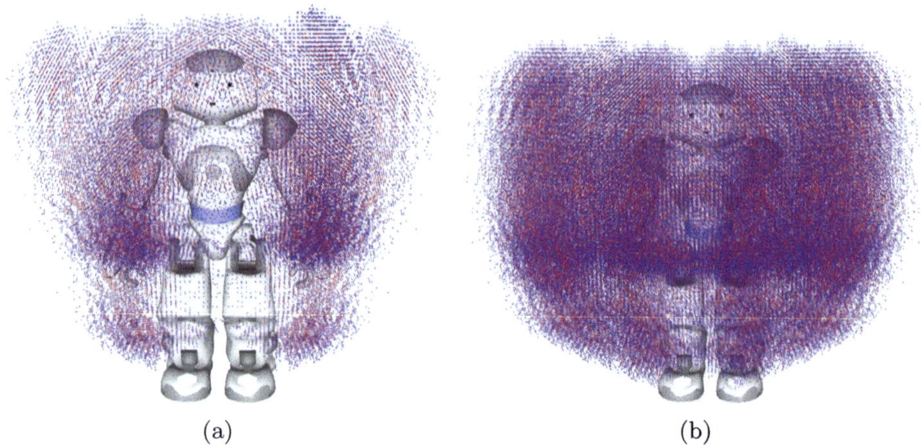

Fig. 8. (a) Depiction of grasp space when using only four fixed positions (b) Depiction of grasp space when body position is calculated online

it offers it by speech and gesture to the human. Once the human touches the object it will be released by the robot.

The evaluation criteria are the time to grasp an object, the number of wrongly detected objects, and the number of successfully grasped objects. All experiments were made on a Nao robot using its Intel Atom (1,6 GHz) processor with 1 GB SDRAM.

6.1 Planning

We evaluated the planning algorithm with the new grasp selection function and compared the results to the results of our previous work. In previous last work we reached a calculation time of 20 ms per frame by using a heuristic and cost function that combined the translation distance between the nodes and the goal with the differences in the lower arm directions per node.

In our new experiment we used the same parameters and the same planner with the extension that the body pose is calculated online. We calculated four possible body positions per arm per frame and tested whether the object can be reached. The average calculation time per frame is 21 ms, which is almost the same as in our previous experiments but the amount of good regions in the overall grasp space could be heavily increased as it can be seen in Fig. 8(a) and Fig. 8(b).

6.2 Motion Path Execution

We also compared the execution of the smoothed paths. Thereby we recorded the commanded hand position (cX, cY, cZ) and the measured hand position (mX, mY, mZ) per frame.

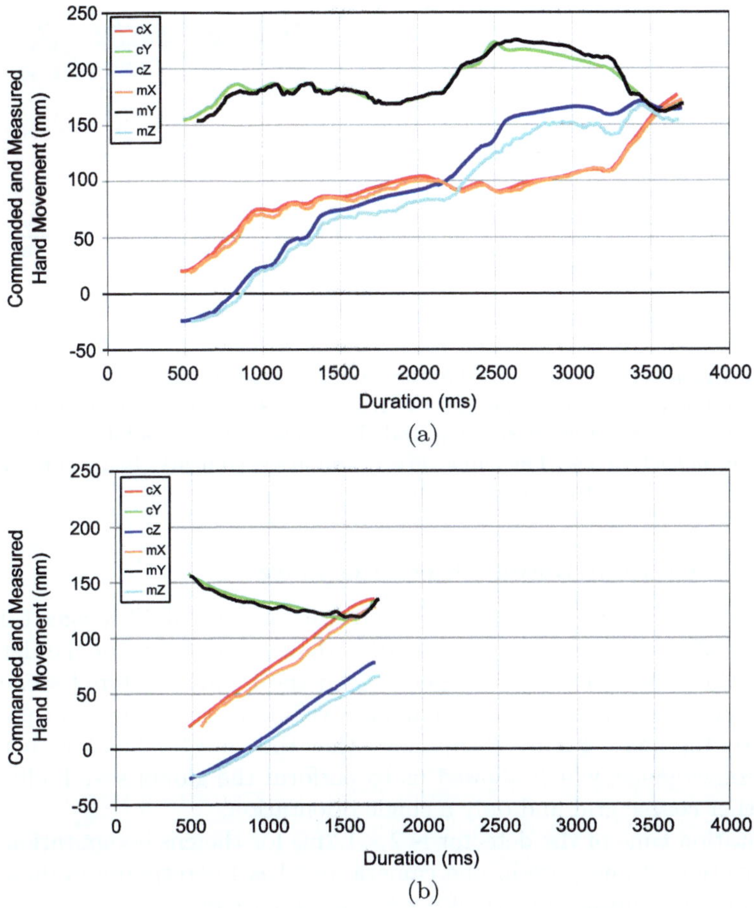

Fig. 9. (a) Comparision of commanded and measured hand motion using our old smoothing method and (b) using the presented least square Bezier fit method

Figure 9(a) depicts a motion path produced by the planner with the smoothing method of our previous work. Although that path was converted into a Bezier path there are a lot of passages where the commanded direction changes a lot.

Figure 9(b) depicts the commanded and measured hand motion of a similar but slightly shorter motion path smoothed by our new method. The commanded path is clearly straighter and the error between commanded and measured hand position on the y-axis is smaller. The deviation on the x- and z-axis results from the backlash of approximately $\pm 3°$ of NAO's arm motors in combination with the weight of the arm.

Due to the much straighter motion paths, we also increased the execution speed. This reduces the duration of the whole grasp.

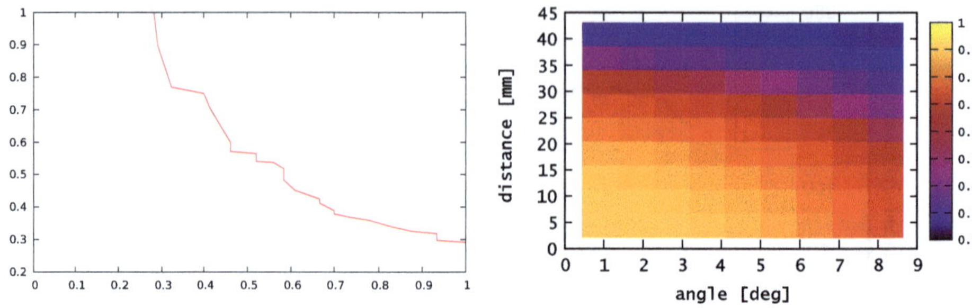

(a) ROC curve (b) Range of Refinement

Fig. 10. (a) Precision over recall for the stereo image based cup detector (b) Probability of the local refinement to converge into the true pose as a function of the angular and translational distance between starting and final pose. The probability is computed with 100 tries in each image. The cup is $95 \times 75mm$ large, so nearly half a cup-diameter in the initial guess leads to a good final pose.

6.3 Stereo Contour-Based Object Response

We evaluated the contour-based stereo detector on a set of 53 images taken in a cluttered office environment. Figure 10(a) shows a roc-curve of the detector. In our opinion the performance is good given the highly cluttered scenes and the fact that often the cup is only partially visible in the image and partially occluded by the hand. Figure 10(b) shows that the detector has a rather large range of convergence, which allowed us to perform the global search efficiently with a rather coarse grid and only a single orientation.

Computation time of the detector is 2×1.2ms for the cns computation, 28μs for rasterization of one pose in one camera and 1μs for response evaluation of one contour, when always blocks of 8×8 contours are evaluated.

6.4 System Level Experiments

Since we could decrease the time to execute a grasping motion, we repeated the system level experiment. In our previous work we did 30 experiments and tested whether a cup could be grasped and how long it took. In doing so, we recorded the time between the first detection and the successful grasp in a normally illuminated simple office environment. We discovered that the average grasping time is now 7.11s instead of 10.026s. In addition, each try was successful.

During the trials we also measured the timings of the object detector. Both cns images are calculated with an average time of 2.5ms per frame, the global search with 21.6ms per frame and the refinement in 1.8ms per frame.

7 Conclusion

We successfully improved our grasp planner by increasing the grasp space but kept the calculation time constant. The whole system operates at 30Hz. We also were able to improve the motion execution by decreasing the duration of the full execution and achieving a smoother movement. In addition, we introduced a releasing function, which completes the whole system and converts it into a prototypical application.

In future work we are planning to include the possibility to grasp an object two-handed. The variety of graspable objects would increase instead of being limited to objects with a handle, because this would enable the robot to perform force-closure grasps. Another point we are planning to investigate is to measure the weight of grasped objects and the impact on the robot's walk. The goal will be that NAO carries an object around in order to bring it to someone or somewhere.

Acknowledgements. This work has been funded by the European Commission in the 7th Framework Programme for RTD in the context of the project "ECHORD – European Clearing House for Open Robotics Development" under the contract number FP7-231143 and by DFG under grant FR2620/1-1 and SFB/TR 8 Spatial Cognition.

References

1. Gouaillier, D., Hugel, V., Blazevic, P., Kilner, C., Monceaux, J., Lafourcade, P., Marnier, B., Serre, J., Maisonnier, B.: Mechatronic design of NAO humanoid. In: IEEE International Conference on Robotics and Automation, ICRA 2009, pp. 769–774 (May 2009)
2. Müller, J., Frese, U., Röfer, T.: Grab a mug - object detection and grasp motion planning with the Nao robot. In: IEEE-RAS International Conference on Humanoid Robots (2012)
3. Azad, P., Asfour, T., Dillmann, R.: Stereo-based 6D object localization for grasping with humanoid robot systems. In: IEEE/RSJ International Conference on Intelligent Robots and Systems, IROS 2007, October 29-November 2, pp. 919–924 (2007)
4. Cotugno, G., Mellmann, H.: Dynamic motion control: Adaptive bimanual grasping for a humanoid robot. In: Proceedings of the Workshop on Concurrency, Specification, and Programming CS&P 2010, Börnicke (near Berlin), Germany, vol. 2 (September 2010)
5. Stuckler, J., Holz, D., Behnke, S.: Robocup@home: Demonstrating everyday manipulation skills in robocup@home. IEEE Robotics Automation Magazine 19(2), 34–42 (2012)
6. Hegger, F., Mueller, C.A., Jin, Z., Alvarez Ruiz, J., Giorgana, G.R.G., Hochgeschwender, N., Reckhaus, M., Paulus, J., Ploeger, P.G., Kraetzschmar, G.K.: The b-it-bots robocup@home 2011 team description paper. In: Proc. RoboCup Symp. (2011)
7. Csurka, G., Dance, C.R., Fan, L., Willamowski, J., Bray, C.: Visual categorization with bags of keypoints. In: Workshop on Statistical Learning in Computer Vision, ECCV, pp. 1–22 (2004)
8. Stückler, J., Steffens, R., Holz, D., Behnke, S.: Efficient 3d object perception and grasp planning for mobile manipulation in domestic environments. Robotics and Autonomous Systems (2012)

9. Fischler, M.A., Bolles, R.C.: Random sample consensus: a paradigm for model fitting with applications to image analysis and automated cartography. Commun. ACM 24(6), 381–395 (1981)
10. Rusu, R., Blodow, N., Marton, Z., Beetz, M.: Close-range scene segmentation and reconstruction of 3d point cloud maps for mobile manipulation in domestic environments. In: IEEE/RSJ International Conference on Intelligent Robots and Systems, IROS 2009, pp. 1–6 (2009)
11. Hsiao, K., Chitta, S., Ciocarlie, M., Jones, E.: Contact-reactive grasping of objects with partial shape information. In: 2010 IEEE/RSJ International Conference on Intelligent Robots and Systems (IROS), pp. 1228–1235 (2010)
12. Stückler, J., Behnke, S.: Following human guidance to cooperatively carry a large object. In: IEEE-RAS Int. Conf. Humanoid Robots (Humanoids), Bled, Slowenia, pp. 218–223 (2011)
13. Kuffner Jr., J.J., Kagami, S., Nishiwaki, K., Inaba, M., Inoue, H.: Dynamically-stable motion planning for humanoid robots. Autonomous Robots 12(1), 105–118 (2002)
14. Burget, F., Hornung, A., Bennewitz, M.: Whole-body motion planning for manipulation of articulated objects. In: Proceedings of the IEEE International Conference on Robotics and Automation, ICRA (2013)
15. Kavraki, L., Svestka, P., Latombe, J.C., Overmars, M.: Probabilistic roadmaps for path planning in high-dimensional configuration spaces. IEEE Transactions on Robotics and Automation 12(4), 566–580 (1996)
16. Harada, K., Kaneko, K., Kanehiro, F.: Fast grasp planning for hand/arm systems based on convex model. In: IEEE International Conference on Robotics and Automation, ICRA 2008, pp. 1162–1168 (May 2008)
17. Vahrenkamp, N., Asfour, T., Dillmann, R.: Simultaneous grasp and motion planning: Humanoid robot armar-iii. IEEE Robotics Automation Magazine 19(2), 43–57 (2012)
18. Zacharias, F., Borst, C., Hirzinger, G.: Object-specific grasp maps for use in planning manipulation actions. In: Kröger, T., Wahl, F.M. (eds.) Advances in Robotics Research, pp. 203–213. Springer, Heidelberg (2009)
19. Zacharias, F., Borst, C., Hirzinger, G.: Capturing robot workspace structure: representing robot capabilities. In: IEEE/RSJ International Conference on Intelligent Robots and Systems, IROS 2007, October 29-November 2, pp. 3229–3236 (2007)
20. Lalibertže, K., Birglen, L., Gosselin, C.M.: Underactuation in robotic grasping hands. Journal of Machine Intelligence and Robotic Control 4, 77–87 (2002)
21. Borst, C., Fischer, M., Hirzinger, G.: Grasping the dice by dicing the grasp. In: Proceedings. 2003 IEEE/RSJ International Conference on Intelligent Robots and Systems, IROS 2003, vol. 4, pp. 3692–3697 (October 2003)
22. Kragten, G., Kool, A., Herder, J.: Ability to hold grasped objects by underactuated hands: Performance prediction and experiments. In: IEEE International Conference on Robotics and Automation, ICRA 2009, pp. 2493–2498 (May 2009)
23. Aptina, `http://www.aptina.com/products/soc/mt9m114/` (accessed: June 03, 2013)
24. DeRose, T.D., Barsky, B.A.: Geometric continuity, shape parameters, and geometric constructions for Catmull-Rom splines. ACM Trans. Graph. 7(1), 1–41 (1988)
25. Müller, J., Laue, T., Röfer, T.: Kicking a ball – modeling complex dynamic motions for humanoid robots. In: Ruiz-del-Solar, J., Chown, E., Ploeger, P.G. (eds.) RoboCup 2010. LNCS (LNAI), vol. 6556, pp. 109–120. Springer, Heidelberg (2010)

26. Itoh, K., Ohno, Y.: A curve fitting algorithm for character fonts. Electronic Publishing 6(3), 195–198 (1993)
27. Herold, J.: Least squares Bezier fit, `http://jimherold.com/2012/04/20/least-squares-bezier-fit/` (accessed: June 03, 2013)
28. Press, W.H., Teukolsky, S.A., Vetterling, W.T., Flannery, B.P.: Numerical recipes in C the art of scientific computing, 2nd edn. Cambridge University Press, New York (1992)

HANDS.DVI: A Device-Independent Programming and Control Framework for Robotic Hands

Gionata Salvietti[1], Guido Gioioso[1], Monica Malvezzi[1], Domenico Prattichizzo[1], Alessandro Serio[2], Edoardo Farnioli[2], Marco Gabiccini[2], Antonio Bicchi[2], Ioannis Sarakoglou[3], Nikos Tsagarakis[3], and Darwin Caldwell[3]

[1] Dept. of Information Engineering and Mathematics, University of Siena, Siena, Italy
{salviettigio,gioioso,malvezzi,prattichizzo}@dii.unisi.it
[2] Centro E. Piaggio, University of Pisa, Pisa, Italy
{alessandro.serio,edoardo.farnioli, gabiccini,bicchi}@centropiaggio.unipi.it
[3] Dept. of Advanced Robotics, Istituto Italiano di Tecnologia, Genova, Italy
{ioannis.sarakoglou,nikos.tsagarakis,darwin.caldwell}@iit.it

Abstract. The scientific goal of HANDS.DVI consists of developing a common framework to programming robotic hands independently from their kinematics, mechanical construction, and sensor equipment complexity. Recent results on the organization of the human hand in grasping and manipulation are the inspiration for this experiment. The reduced set of parameters that we effectively use to control our hands is known in the literature as the set of synergies. The synergistic organization of the human hand is the theoretical foundation of the innovative approach to design a unified framework for robotic hands control. Theoretical tools have been studied to design a suitable mapping function of the control action (decomposed in its elemental action) from a human hand model domain onto the articulated robotic hand co-domain. The developed control framework has been applied on an experimental set up consisting of two robotic hands with dissimilar kinematics grasping an object instrumented with force sensors.

Keywords: Robotic hand, grasping, object-based mapping, human hand synergies.

1 Introduction

The HANDS.DVI experiment deals with the development of a unified structure for programming and controlling robotic hands based on a number of fundamental primitives, and abstracting, to the possible extent, from the specifics of their kinematics, mechanical construction, sensor equipment. HANDS.DVI hinges on the study of how the embodied characteristics of the human hand and its sensors, the sensorimotor transformations, and the many constraints they impose, affect and determine the learning and control strategies we use for such fundamental cognitive functions as exploring, grasping and manipulating. The ultimate goal is to learn how to devise simplified and device independent control system architectures for robotic hands from human data available

F. Röhrbein et al. (eds.), *Gearing Up and Accelerating Cross-Fertilization between Academic and Industrial Robotics Research in Europe*, Springer Tracts in Advanced Robotics 94,
DOI: 10.1007/978-3-319-03838-4_10,

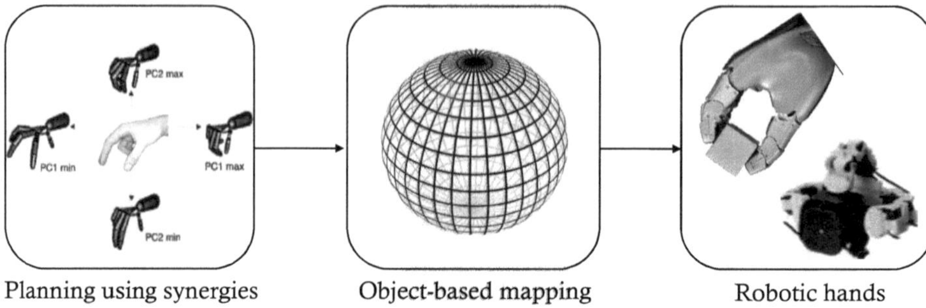

Fig. 1. Idea of the project. The synergies defined for a paradigmatic hand model are used as a simplified language to describe different grasping and manipulation tasks. The arising motion is mapped through and object based approach to different robotic hands without considering the specific kinematics.

in the literature and hypotheses driven simulations. The acronym HANDS.DVI was chosen in analogy to device-independent files that are generated by the TeX typesetting language.

The experiment has been separated into three different phases called SYN, DVI and EXP. The phase SYN dealt with the development of strategies for grasp force control based on studies in neuroscience concerning the sensorimotor organization of the human hand [1]. These studies demonstrated that, notwithstanding the complexity of the hand, a few variables are able to account for most of the variance in the patterns of configurations and movements. The reduced set of parameters that humans effectively use to control their hands, known in the literature as *synergies*, represents a possible set of words for a unified control language for robotic hands. In the phase DVI, we focused on designing control algorithms based on synergies for robotic hands with a generic kinematic structure, not necessarily bio-inspired. We developed on object-based approach to map human synergies onto several robotic hand types including non-anthropomorphic hands. The main idea of the mapping algorithm was to involve a virtual object as a mediator between human and robotic hand. Finally, in phase EXP, we implemented and evaluated the synergy based approach on two devices: the ModHa 39p hand, a modular hand developed at University of Siena, and the DLR-HIT II hand. For the evaluation we also used an instrumented object developed within the experiment that allows to measure contact point positions and the relative contact forces. The general framework of the experiment is pictorially represented in Fig. 1.

The rest of the chapter is organized as it follows. Section 2 deals with the definition of a paradigmatic hand able to capture the synergistic organization of the human hand. In Section 3 a description of the object-based mapping, that represents the algorithmic core of HANDS.DVI, is given. In Section 4 the setup and relative results for the experiments in motion and force control evaluation are shown. Finally in Section 5 conclusions and future possible applications are outlined.

2 The Paradigmatic Hand

A deeper understanding of the way humans use their hands enable an approach to programming hands that allows users to more easily control the different devices that may be used in a robotic system, by encapsulating the hand hardware in functional modules, and ignoring the implementation-specific details. In the HANDS.DVI experiment, such new methodology for grasping analysis has been based on the concept of synergies. The results on the organization of the human hand in grasping and manipulation presented in [1, 2] were based on experimental tests in which subjects were asked to perform grasping actions on a wide variety of objects. Data were recorded by means of data gloves and were analysed with principal component analysis (PCA) techniques. The results showed that the first two principal components account for most of the variability in the data, more than 80% of the variance in the hand postures. In this context the principal components were referred to as synergies to capture the concept that, in the sensorimotor system of the human hand, combined actions are favoured over individual component actions, with advantages in terms of simplification and efficiency of the overall system. Further, synergies where shown to exist not only in movement of hand configurations while preparing to grasp but also in force control. In [2] the authors suggested an explanation of the coordination of isometric forces exerted during grasping, namely that there exist a few basic patterns, the synergies, which are suitable for coarse control of force and that these synergies can be modified by superimposing a finer control. To summarize the results of the recent neuroscience studies: notwithstanding the human hand is characterized by a complex mechanical structure with many degrees of freedoms (DoFs) it results that most of the actions of the human hand can be represented as a combination of a relatively small number of basic primitives of motion, referred to as synergies. In other words, even if the human hand has a complex kinematic structure, with many degrees of freedom, in most everyday tasks it appears to be controlled by a much smaller set of knobs, corresponding to compound actions, movements and variables. Our idea is to use this few knobs to control different robotic hands without considering the specific kinematic. However the development of such control framework passes through the definition of a generic human hand model, i.e. a *paradigmatic* hand, where the synergies can be easily defined.

The paradigmatic hand is a kinematic model inspired by the human hand that represents a trade–off between the complexity of the human hand model accounting for the synergistic organization of the sensorimotor system and the simplicity, and accessibility, of the models of robotic hands available on the market. Examples of human hand biomechanical models are available in the literature [3, 4]. The fingers are usually modelled as kinematic chains independent from each other, sharing only their origin in the hand palm. In absence of disabilities or handicaps, the ratios between the bones lengths of each finger are almost constant [5]. Hence, in Tab. 1 the bones length ratios defined with respect to the length of the distal phalanges bone (dp) of each finger [6] are reported (for bone acronyms please refer to the table caption).

The human hand joints can mainly be divided into 1-DoF and 2-DoF joints. The 1-DoF joints in the hand can be represented as revolute joints; the 2-DoF joints can be divided into two types. The *trapeziometacarpal* joint of the thumb is a saddle joint with non-orthonormal axes, the *metacarpophalangeal* joints of the fingers are condyloid.

Table 1. Table of bone-to-bone length ratios (Bone names: distal phalanx (dp), proximal phalanx (pp), middle phalanx (mp), metacarpal (mc))

Finger		mp/dp	pp/dp	mc/dp
Thumb	right	–	1.37	2.09
	left	–	1.36	2.08
Index	right	1.41	2.45	4.17
	left	1.41	2.44	4.10
Middle	right	1.60	2.54	3.71
	left	1.59	2.54	3.71
Ring	right	1.50	2.33	3.25
	left	1.49	2.31	3.22
Pinky	right	1.15	2.04	3.32
	left	1.16	2.04	3.32

The main difference between saddle and condyloid joints is that condyloid joints have approximately intersecting axes while saddle joints do not. For the thumb, the axes of the metacarpal are non-orthogonal screw. Therefore, the *metacarpophalangeal* joint of the index, middle, ring and little fingers are usually modelled as a two DoFs joint (one for adduction/abduction and another flexion/extension). The *proximal interphalangeal* and *distal interphalangeal* joints of the other fingers can be modelled as a one DoF (revolute) joint. The thumb has at least 5 DoF: 2 DoF in *trapeziometacarpal* joint, 2 DoF in *metacarpophalangeal* joint, and 1 DoF in *interphalangeal* joint. Anyway, the range of deviation of metacarpophalangeal joint is so small that generally can be modelled as a single DoF joint, while the trapeziometacarpal joint is more important in the analysis of the thumb kinematics [5]. In order to avoid unnatural finger positions, the set of angle constraints reported in Tab. 2 is taken into account [7]. Our model, showed in Fig.2, has therefore 20 DoFs corresponding to 4 DoFs for the thumb (TR, TA, TM, TI

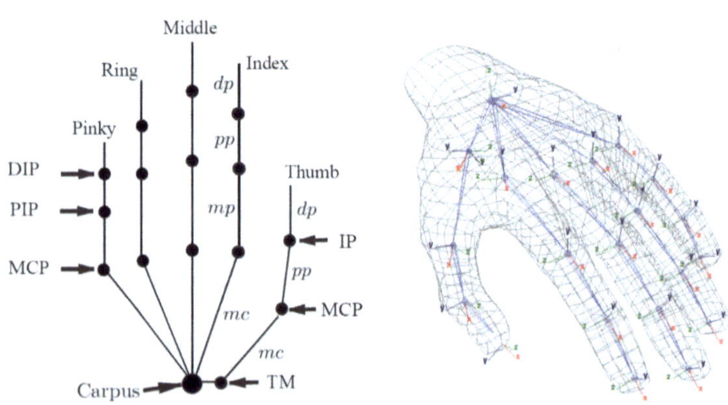

Fig. 2. 20 DoFs kinematic model of the paradigmatic hand

Table 2. Allowed ranges for the joint angle variables

Finger	q_1	q_2	q_3	q_4
Thumb	-10°, 80°	0°, -55°	0°, -55°	0°, -40°
Index	0°, 90°	-15°, 15°	0°, 110°	0°, 90°
Middle	0°, 90°	-12°, 12°	0°, 110°	0°, 90°
Ring	0°, 90°	-10°, 10°	0°, 110°	0°, 90°
Pinky	0°, 90°	-12°, 12°	0°, 110°	0°, 90°

- Thumb Rotation, Abduction, Metacarpal, Interphalangeal) and 4 DoFs for the index, middle, ring and pinky (Index Abduction, Metacarpal, Proximal interphalangeal, Distal interphalangeal).

3 The Mapping Algorithm

In this Section the mapping algorithm developed in the DVI phase is revised. For further details, the reader is referred to [8,9]. The proposed mapping algorithms tries to match the effects of the manipulation tasks performed by the paradigmatic hand, and the real robotic hand. In other terms the paradigmatic hand drives the real robotic hand through this mapping. The proposed approach is not specific for a given task or a given grasped object but can be extended to most of the manipulation tasks. Such a generality is gained considering that the principal actions in manipulation are to guarantee the stability of the grasp and to move the grasped object along planned trajectories. Other mapping methods have been proposed in the literature which generally are based on a joint-to-joint mapping [10] or on a fingertip position mapping [11]. We used a virtual object approach to capture the generality of the many possible objects to be manipulated in our model. Two virtual spheres are used, one for the paradigmatic hand and the other for the robotic hand. These are defined by the hands' posture and change during the task. Note that the use of a spherical virtual object does not restrict the use of this algorithm to spherical objects [8]. The main idea is to reproduce movements and deformations exerted by the paradigmatic human-like hand, controlled by synergies, on the virtual sphere computed as the minimum sphere containing a set of reference points that can be arbitrarily placed on the hands (see Fig. 3).

In the following we describe how the map is obtained. Let the paradigmatic hand be described by the joint variable vector $q_h \in \Re^{n_{qh}}$ and assume that the subspace of all configurations can be represented by a lower dimensional input vector $z \in \Re^{n_z}$ (with $n_z \leq n_{qh}$) which parametrizes the motion of the joint variables along the synergies $q_h = S_h z$ being $S_h \in \Re^{n_{qh} \times n_z}$ the synergy matrix. In terms of velocities one gets

$$\dot{q}_h = S_h \dot{z}. \tag{1}$$

The ultimate goal of this mapping is to find a way of controlling the reference joint variables $\dot{q}_r \in \Re^{n_{qr}}$ of the robotic hand in a synergistic way using the vector of synergies

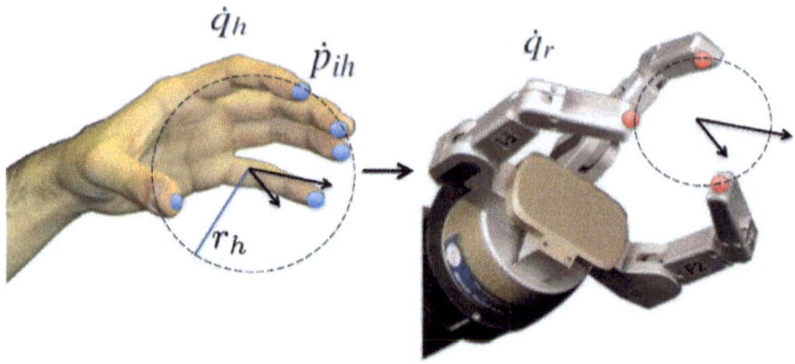

Fig. 3. Mapping synergies from the paradigmatic human hand to the robotic hand: the reference points on the paradigmatic hand p_h (blue dots) allows to define the virtual sphere. Activating the human hand synergies, the sphere is moved and strained; its motion and strain can be evaluated from the velocities of the reference points $\dot{p}_h$. This motion and strain, scaled by a factor depending on the virtual sphere radii ratio, is then imposed to the virtual sphere relative to the robotic hand, defined on the basis of the reference points p_r (red dots).

z of the paradigmatic hand. In other terms we want to design a map S_r to steer the robotic joint reference variables as follows

$$\dot{q}_r = S_r \dot{z} \tag{2}$$

where map S_r depends on the synergy matrix S_h and other variables as explained in the following.

To define the mapping we assume that both the paradigmatic and the robotic hands are in given configurations q_{0h} and q_{0r} (Fig. 3). A set of reference points p_h are chosen on the paradigmatic hand. We have chosen the fingertip points as reference points. Other choices are possible as, for example, in the intermediate phalanges or in the hand palm since the number of reference points can be arbitrary set [8,9].

The virtual sphere object is then computed as the minimum sphere containing the reference points in p_h (Fig. 3). Note that these points in general do not lie on the sphere surface. Let us parametrize the virtual sphere by its center o_h and radius r_h. The motion imposed to the hand reference points moves the sphere and changes its radius.

The motion of the hand due to synergies could be described using a large set of parameters, in this algorithm we simplify the problem assuming the following transformation for the virtual sphere:

- a *rigid-body* motion, defined by the linear and angular velocities of the sphere center $\dot{o}_h$ and ω_h, respectively
- a *non-rigid* strain represented by the radius variation $\dot{r}$ of the sphere.

Representing the motion of the hand through the virtual object, the motion of the generic reference point p_{ih} can be expressed as

$$\dot{p}_{ih} = \dot{o}_h + \omega_h \times (p_{ih} - o_h) + \dot{r}_h (p_{ih} - o_h) . \tag{3}$$

Grouping all the reference point motions, one gets

$$\dot{p}_h = A_h \begin{bmatrix} \dot{o}_h \\ \omega_h \\ \dot{r}_h \end{bmatrix}, \tag{4}$$

where matrix $A_h \in \Re^{n_{ch} \times 7}$ is defined as follows

$$A_h = \begin{bmatrix} I & -S(p_{1h} - o_h) & (p_{1h} - o_h) \\ \cdots & \cdots & \cdots \\ I & -S(p_{ih} - o_h) & (p_{ih} - o_h) \\ \cdots & \cdots & \cdots \end{bmatrix} \tag{5}$$

and $S()$ is the skew operator. Matrix A_h depends on the type of motion that we decide to reproduce on the robotic hand and then it depends on the task. From these equations we can evaluate the virtual sphere motion and deformation as a function of the synergy vector velocity $\dot{z}$ of the paradigmatic hand

$$\begin{bmatrix} \dot{o}_h \\ \omega_h \\ \dot{r}_h \end{bmatrix} = A_h^{\#} \dot{p}_h = A_h^{\#} J_h S_h \dot{z}, \tag{6}$$

where $A_h^{\#}$ denotes the pseudo-inverse of matrix A_h. We now need to map these motions and deformations on the robotic hand. The robotic hand is in a given configuration $q_{0r} \in \Re^{n_{qr}}$ with resulting reference point location vector $p_r \in \Re^{n_{cr}}$. Note that no hypothesis were imposed on the number of reference points on the paradigmatic human and robotic hands, in general we can consider $n_{ch} \neq n_{cr}$, neither on their locations, and neither on the initial configuration of the two hands. The same use of the virtual sphere is applied here: find the minimum sphere enclosing the reference points and indicate with o_r its center coordinates and with r_r its radius (Fig. 3). Let us thus define the *virtual object scaling factor* as the ratio between the sphere radii $k_{sc} = \frac{r_r}{r_h}$. This factor is necessary to scale the velocities from the paradigmatic to the robotic hand workspaces. Note that the scaling factor depends on the hand dimensions, but also on their configuration.

Then, the motion and deformation of the virtual sphere generated by the paradigmatic hand are scaled and tracked by the virtual sphere referred to the robotic hand

$$\begin{bmatrix} \dot{o}_r \\ \omega_r \\ \dot{r}_r \end{bmatrix} = K_c \begin{bmatrix} \dot{o}_h \\ \omega_h \\ \dot{r}_h \end{bmatrix} \tag{7}$$

where the scale matrix $K_c \in \Re^{7 \times 7}$ is defined as

$$K_c = \begin{bmatrix} k_{sc} I_{3,3} & 0_{3,3} & 0_{3,1} \\ 0_{3,3} & I_{3,3} & 0_{3,1} \\ 0_{1,3} & 0_{1,3} & 1 \end{bmatrix}. \tag{8}$$

According to eq. (4) and (5), the corresponding robot reference point velocity is given by

$$\dot{p}_r = A_r \begin{bmatrix} \dot{o}_r \\ \omega_r \\ \dot{r}_r \end{bmatrix}, \tag{9}$$

where matrix $A_r \in \Re^{n_{cr} \times 7}$ is defined as follows

$$A_r = \begin{bmatrix} I & -S(p_{1r} - o_r) & (p_{1r} - o_r) \\ \cdots & \cdots & \cdots \\ I & -S(p_{ir} - o_r) & (p_{ir} - o_r) \\ \cdots & \cdots & \cdots \end{bmatrix}. \tag{10}$$

Recalling eq. (6) and (7) we can express the robotic hand reference point velocities $\dot{p}_r$ as a function of the synergy velocities $\dot{z}$

$$\dot{p}_r = A_r K_c A_h^{\#} J_h S_h \dot{z} \tag{11}$$

and, considering the robot hand differential kinematics $\dot{p}_r = J_r \dot{q}_r$, where $J_r \in \Re^{n_{cr} \times n_{qr}}$ is its Jacobian matrix, the following relationship between robot hand joint velocities and synergy velocities is defined

$$\dot{q}_r = J_r^{\#} A_r K_c A_h^{\#} J_h S_h \dot{z}. \tag{12}$$

Finally the synergy mapping S_r in (2) for the robotic hand is computed as

$$S_r = J_r^{\#} A_r K_c A_h^{\#} J_h S_h, \tag{13}$$

where $J_r^{\#}$ is the pseudoinverse of the Jacobian of the robotic hand and J_h is the Jacobian of the paradigmatic hand. Note that matrix $J_r^{\#} A_r S_c A_h^{\#} J_h$ depends on

- paradigmatic and robotic hand configurations q_{0h} and q_{rh};
- location of the reference points for the paradigmatic and robotic hands, p_h and p_r.

4 Experimental Result

The object-based mapping able to transfer human hand synergies onto robotic hand has been validated in an experimental setup where the capability of reproducing object motions and exerted grasping forces were considered. All the experiments have been performed on two model of robotic hands: the DLR-HIT II hand [12] and the ModHa39p hand [8]. The DLR-HIT II hand has an anthropomorphic structure with 5 fingers, 15 DoFs and 12 actuated joints. The ModHa39p hand is a fully-actuated robotic hand with a modular structure. Each module ($42 \times 33 \times 16$mm) has one DoF and it can be easily connected to the others obtaining kinematic chains that we can consider as fingers. These chains are connected to a common base that can be thought as a palm. In the proposed configuration each finger has 3 DoFs, thus the hand has globally 9 DoFs. In the rest of the Section, the obtained results are presented.

4.1 Object Motion Evaluation

In this experiment we compared the trajectory of the center of a virtual grasped object moved by the paradigmatic hand and the same trajectory of a real object grasped and moved by the robotic hands. A tracking system was thus necessary. We decided to use a cheap and efficient optical system since that met our specifics on precision. In particular,

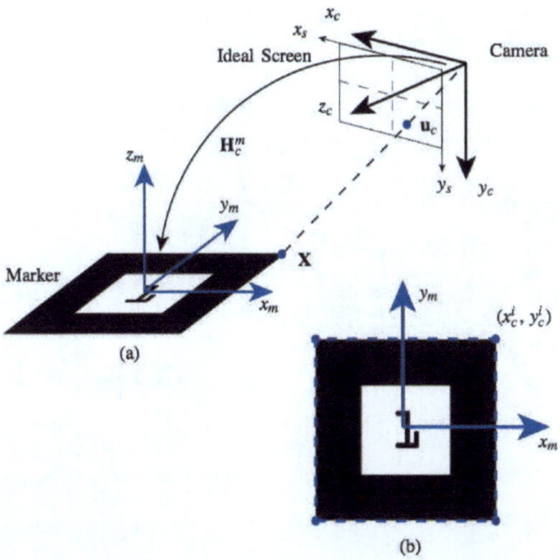

Fig. 4. *ARToolkit coordinates frames*: (a) camera frame $\langle x_c, y_c, z_c \rangle$, marker frame $\langle x_m, y_m, z_m \rangle$ and ideal screen frame $\langle x_s, y_s \rangle$; (b) marker line contour (dotted lines) and corners (x_c^i, y_c^i), $i = 1,2,3,4$

we used the ARToolkit library [13, 14] to track the motion of a marker placed on the grasped object through a camera.

ARToolKit is a C and C++ language software library that lets programmers easily develop Augmented Reality applications. It uses computer vision techniques to compute the real camera position and orientation relative to marked cards, allowing the programmer to overlay virtual objects onto these cards. Given square markers are used as a base of the coordinates frame where virtual objects are represented.

Let us consider the setup reported in Fig. 4 where a perspective camera $\langle c \rangle$ is observing a marker $\langle m \rangle$. If we call the 3-D coordinates of a point $\mathbf{X}$ relative to the camera and marker reference frame $\mathbf{P}_c \in \mathbf{R}^3$ and $\mathbf{P}_m \in \mathbf{R}^3$, they are related by a rigid-body transformation

$$\mathbf{P}_c = \mathbf{R}_c^m \mathbf{P}_m + \mathbf{t}_c^m, \quad (14)$$

where $\mathbf{R}_c^m \in \mathbf{SO}(3)$ is the rotation matrix which relates the camera and marker reference frame and $\mathbf{t}_c^m$ is the corresponding translation vector. Considering $\widetilde{\mathbf{P}}_c$, $\widetilde{\mathbf{P}}_m$ the relative extension in homogeneous coordinates, eq. (14) can be written as

$$\widetilde{\mathbf{P}}_c = \mathbf{H}_c^m \widetilde{\mathbf{P}}_m \quad (15)$$

where

$$\mathbf{H}_c^m = \begin{bmatrix} \mathbf{R}_c^m & \mathbf{t}_c^m \\ \mathbf{0}_{1\times 3} & 1 \end{bmatrix}. \quad (16)$$

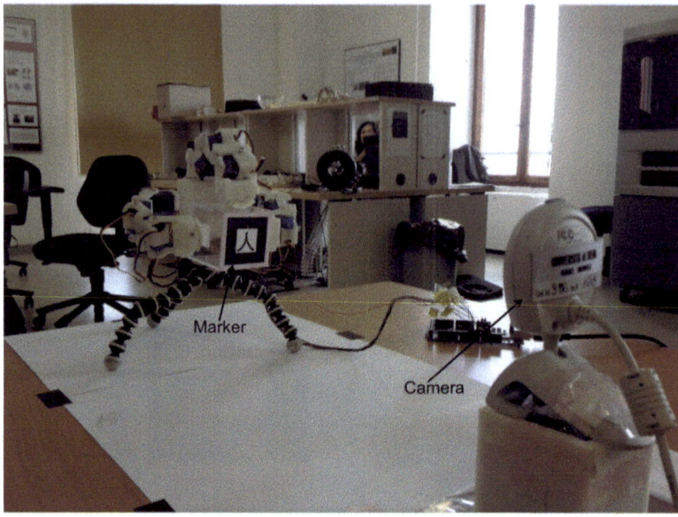

Fig. 5. The experiment setup for object motion evaluation. The camera measures the movement of the marker placed on the grasped object.

Let us assume that the intrinsic camera calibration matrix is given by,

$$\mathbf{K} = \begin{bmatrix} f_x & s & u_0 \\ 0 & f_y & v_0 \\ 0 & 0 & 1 \end{bmatrix}$$

where f_x, f_y (pixels) denote the focal lengths of the camera along the x and y directions, s is the skew factor and (u_0, v_0) (pixels) is the principal point of the CCD. According to perspective projection models, the projection $\widetilde{\mathbf{u}}_c \triangleq [x_c\, y_c\, 1]^T$ of $\widetilde{\mathbf{P}}_c$ on the camera is given by

$$\widetilde{\mathbf{u}}_c = \mathbf{K}[\mathbf{I}\ \mathbf{0}]\,\widetilde{\mathbf{P}}_c, \tag{17}$$

where $\mathbf{I} \in \mathbf{R}^{3\times3}$ represents the identity matrix. By putting together eq. (15) and eq. (17) we obtain,

$$\widetilde{\mathbf{u}}_c = \mathbf{K}[\mathbf{I}\ \mathbf{0}]\,\mathbf{H}_c^m\,\widetilde{\mathbf{P}}_m$$

which describes the projection on the camera image plane of a 3-D point expressed in the marker reference frame $\langle m \rangle$. The marker pose and position ($\mathbf{H}_c^m$) can be obtained by minimizing the reprojection error

$$err = \frac{1}{4}\sum_{i=1,2,3,4} (\hat{x}_c^i - x_c^i)^2 + (\hat{y}_c^i - y_c^i)^2 \tag{18}$$

where $\hat{x}_c^i$, $\hat{y}_c^i$ are noisy measurements of the i-th marker corner and x_c^i, y_c^i are the ideal corresponding points (see Fig. 4), [15, ch.6], [16, ch.4].

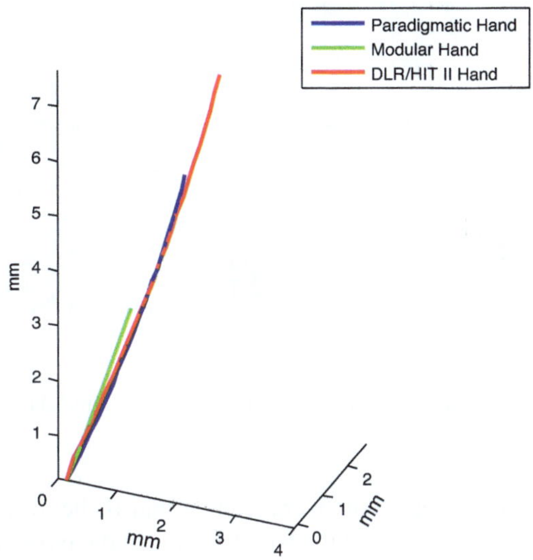

Fig. 6. Object trajectories obtained during the experiments

In Fig. 5 the setup of the experiment is showed. We used the obtained measures to estimate the motion of the grasped object. We considered a cube as grasped object. Only the first four synergies were activated on the paradigmatic hand model. Given this underactuation condition, for each configuration of the hand, only one feasible rigid body motion of the cube exists, corresponding to a particular combination of the four synergies [17]. This particular combination were activated on the hand at each time step, resulting in an object movement that is represented by the blue line in Fig. 6.

The synergistic movement of the paradigmatic hand were mapped on the two considered robotic hands. The mapped movement produced an object displacement in the two cases and the resulting trajectories are represented in Fig. 6. We performed 20 trials for each robotic hand and the plotted trajectories that we considered to analyse the performances have been computed as the average of the 20 obtained trajectories. We can observe that the three paths have different lengths. This is due to the scaling factor that we introduced in the mapping algorithm. In particular, the paradigmatic hand performed a 5.9 mm movement. The ModHa39p performed a 3.61 mm movement corresponding to the scaling factor 0.61 computed by the mapping algorithm. The DLR-HIT II Hand, with a scaling factor of 1.3, produced an object displacement of 7.69 mm.

Note that the DLR-HIT II hand obtained better results in terms of object motion trajectory. This is due to its higher redundancy and thus dexterity. It was not possible for the ModHa39p to reproduce the movement, given its simple kinematic structure. However the mapping algorithm, with its pseudoinverse computation, produced the closest feasible trajectory for this hand.

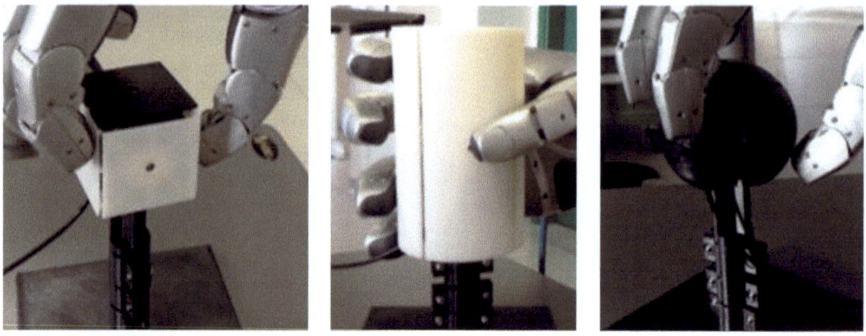

(a) (b) (c)

Fig. 7. The instrumented object: in the cube (a), the cylinder (b) and the sphere (c) configuration

In the two cases we computed an average error that is the average angular distance between the linear velocity vector of the paradigmatic and the robotic hand at each time step. For the ModHa39p this error was 7.4 degrees while for the DLR-HIT II hand it was practically zero.

4.2 Internal Forces Evaluation

Within the project we developed an instrumented object in order to measure the forces exerted by the robotic hands during grasping and manipulation and to compare them with those exerted by the paradigmatic hand. The instrumented core of the object is a cube with each of its faces being a 6 axis force torque sensor. An external interaction face is mounted at the centre of each face. The external faces are interchangeable and by attaching the appropriate face it is possible to create a cube, a sphere and a cylinder, as shown in Fig. 7. Mounting of the external faces is easy with a single locating screw per face. Therefore change between different objects shapes are relatively fast. With the current size of the instrumented core the smallest object size that can be assembled are:

- a cube of 50mm side;
- a sphere of 70mm diameter;
- a cylinder of 70mm diameter and 120mm length.

The size of the object can be increased by attaching larger external faces while respecting the sensing element torque limits.

All electronics and sensitive parts are covered inside the instrumented core. A single flexible cable provides power to the electronics and a fast TCP/IP-UDP connection. The main board uses a Texas LM3S8962 Microcontroller running at 50MHz. The system uses a 16bit DACs to read 36 multiplexed channels. The circuit also uses a 16bit DAC to eliminate the offsets between the channels. The Ethernet is a 100Mb/s link which allows a frame rate of 1kHz.

The force/torque elements have been designed based on a FEM analysis in order to ensure a reasonable loading range of the material. For the intended loads material stress must be kept low enough to lie within the elastic region of the material. At the same time the strain induced on the measuring gauges should be large enough to allow appropriately low amplification gains ensuring a high signal to noise ratio. In this way fatigue free operation of the material can guarantee long life and repeatable measurement, while high signal to noise ratio should enable high force/torque resolution. To achieve a good signal to noise ratio the amplifier gain has been specified to be in the region of 20. The maximum continuous load (pure force in the middle of the face with no torques) was specified at 100N. The chosen sensing material was 39NICRMO3 (steel) with a Yield stress between 600-900 MPa. Based on the above specifications a FEM and an electronics analysis was conducted leading to a design with the following predicted characteristics.

For the maximum continuous load the stress of the material was designed to lie approximately in the $1/3\times$Yield stress range where plastic deformation and fatigue are avoided. Based on the selected gain of 20 the 16Bit DAC provides a force resolution of 3mN and a force range of $\pm$100N. The peak load (pure force in the middle of the face with no torques) for a $2/3\times$Yield-Stress is in the region of $\pm$200N. This load should not be exceeded to avoid any possibility of plastic deformation and fatigue. The operating range is near $\pm$100N for pure force (pure force in the middle of the face with no torques). The sensor has not been designed for continuous cyclic loading. Cyclic loading should be kept sufficiently lower than the maximum continuous load so as to avoid fatigue.

This instrumented object allows 6 wrench vectors (1 per external face) to be measured. This limits interactions to one contact point per face. Assuming accurate readings in all 6 DoF (forces *xyz* and torques *xyz*) an arbitrary force vector applied on an external face could be resolved, Fig. 8. A soft finger model at the contact points is computed using the 6 measures and the intrinsic contact sensing algorithm presented in [18].

Another important aspect is that the object can be grounded by mounting one of its faces on a table or other grounded surfaces. This allows experiments with the hand/arm control while a total force is needed to be applied to the environment through the particular object geometry. The grounded object would directly provide forces and torques with respect to the world frame measured at the grounded face as well as those at the robot interaction points.

In order to evaluate the performances of the mapping algorithm in grasping and manipulation tasks we focused on the computation of exerted internal forces. Internal forces are those forces that do not move the contact points playing an important role in the grasp stability [19]. We performed experiments taking advantage of the instrumented object previously described. For the sake of simplicity, only results obtained with the cubic and the spherical configuration of the object are reported in this chapter. The target of the experiment was to compare the internal forces exerted by the paradigmatic hand on the object in simulation environment with those reproduced by the robotic hand on the real instrumented object. A direct comparison between the two hands was not possible due to the fact that the "driven" robotic hand could have a different number of contact points (for instance, the ModHa39p hand can have maximum 3 contact points if fingertip grasping is considered). For this reason, we adopted a measure of the whole object deformation energy produced by the activation of a

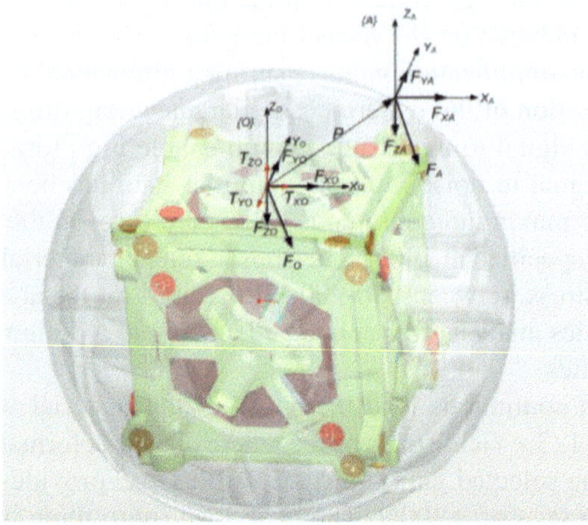

Fig. 8. The vector of a single force F_A acting on a surface can be solved from the force and torque vectors measured at the centre of the respective load cell

combination of synergies to evaluate and compare the performance of the analysed mapping procedures. A variation of the internal forces $\delta\lambda$ is obtained activating a combination of synergies on the paradigmatic hand.

Considering the model of soft synergies described in [17], the contact force variation can be computed considering the associated compliance. Let indicate with δx the vector containing the deformation components of each contact point evaluated as $\delta x = K_s^{-1}\delta\lambda$. The elastic energy variation produced by the activation of synergies can be computed as

$$\delta E_{el} = \frac{1}{2}K_s\|(\delta x)^2\| = \frac{1}{2}K_s^{-1}\|(\delta\lambda)^2\|, \tag{19}$$

where K_s is a contact stiffness matrix. The $\delta\lambda$ values were computed in simulation environment for the paradigmatic hand while they were read directly from the force sensors of the object for the robotic hands (see Fig. 9). We considered the same K_s value for the two hands. We computed the energy variation index expressed in (19) considering the activation of the first three synergies and we evaluated the percentage difference between the paradigmatic and the robotic hand.

The exerted forces and the consequent energy variations due to the synergy activations for the paradigmatic hand were evaluated in simulation environment using the Matlab SynGrasp Toolbox [20]. In Fig. 10 is represented the Matlab model of the paradigmatic hand used for the simulations.

Values obtained in simulation activating separately the first three synergies were compared with energy variations obtained by the robotic hands. The object was kept in a fixed reference position. This allows to compensate the gravity effects. A specific calibration procedure has been developed for this porpoise. In Fig. 11 the two robotic

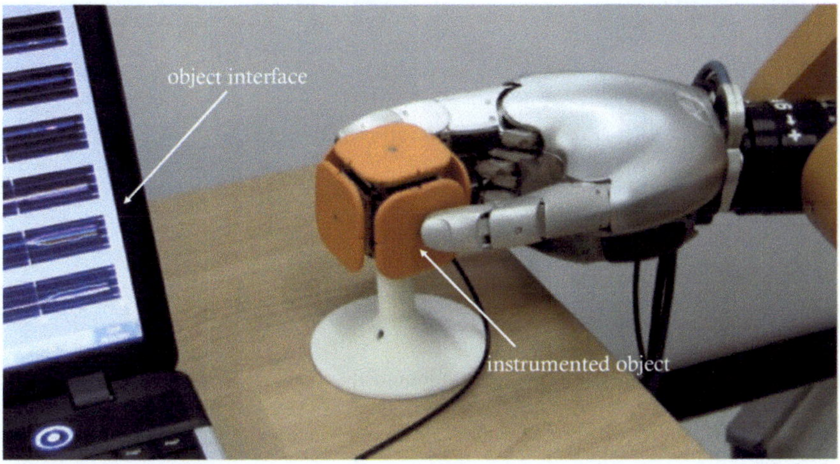

Fig. 9. Internal force evaluation setup. The six boxes on the object interface (on the left) represent the three forces and the three torques applied on a single face of the cube.

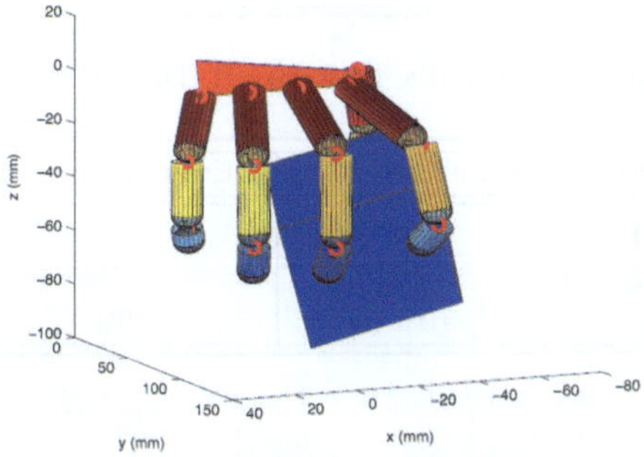

Fig. 10. The Paradigmatic Hand model grasping the cube in SynGrasp

hands grasping the instrumented object in "cube-configuration" are shown, while in Fig. 12 the DLR-HIT II hand grasping the object in "sphere-configuration" is reported.

We performed 20 trials with both the two robotic hands and the two object configuration and we computed the average percentage error obtained by the robotic hands with respect to the total amount of energy variation produced by the paradigmatic hand. Results are summarized in Tables 3 and 4.

It is worth noting that, as expected, the DLR-HIT II hand achieves best performances also in terms of energy. The worst performances that we obtained correspond to the third synergy mapped on the ModHa 39p. This can be explained considering that the third synergy is substantially a movement that constrains the fingers of the paradigmatic hand to spread out mainly using the adduction/abduction joints. This kind of movement can not be reproduced on the ModHa 39p because there are not adduction/abduction joints on that hand and thus also the total energy variation can not be reproduced.

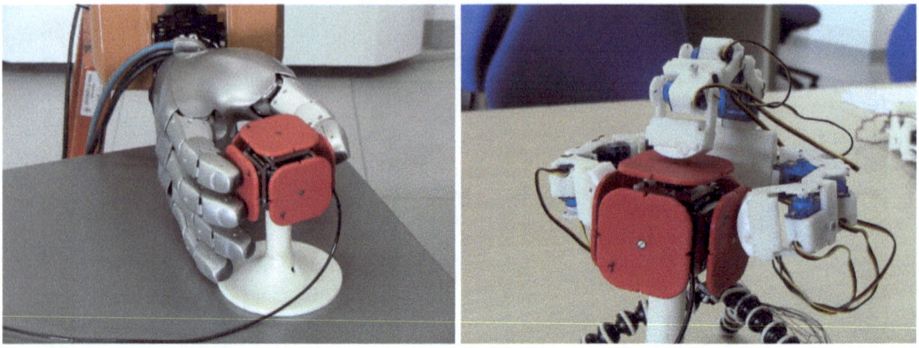

Fig. 11. The two robotic hands grasping the sensorized object in the reference positions.

Table 3. Energy variation error for the ModHa 39p Hand - Cubic and spherical object

Synergies	% Error Cubic Obj.	% Error Spherical Obj.
Syn 1	21%	18%
Syn 2	10%	5%
Syn 3	50%	45%

Table 4. Energy variation error for the DLR-HIT II Hand - Cubic and spherical object

Synergies	% Error Cubic Obj.	% Error Spherical Obj.
Syn 1	1%	1%
Syn 2	26%	22%
Syn 3	10%	8%

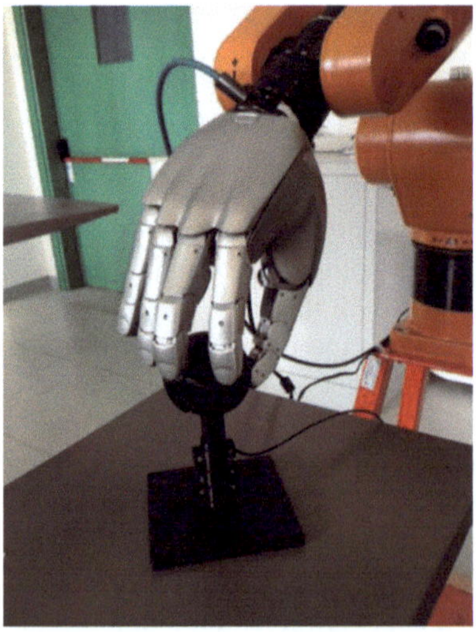

Fig. 12. The DLR-HIT II hand grasping the instrumented object in the sphere configuration

5 Conclusions

The chance of a large diffusion of robotic hands, especially in industrial scenarios, necessarily passes through a simplification of the rules underlying their control laws. In several fields the replacement of well known instruments with new user-friendly and easy-to-control ones has allowed the gaining of new market segments. Close to the simplicity concept, the device independent property is of fundamental importance for a wide integration in the hyper-flexible cell scenario. The middleware approach we developed, will ensure the possibility to perform our control approach, based on synergies, upon different existing robot hands.

This is the target of the HANDS.DVI experiment: hiding to users both the complexity of the robotic hand structure, and making possible the use of different devices with the same abstraction layer, i.e. transparent to programmers. While the quest for a solution to the decades-long problem of trading off simplicity and performance in the programming and control of robot hands has often been solved via empirical methodologies, our goal was replicating in the robotic device an organized set of synergies, ordered by increasing complexity, so that a correspondence can be made between any specified task set (in terms of a number of different grasps and exploration actions) and the least number of synergies whose aggregation makes the task feasible. Such "principled simplification" approach will eventually lead to the possibility of expanding tasks in a basis of synergies correctly defined for the artificial hand to be controlled, thus making it possible to come closer to practical applications of such devices in the industrial hyper-flexible cell scenario.

The main innovation of HANDS.DVI concerns the integration of the methodologies and technologies from neuroscience, robotics and control theory in a coherent set of theoretical tools and a methodology for the integration of the current hands model in a larger context such that of hyper-flexible cells. In particular, possible innovations based on the outcome of this project are:

- to allow robot hands to be programmed more easily and to adapt more robustly to different task/environment conditions;
- to improve the human-robot cooperation inside an industrial assembly chain by exploiting similarities between human and robotic hand behaviours.

Consider as example an operator that has to "teach" to a complex robotic hand the grasp of a new object. In the HANDS.DVI scenario, the operator uses only few knobs (for instance those controlling the first two synergies) to shape the hand around the object decreasing the time and the complexity of the operation. Dually, when virtual artificial hands work as avatars in visuo-haptic rendering of the hyper-flexible cell scenario, exploiting synergies allows to use less sensors to track accurately and in real-time the "principal motion" while performing a manipulation task.

In conclusion, two different experiments to validate the proposed framework were described in this chapter. In the first one, we evaluated the trajectory of a grasped object moved both from the paradigmatic and the robotic hand. We observed that using our mapping method it is possible to reproduce the synergistic movement of a model of the human hand. We further observed that, as imagined, performance decreases according to the limitation of the kinematic structure of the hand. Anyway the generality of the method is preserved. In the second experiment we evaluated the forces exerted by the paradigmatic and the robotic hands over a grasped object. An instrumented object developed within the project has been used to measure these forces. We used an estimation of the total energy used in the grasp to overcame mismatching in the number of contact points that do not consent a direct comparison. Also in this case, our mapping results efficient and the performance decrease with simpler hand.

However, this approach presents some drawbacks. The proposed mapping is based on a heuristic approach: we choose to reproduce a part of the hand motion, which practically corresponds to move and squeeze a spherical object. Although squeezing and moving an object explains a wide range of tasks, many other possibilities exist in manipulating objects which are not modelled with this mapping. Work is in progress to generalize the proposed method enriching the possible motions to be reproduced.

Acknowledgments. This work was supported by the EU Project EC FP7-ICT-231143 ECHORD.

References

1. Santello, M., Flanders, M., Soechting, J.: Postural hand synergies for tool use. The Journal of Neuroscience 18, 10105–10115 (1998)
2. Santello, M., Soechting, J.F.: Force synergies for multifingered grasping. Experimental Brain Research 133, 457–467 (2000)
3. Lin, J., Wu, T.: Modeling the Constraints of Human Hand Motion. Urbana 51(61), 801 (2000)
4. Linscheid, R., An, K., Gross, R.: Quantitative analysis of the intrinsic muscles of the hand. Clinical Anatomy 4(4), 265–284 (1991)
5. Kim, K., Youm, Y., Chung, W.: Human kinematic factor for haptic manipulation: The wrist to thumb. In: Proceedings of the 10th Symposium on Haptic Interfaces for Virtual Environment and Teleoperator Systems, HAPTICS 2002, pp. 319–326 (2002)
6. Youm, Y., Holden, M., Dohrmann, K.: Finger ray ratio study. Tech. Rep. (1977)
7. Chao, E.: Biomechanics of the hand: a basic research study. World Scientific Publishing Company (1989)
8. Gioioso, G., Salvietti, G., Malvezzi, M., Prattichizzo, D.: An object-based approach to map human hand synergies onto robotic hands with dissimilar kinematics. In: Robotics: Science and Systems VIII. The MIT Press, Sidney (2012)
9. Gioioso, G., Salvietti, G., Malvezzi, M., Prattichizzo, D.: Mapping synergies from human to robotic hands with dissimilar kinematics: an approach in the object domain. IEEE Trans. on Robotics (2013)
10. Ciocarlie, M., Goldfeder, C., Allen, P.: Dimensionality reduction for hand-independent dexterous robotic grasping. In: IEEE/RSJ International Conference on Intelligent Robots and Systems, IROS 2007, pp. 3270–3275 (2007)
11. Peer, A., Einenkel, S., Buss, M.: Multi-fingered telemanipulation - mapping of a human hand to a three finger gripper. In: The 17th IEEE International Symposium on Robot and Human Interactive Communication, RO-MAN 2008, pp. 465–470 (August 2008)
12. Butterfass, J., Grebenstein, M., Liu, H., Hirzinger, G.: DLR-hand II: next generation of a dextrous robot hand. In: Proceedings 2001 ICRA IEEE International Conference on Robotics and Automation, vol. 1, pp. 109–114 (2001)
13. Artoolkit library for Augmented Reality, `http://www.hitl.washington.edu/artoolkit/`

14. Kato, H., Billinghurst, M.: Marker Tracking and HMD Calibration for a Video-Based Augmented Reality Conferencing System. In: Proc. IEEE ACM Int. Workshop on Augmented Reality (1999)
15. Ma, Y., Soatto, S., Kosecká, J., Sastry, S.: An Invitation to 3-D Vision: From Images to Geometric Models. Interdisciplinary Applied Mathematics. Springer (2003)
16. Hartley, R., Zisserman, A.: Multiple View Geometry in Computer Vision, 2nd edn. Cambridge University Press (2004)
17. Gabiccini, M., Bicchi, A., Prattichizzo, D., Malvezzi, M.: On the role of hand synergies in the optimal choice of grasping forces. Autonomous Robots 31, 235–252 (2011)
18. Bicchi, A., Salisbury, J.K., Brock, D.L.: Contact sensing from force measurements. The International Journal of Robotics Research 12(3), 249–262 (1993)
19. Bicchi, A.: On the closure properties of robotic grasping. The Int. J. of Robotics Research 14(4), 319–334 (1995)
20. Malvezzi, M., Gioioso, G., Salvietti, G., Prattichizzo, D., Bicchi, A.: Syngrasp: a matlab toolbox for grasp analysis of human and robotic hands. In: Proc. IEEE Int. Conf. on Robotics and Automation (2013)

DEXDEB – Application of DEXtrous Robotic Hands for DEBoning Operation

Guowu Wei[1], Franck Stephan[2], Vahid Aminzadeh[1], Helge Würdemann[1], Rich Walker[3], Jian S. Dai[1], and Grigoré Gogu[2]

[1] Centre for Robotics Research, King's College London,
Strand, London WC2R 2LS, United Kingdom
jian.dai@kcl.ac.uk

[2] Pascal Institute, French Institute for Advanced Mechanics / Blaise Pascal University/CNRS, Campus de Clermont-Ferrand/Les Cézeaux,
CS 20265, 63175 Aubière Cedex, France
Grigore.Gogu@ifma.fr

[3] Shadow Robot Company LTD.,
251 Liverpool Road, London N1 1LX, United Kingdom
rw@shadowrobot.com

Abstract. This paper presents for the first time an application Investigation of using dexterous robotic hands for deboning operation so as to establish a human-robot co-working platform for cutting, deboning and muscle extraction operation in meat industry. By setting up a test rig consisting of a support and a customized knife integrated with force sensors, and utilizing a modified data glove, manual ham deboning operations are carried out providing essential information and background for the robotic hand design, appropriate force/torque and position sensors identification, and human-robot co-working platform trajectory planning. Principle component analysis method is then employed for trajectory mapping and planning associated with the knife peak coordinates, and concept of force cone is introduced leading to an efficient algorithm for trajectory planning. Further, design and kinematics of a metamorphic hand are investigated laying a background for measuring manipulation and grasp quality of the proposed robotic hand. The above experimental, theoretical, hardware and software preparations finally lead to the applications of using two dexterous robotic hands, i.e. one Shadow C6M left hand and one KCL metamorphic hand to replace human left hand in deboning operation. The experiment thus laid background work for the robotization of meat industry and gave insight into the benchmarking of utilizing dexterous hand in deboning operation constructing a human-robot co-working hyper-flexible cell.

Keywords: Dexterous robotic hand, metamorphic hand, deboning operation, robot-human co-working, hyper-flexible cell.

1 Introduction

The meat sector is one of the most important sectors in the European Union (EU) agriculture. Four main meat types, i.e. beef and veal, pork, poultry, and

F. Röhrbein et al. (eds.), *Gearing Up and Accelerating Cross-Fertilization between Academic and Industrial Robotics Research in Europe*, Springer Tracts in Advanced Robotics 94,
DOI: 10.1007/978-3-319-03838-4_11,

lamb/goat-meat, account for one quarter of the total value of agricultural production. Half of all the EU farms have livestock. 90% of farmers with ruminant animals (cattle, sheep and goats) are specialist livestock producers. In the consumer market, the EU customers consume roughly 35 million tonnes of the various types of meat each year, i.e. around 92 kilogrammes per person per year on average. In addition, the EU is a major meat producer occupying over 16% of global meat production and plays an important role in meat trading.

Just in France the meat sector is the largest section in the AFI (Agriculture and Food Industries) regarding employment (122,049 employees in 2005) and sales (31.1 billion Euros in 2005). In UK, there were around 112,000 employees in 2005 and the sales reached over 11 billions Pounds according to Department of Food and Rural Affairs (Defra). In the EU meat industry, three crucial issues, i.e. labour shortages, dangerous and strenuous work condition, and competition from meat exported from countries with low labour costs, are encountered and needed to be solved so as to keep the market moving.

Today, companies working in meat activities, especially those in the slaughtering/quartering sector, are having increasing difficulties in finding qualified labour. This loss of interest in the meat sector by skilled workers and young people is largely due to a devalued image of the job, difficult work conditions (unsociable working hours, cold temperatures, etc), and the dangerous and strenuous work involved (severe injuries, repetitive tasks, heavy loads to carry and handle, etc). Therefore, companies are faced with a disparity between hiring unskilled workers and the qualifications that are necessary to carry out meat quartering tasks. This problem has a profound impact on productivity and consequently on the companies profitability.

Therefore since the last decade, robotization of these tasks has become an important objective and a crucial challenge for companies working on meat transformation and meat product sector who seek to improve the health and safety of workers in this sector, as well as to find new solutions concerning the increasing production costs linked to current and future labour shortages. Commonly, robots have long been imagined as mechanical workers, helping us with the chores of our daily life. Robots are expected to work alongside us in our homes and workplaces, to extend the time an elderly person can live at home, to provide physical assistance to a worker on an assembly line, and to help with household chores. Since the first industrial robot proposed by Dr. Engelberger [1] in the early 1960s, robots are widely used and working in factories around the world, performing manipulation tasks with remarkable speed and precision. Most robotic work cells consist of a commercially available industrial robot, a specialized end effector, and a control scheme, integrated with a sensing system.

The main challenge in the meat industrial field is to establish robotic system for cutting and deboning tasks that have all constraints and which needs to be capable of adapting to all types of beef carcasses or pork ham in a hyper-flexible cell. Relevant research has been carrying out to developing robotic systems for poultry/beef/pork deboning [2–8] or to replacing the right-hand of an operator with robotic hand for deboning operation [9–13]. However, no report presents the application of using robotic hand in deboning operation to replacing left-hand of a human operator.

Grasping and manipulation of soft and deformable objects is still a challenge to robotics community and industrial handling. Study has been made in a number of areas, including textile handling for manipulating clothes [14], food handling for disarray-to-order processing [15], flexible material handling including papers and cartons for packaging and soft tissue grasping for surgical robotics [16]. Meat grasping and handling has been topics of recent research [17].

In order to bridge the research gap, this paper presents the application of two dexterous robotic hands, one commercialized Shadow C6M dexterous anthropomorphic hand and one four-fingered metamorphic hand, for deboning operation shedding light on using robotic hands for meat handling in deboning automation. This fully utilizes the human-robot co-working environment and leads to setting up a hyper-flexible meat operating cell for deboning operation and muscle extraction.

The rest of this paper is organized as follows. Section 2 presents experiments on deboning operation conducted by a skilled human operator providing data for robotic hand design, and operating gaits and motion trajectory planning. Section 3 delivers the design of the two dexterous robotic hands with focus on the design and kinematics investigations of a four-fingered metamorphic hand. Section 4 demonstrates the scenario of human-robot co-working system in deboning operation, and Section 5 presents the conclusions and benchmark for robotised deboning operation.

2 Experiments of Manual Deboning Operation

Since the main purpose of the investigation carried out in this paper is to use dexterous robotic hands to replace left-hand of a human operator so as to form a human-robot co-working platform for deboning operation, the information involved in deboing operation carried out by a human operator, including left hand motion trajectory along with the trajectory of each finger, and force applied for cutting and grasping, needs to be obtained and analysed providing background for the design of left-hand robotic hand and for the robotic hand to emulate the operations performed by skilled human operator in handing, pulling, pushing and twisting meat to facilitate and assist cutting by the cutter manipulated by the right hand of the human operator.

2.1 Cutting Force Measurement with Kistler Based Test Rig

In order to carry out the experiments on measuring cutting force exerted by a skilled human operator during deboning operation, a test rig was set up integrated with Kistler force sensor as illustrated in Fig. 1. In this test rig, hams were fixed on the support equipped with the Kistler sensor and all Kistler sensors were related to a charge amplifier with 4 channels using a connection cable. The signals were collected and transmitted to a PC and were processed with respect

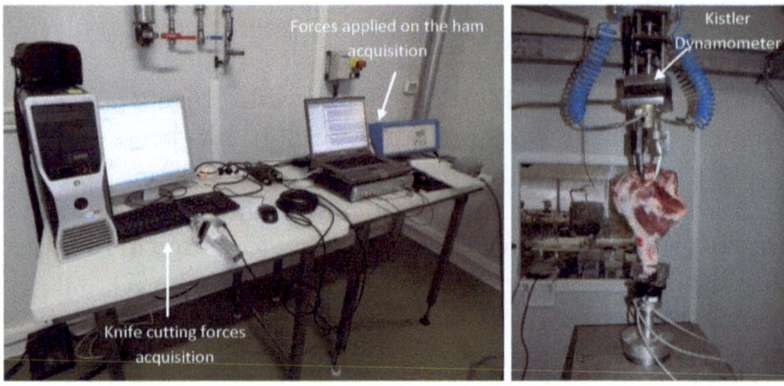

Fig. 1. Test bench for manual ham deboning operation

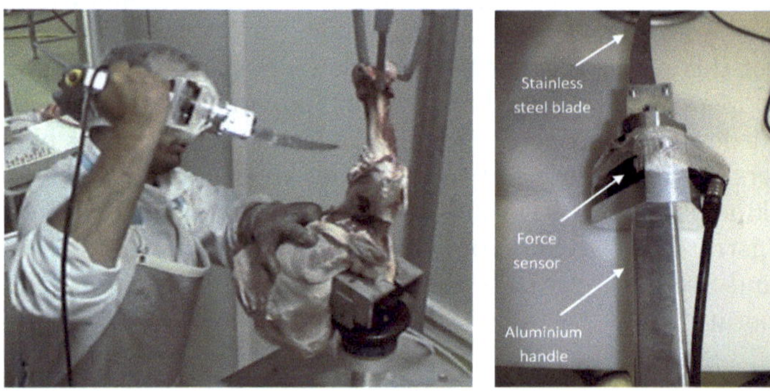

Fig. 2. Knife developed for estimating cutting forces during ham deboning operation

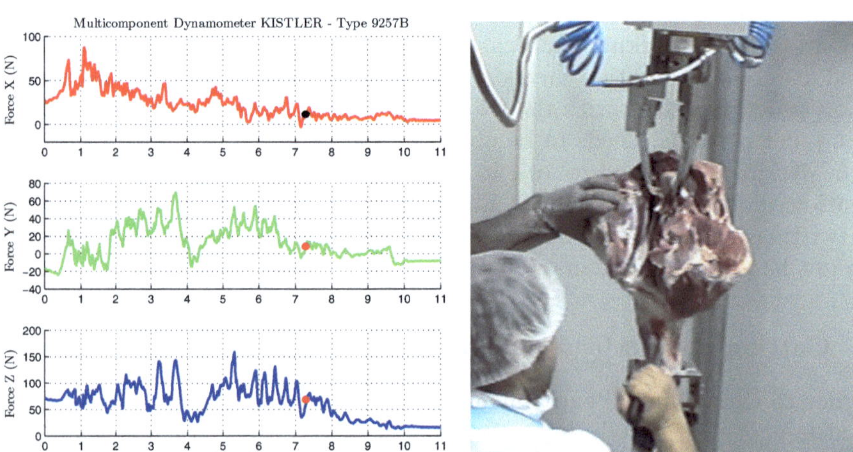

Fig. 3. Force information for manual deboning of a ham held vertically

to time. Further, a knife prototype was developed to obtain a first estimate of the cutting force during ham deboning. This knife was equipped with a 6-axis force sensor and consists of (see Fig. 2) a specifically designed stainless steel blade which was fixed on the support of the force sensor i.e., a 6-axis force sensor (ATI FT-Gamma SI-32-2.5) with an aluminium handle fixed under it. The sensor was connected to computer via a 10m length cable.

Based on the above test rig, with a force sensor ATI FT-Omega160 mounted on an ABB robot arm, a series of experiments on manual ham deboning operation were carried out and data containing force and torque information during the operation was collected and stored. Figure 3 shows the force information for a cutting step during the deboning operation and the experiment indicates that the maximum force applied along the z-axis during the deboning does not exceed 170N.

2.2 Identification of the Left Hand Movements with a Data Glove

Aiming to get the motion and trajectory information of the left hand during deboning operation, a commercialized data glove was adapted and equipped with tactile and flex sensors in order to record the movements of the butcher's left hand during manual ham deboning operation. As illustrated in Fig. 4, the modified data glove was integrated with six tactile sensors (FSR^{TM}) allowing the glove to measure forces applied by the fingers and the palm of the left hand (the palm and each fingertips are equipped with a sensor), and ten flex sensors (2 flex sensors per finger) in order to measure the joints angles value of the proximal interphalangeal (PIP) and metacarpophalangeal (MCP).

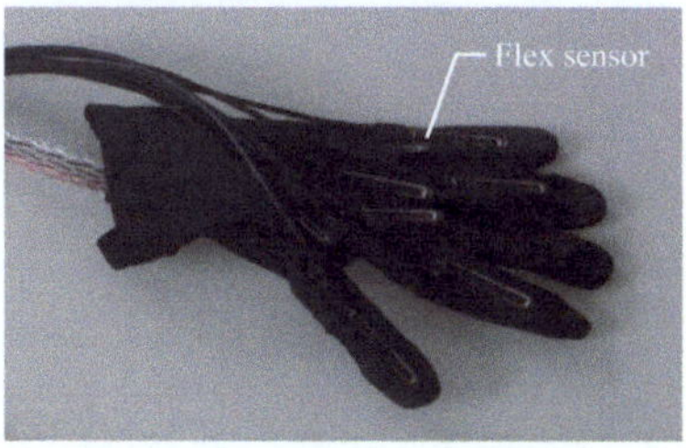

Fig. 4. A modified data glove

Using the modified data glove, several sets of experiments were carried out aiming to measure forces applied by the butcher's left hand fingertips on the meat as well as to record the joints angular coordinates of the proximal interphalangeal and metacarpophalangeal joints, in order to define the butcher's left hand poses during the several steps of hams deboning. Figure 5 and Fig. 6 show respectively a segment of force information and joint displacement information obtained in the experiments.

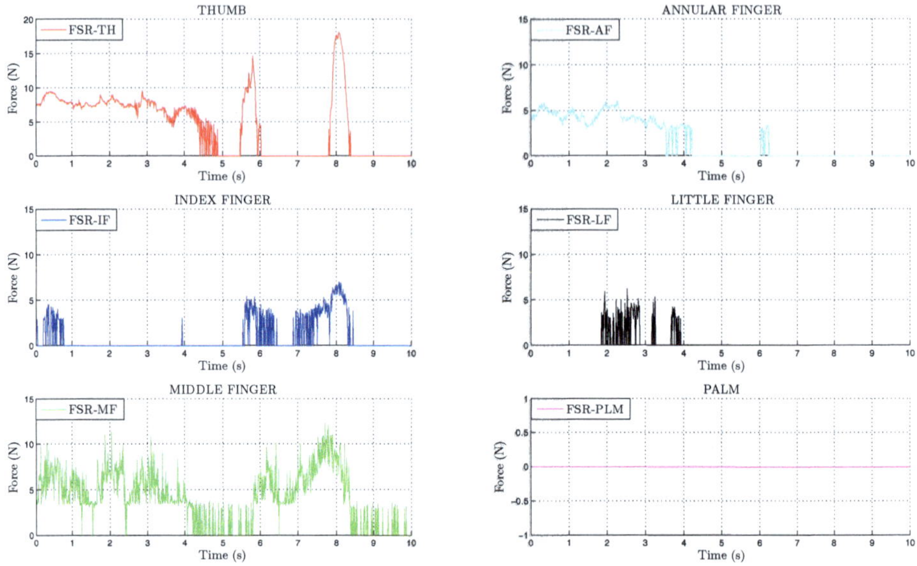

Fig. 5. FSR sensor signals during manual ham deboning

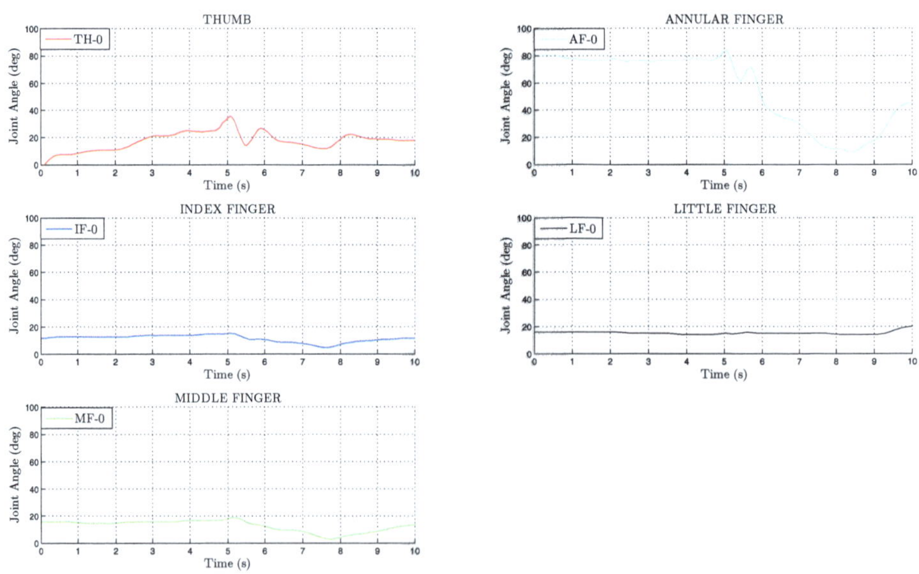

Fig. 6. MCP joints angular coordinates during manual ham deboning

2.3 An Efficient Algorithm for Trajectory Planning

During the ham deboning actions carried out by a skilled worker, coordinates of the knife peak were recorded over time and based on the real time coordinates, principle component analysis over time method [18] was employed such that multi-variant statistical technique is used to analyse a list of data arrays and describe the data as a set of orthogonal principal components to find the direction of variation with time component. Further, in order to map the experiment data to the robotic hand, a trajectory planning algorithm is developed based on the assumptions that: a) movement should be away from the cut surface; b) the absolute value of the force applied is meant to be limited; c) while applying the force, the left hand should travel a certain distance. Based on the above assumption, left-hand force cone is constructed (see Fig. 7) and the associated results are obtained based on the control laws and the vector field according to the gradient of a potential function.

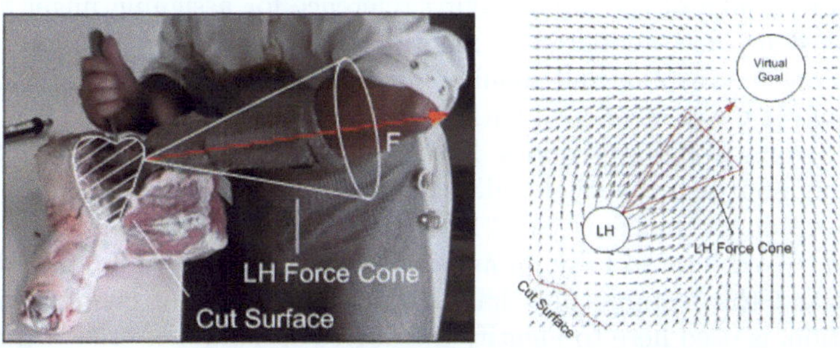

Fig. 7. Left-hand force cone construction

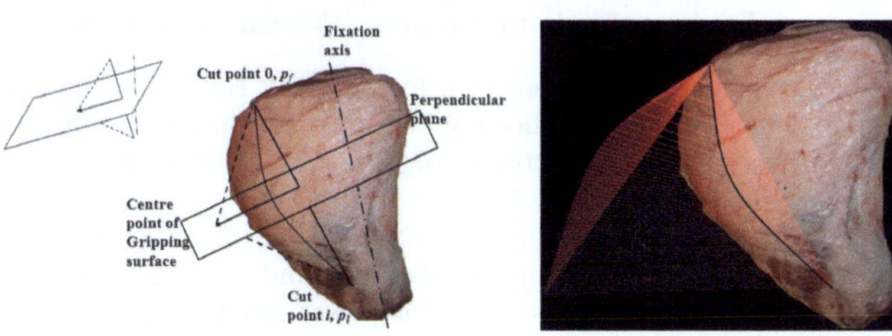

a) Grasping planning b) Grasping trajectory

Fig. 8. Trajectory planning algorithm and simulation

Further, based on the above assumptions and analysis, a trajectory planning algorithm is developed by defining fixation axis, cutting point perpendicular plane and griping surface in a ham as illustrated in Fig. 8a and computational simulation results are obtained and graphically demonstrated in Fig. 8b.

The manual deboning operation force and trajectory information achieved and analysed in this section provided essential background not only for mapping the deboning operation task workspace to the robotic hand joint space so as to design and develop dexterous left-hand robotic hand for the proposed tasks, but also for identifying appropriate force/torque and position sensors to be integrated in the robotic hand supplying real-time feedback and control for performing human-robot co-working deboing operation.

3 Development of Dexterous Robotic Hands

In this investigation, two dexterous robotic hands, i.e. one Shadow robotic hand and one KCL metamorphic robotic hand, are to be developed so as to examine the feasibility of using the mechanical hand in deboning operation as an initial step towards setting up a hyper-flexible human-robot co-working cell. The trajectory and force information obtained in Section 2 is respectively considered for structure design of the hands giving reference for assigning finger and phalanx dimensions, and for actuator identification in order to provide sufficient torque for meat grasping and manipulation. Although a left-hand Shadow^{TM} robotic hand [19] was developed within the framework of the DEXDEB project, it is now a commercialized robotic hand, thus only the development of a four-fingered metamorphic robotic hand, which was invented based on the principle of metamorphic mechanisms [20, 21], is presented in this paper. The Shadow hand is employed here to fully mimic the left-hand of human hand in deboning operation and the metamorphic hand, which is capable of changing its palm configuration, is used here to identify the degree of hand dexterity demanded in deboning operation.

3.1 Structure Design of a Four-Fingered Metamorphic Hand

Based on the previous development of a five-fingered right-hand metamorphic robotic hand [22–24], for the specified application of the hand for deboning operation, a four-fingered left-hand dexterous metamorphic robotic hand is developed in this paper as illustrated in Fig. 9.

The hand consists of a reconfigurable palm and four fingers, i.e. a four-DOF thumb, and three-DOF index finger, middle finger and ring finger. Since in most case the little finger is used in assisting for power grasp, it is omitted in this design. The reconfigurable palm is a spherical five-bar linkage containing five links 1 to 5 with the base link 5 connected to a wrist. The palm itself has two degrees of freedom and it is actuated by drive 1 and drive 2 as illustrated in Fig. 9. The two drives are utilized to adjust configurations of the palm and drive 1 is in particular used to change the structure of the reconfigurable palm by rotating the crank link 1 so as to form a four-bar linkage in an innate metamorphic phase. In this configuration link 1 overlaps link 5 such that locking drive 1, the palm evolves into a one degree of freedom four-bar mode as shown in Fig. 10. A 3-phalanxed thumb is mounted to link 2 connected by an additional link

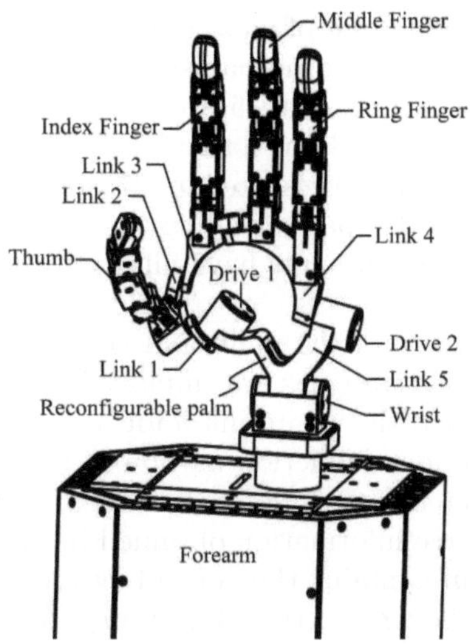

Fig. 9. Structure of a four-fingered metamorphic robotic hand

and revolute joint providing a fourth degree of freedom, a 3-phalanxed index finger is mounted at link 3, and the 3-planlaxed middle finger and ring finger are fixed at link 4. It should be pointed out that except for the thumb, each of the index finger, middle finger and ring finger only contains three parallel joints that provide only flexion/extension motions but no adduction/abduction motion. It is expected that the introduction of the reconfigurable palm will compensate the absence of adduction/abduction motions of the three fingers and provide dexterous manipulation and grasp by adapting configurations of the palm for various tasks and different environment.

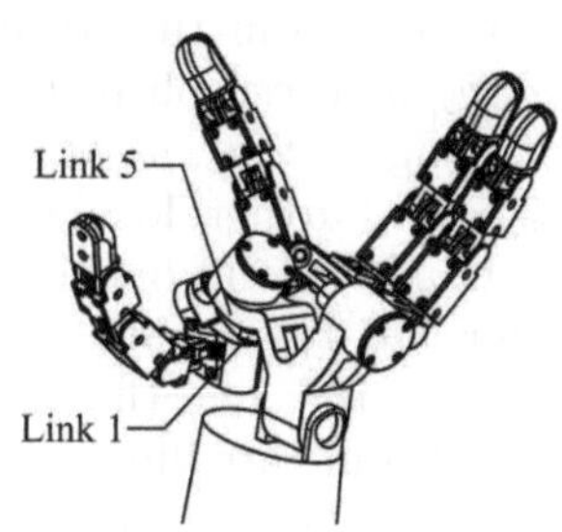

Fig. 10. Metamorphic hand in an innate metamorphic phase

Further, in accordance with experiment data obtained in Section 2 and the size of an adult's hand, the maximum radius of the spherical linkage palm is set as 50mm. In order to increase the dexterity of the palm, both human hand structure from the humanoid point of view and rotatability criterion of spherical linkage [27] from the kinematics of mechanism point of view are considered and the angles corresponding to links 1 to 5 are assigned as $\alpha_1 = 25°$, $\alpha_2 = 40°$, $\alpha_3 = 70°$, $\alpha_4 = 112°$ and $\alpha_5 = 113°$ such that $\alpha_1+\alpha_2+\alpha_3+\alpha_4+\alpha_5 = 360°$. The angles comply with the condition presented in [27] so that $\alpha_5+\alpha_1+\alpha_2 = \alpha_3+\alpha_4$ leading to the desired property that excluding the indeterminate position where all links fall on a great circle, all joints have full rotatability except for the joint between link 4 and link 5.

In order to reduce the size of the palm and the weight of the hand, tendon driven and under-actuation strategies are employed in the design. Three pulleys are embedded in the palm with two in link 5 for actuating the palm and one in link 2 for realizing adduction/abduction motion of the thumb. The pulleys are driven by tendons connected to DC motors installed in the forearm. Motors are identified based on the force information obtained in manual deboing operation in Section 2. The forearm contains the control system of the hand. The bases of the index finger, middle finger, ring finger and little finger are distributed on the palm in such an arrangement that, when all the palm links lie on a same plane (i.e. all links fall on a same great circle), the MCP joints of these fingers are perpendicular to a plane passing through the centre of the palm and perpendicular to twist joint. While, the finger base of the thumb is fixed to an additional link connecting to the pulley mounted in link 2, leading to an additional degree of freedom of the thumb. Dimensions of the fingers are determined according to the motion information of the human operator collected in Section 2.

3.2 Kinematics of the Metamorphic Robotic Hand

In this design, the four fingers are connected to the aforementioned reconfigurable palm via finger bases that are attached at points B_1, B_2, B_3, and B_4 as shown in Fig. 11. Based on the geometric constraint of the reconfigurable palm investigated in [23], in order to transform the palm motion to finger motion, moving coordinate frame B_i-$x_{b_i}y_{b_i}z_{b_i}$ is established at point B_i with z_{b_i}-axis directed along OB_i, y_{b_1} directed along $\boldsymbol{z}_{b_1} \times \boldsymbol{z}_2$, y_{b_2} directed along $\boldsymbol{z}_{b_2} \times \boldsymbol{z}_3$, and y_{b_i} ($i = 3$ and 4) directed along $\boldsymbol{z}_{b_i} \times \boldsymbol{z}_4$. Further, local coordinate frame F_i-$x_{i_1}y_{i_1}z_{i_1}$ ($i = 1$, 2, 3 and 4) is set up at the MCP joint of each finger with z_{i_1}-axis aligned with MCP joint of the ith finger and x_{i_1}-axis directed along B_iF_i. For $i = 2$, 3 and 4, the angle between z_{b_i} and x_{i_1} is γ_i and the distance between B_i and F_i is a_i. In addition, parameters of the thumb are illustrated in Fig. 11 with γ_1 denoting the angle between OB_1 and joint G, and a_1 denoting the distance between B_1 and F_1. The angle between z_{11} and z_0 is given as β. Based on the above geometric relationship and coordinate systems, the positions and orientations of the finger bases can be expressed in the global coordinate frame established in the palm by firstly calculating

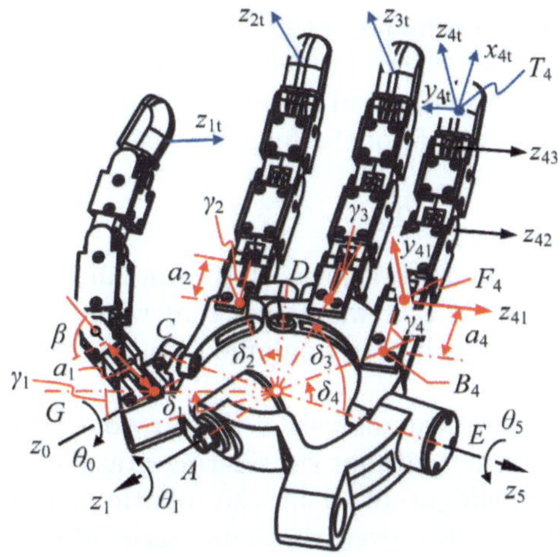

Fig. 11. Transformation from the palm to finger bases

$$\mathbf{R}_{ob_i} = \begin{cases} \mathbf{R}(z_1, \theta_1)\mathbf{R}(y_1, -\alpha_1)\mathbf{R}(z_2, \theta_2)\mathbf{R}(y_2, -\delta_1) & \text{if } i = 1 \\ \mathbf{R}(y, \alpha_5)\mathbf{R}(z_5, \theta_5)\mathbf{R}(y_4, \alpha_4)\mathbf{R}(z_4, \theta_4)\mathbf{R}(y_3, \delta_2) & \text{if } i = 2 \\ \mathbf{R}(y, \alpha_5)\mathbf{R}(z_5, \theta_5)\mathbf{R}(y_4, \delta_i) & \text{if } i = 3 \text{ and } 4 \end{cases} . \tag{1}$$

Subsequently, transformation from the finger base coordinate frame to the global coordinate frame can be given as

$$\mathbf{T}_{ob_i} = \begin{bmatrix} \mathbf{R}_{ob_i} & \mathbf{R}_{ob_i}\boldsymbol{k} \\ 0 & 1 \end{bmatrix} (i = 1,\, 2,\, 3 \text{ and } 4), \tag{2}$$

where $\boldsymbol{k} = [0, 0, \mathrm{R}]^T$ and $\mathbf{R}_{ob_i}\boldsymbol{k}$ gives the position vector of point B_i in the global coordinate frame. R is the theoretical radius of a virtual sphere on which all the links lie.

Further, the homogeneous transformation from coordinate frames of the MCP joints to the global coordinate frame can be derived according to Fig. 11 as

$$\mathbf{T}_{of_i} = \mathbf{T}_{ob_i}\mathbf{T}_{bf_i} = \begin{bmatrix} \mathbf{R}_{of_i} & \boldsymbol{p}_{of_i} \\ 0,0,0 & 1 \end{bmatrix}, \tag{3}$$

for $i = 1$, it has

$$\mathbf{T}_{bf_i} = \begin{bmatrix} -c\gamma_1 c\theta_0 & c\gamma_1 s\theta_0 & s\gamma_1 & -a_1 c\gamma_1 c\theta_0 \\ s\theta_0 & c\theta_0 & 0 & a_1 s\theta_0 \\ -s\gamma_1 c\theta_0 & s\gamma_1 s\theta_0 & -c\gamma_1 & -a_1 s\gamma_1 c\theta_0 \\ 0 & 0 & 0 & 1 \end{bmatrix},$$

and for $i = 2$, 3 and 4, it has

$$\mathbf{T}_{bf_i} = \begin{bmatrix} c\gamma_i & 0 & -s\gamma_i & a_i c\gamma_i \\ 0 & 1 & 0 & 0 \\ s\gamma_i & 0 & c\gamma_i & -a_i c\gamma_i \\ 0 & 0 & 0 & 1 \end{bmatrix}.$$

Eventually, given the coordinates of the fingertips in the finger base coordinate frames as $\mathbf{T}_{ft} = e^{[\boldsymbol{S}_{i1}]\theta_{i1}} e^{[\boldsymbol{S}_{i2}]\theta_{i2}} e^{[\boldsymbol{S}_{i3}]\theta_{i3}} \mathbf{M}$, with $\mathbf{M} = (\mathbf{I}, \boldsymbol{p}_{bt})$ and $\boldsymbol{p}_{bt} = [a_{i1} + a_{i2} + a_{i3},\ 0\ ,\ 0]^T$, where a_{i1}, a_{i2}, and a_{i3} are the lengths of the three phalanxes of the ith finger, the coordinate of the fingertips can be expressed in the global coordinate frame as

$$\mathbf{T}_{ot_i} = \mathbf{T}_{of_i} \mathbf{T}_{ft_i}, \tag{4}$$

From (4), the augmented workspace of the metamorphic hand can be obtained and as indicated in [23], it was found that with the reconfigurable palm, workspace of the hand is significantly enlarged leading to various hand configurations and poses.

Further, from the theoretical analysis of the previous version of an anthropomorphic metamorphic hand, without considering grasp stability and rigidity, it is shown that with a reconfigurable palm the robotic hand has better dexterity and manipulability [23–26]. However, through a series of tests, it is revealed that the involvement of the articulated palm increases complexity of the hand which not only requires the use of feedback control but also results in an overall decrease of grasp rigidity. Therefore, in order to provide a theoretical background for optimizing the design and control of the metamorphic hand, and give some insight into grasp planning, grasp kinematics and grasp constraints of the metamorphic hand are investigated in this project [28] based on multifingered hand kinematics [29]. Grasp map and grasp constraint derived in this investigation can be used to evaluate the quality of grasps performed by the metamorphic hand. The grasp map helps to reveal whether a grasp performed by the metamorphic hand is force-closure grasp and the grasp constraint helps to determine whether a grasp is manipulable. If for any object motion $\boldsymbol{V}^b_{op}$, which denotes the body velocity of the object expressed in the global coordinate frame, there exists joint velocity vector $\dot{\theta}$ satisfying the fundamental grasping constraint of a specified robotic hand, the grasp executed by the metamorphic hand is supposed to be manipulable. The grasp constraint presented in this investigation can be adapted and applied to the grasping analysis of the Shadow hand as well as the other dexterous robotic hands.

3.3 Prototype of a Metamorphic Robotic Hand

Considering the specified application of the robotic hand in meat deboning operation, based on the kinematic analysis, computer simulation and feedback received from the tests of the previous version hand, a left-hand metamorphic hand as illustrated in Fig. 12 was designed and fabricated. The robotic hand was subsequently assembled and tendon driven method was employed for actuating the reconfigurable palm and the fingers. Underactuation scheme was used for the fingers such that each finger only involves two actuators one for actuating the MCP joint and the other one for actuating the DIP and PIP joints which are coupled through torsion springs. Steel tendon was used with sheath guiding the tendons passing through the wrist and the tensor units into the forearm where motors and control boards are settled. Miniature potentiometers were embedded

Fig. 12. A four-fingered metamorphic robotic hand

in the finger joints serving as position sensors for measuring the motion of the fingers. NI^{TM} labview was chosen as user interface and CAN bus with CANOpen interface was employed for robust communication of the setpoints between Labview and embedded motor control boards. Since the communication distance was short, the communication speed was set to 1Mbps to maximize the performance.

4 Experiment of Deboning Operation by Human-Robot Co-working Platform

4.1 Definition of the Robotic Hand Poses

In order to achieve grasp of ham muscles by using robotic hand, at the beginning of each deboning step, initial poses of the hand must be identified. By taking the Shadow Hand as an example, this identification was done in collaboration with a skilled butcher of ADIV[1] in France. The butcher's demonstration, advises and expertise helped to define several initial poses for the Shadow hand, which directly emulate the movements generated by the left hand of the butcher during a manual hams deboning process.

As illustrated in Fig. 13, the defined poses were then applied to the Shadow hand for each step of deboning operation. As soon as the hand is in contact with the ham, the hand open or close its fingers to adapt its shape with respect to the muscle shape. The tactile sensors mounted on the Shadow hand fingertips allow it to adjust - through their measurements - the positioning of each finger of the hand, to ensure a stable grasp of the muscle before realizing the robotised ham deboning operation.

For each step of deboning, two poses are defined, one with the opened hand, the other with the closed hand. The pose with the opened hand is used during the approach phase of the hand towards the ham. As long as the hand palm is in contact with the ham, the fingers are closed to perform the grasp of the ham.

[1] http://www.adiv.fr/

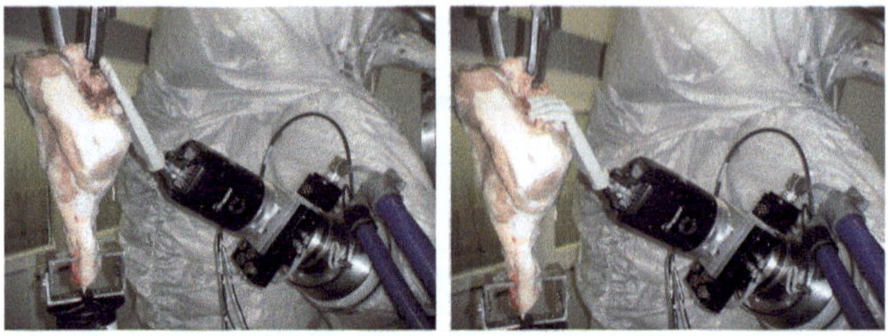

Fig. 13. Initial pose of the Shadow hand for ham deboning

4.2 Formalization of ABB Robot and Robotic Hand Movements

In order to identify the various steps of hams deboning modelling and define the control strategy of the arm-hand system used to perform manipulation and grasping tasks in deboning operation, by integrating the Shadow hand with a 6-DOF ABB robot arm, kinematics of the arm-hand system can be formulated as

$$\boldsymbol{p} = \prod \mathbf{T}\left(\theta_{Ai},\, \mu_{Ai}\right) \prod \mathbf{T}\left(\theta_{Hj},\, \mu_{Hj}\right), \tag{5}$$

where $\mathbf{T}\left(\theta_{Ai},\, \mu_{Ai}\right)$ denotes coordinate transformation matrix of the arm with respect to the joint variable θ_{Ai} and geometric parameter μ_{Ai} of the arm, and $\mathbf{T}\left(\theta_{Hj},\, \mu_{Hj}\right)$ denotes coordinate transformation matrix of the hand related to the joint variable θ_{Hj} and geometric parameter μ_{Hj} of the hand, Based on the kinematics of the arm-hand system integrating with force to be exerted, the arm-hand system composed of a 6-DOF ABB robot arm and a 24-DOF Shadow hand is impedance controlled.

The arm-hand manipulator is consequently used in the meat deboning operation task to help a butcher aiming to replace butchers left arm-hand with this robotic manipulator so as to prevent the butcher from unexpected injury. This cooperative meat cutting task is shared to the manipulator and to the butcher holding a knife. Manipulator has three sequential sub-tasks to do: (1) approaching to the meat; (2) grasping a piece of meat; and (3) pulling accompany with grasping. With such a assistance provided by the left-hand robotic hand, it is expected that the butcher can cut the stretched meat easily and safely.

4.3 Human-Robot Co-working Experiments for Deboning Operation

The aforementioned poses definition and movement formulation are then applied to both the Shadow robotic hand and the metamorphic robotic hand. Separately integrating the two robotic hands into an ABB robot arm in the lab aiming for the robotization of meat processing at ADIV in France, experiments on human-robot co-working ham deboning were carried out and Fig. 14 and Fig. 15 show respectively the experiments performed by the Shadow robotic hand and the metamorphic robotic hand.

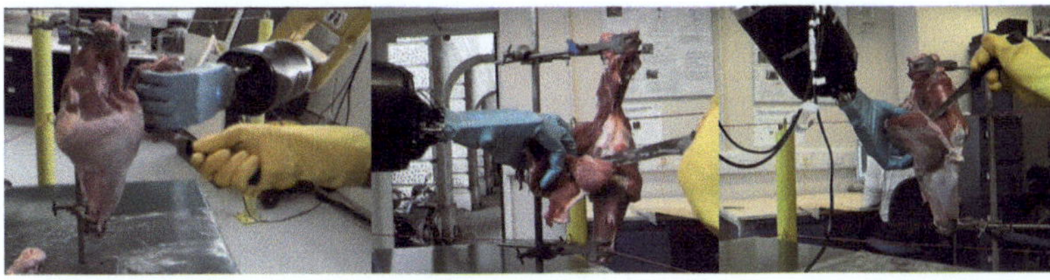

(a) Shadow hand integrated with a Fanuc robot arm at KCL in UK

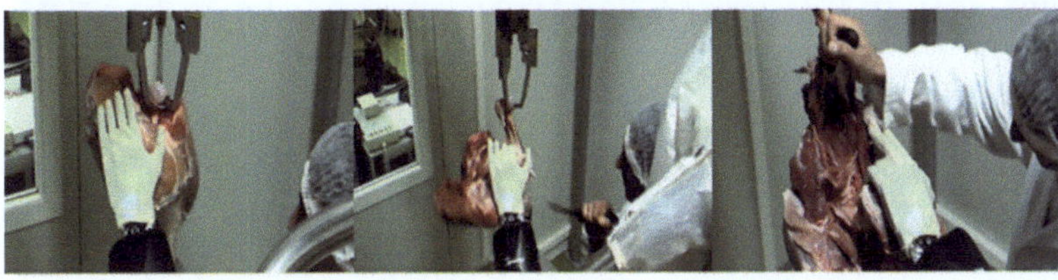

(b) Shadow hand integrated with an ABB robot arm at ADIV in France

Fig. 14. Shadow robotic hand applied in meat grasping for deboning operation

In a meat cutting task, usually butcher's left hand manipulates (grasps, pulls, pushes) one part of the fixed meat while his right hand is cutting the meat with a knife. In such manipulations, the butcher often severely hurts his left arm-hand with unexpected sharp knife strokes. Therefore, in this experimental scenario, we replace butchers left arm-hand with a arm-hand manipulator. The manipulator first reaches to the fixed meat, then grasps a part of it and finally pulls while grasping, so that the butcher can cut the stretched meat easily and without any self-injury risk. Thus, during the operation, the arm-hand manipulator has three sequential tasks to do: reach, grasp, and grasp and pull.

The reach task was performed with teach-by-show method. A skilled butcher taught the manipulator how to reach to the meat for grasping, this reach trajectory is defined as X_{reach}. In autonomous replay of the reach task, manipulator looks for the meat along the learnt trajectory until it feels a resistance force f_{reach} receiving from the wrist force sensor.

Consequently, both Shadow robotic hand and KCL (King's College London) metamorphic robotic hand were partially used to replace human left-hand in deboning operation for meat grasping and manipulation by following the predefined trajectory and qualitative information that was obtained calibrated with the position and force sensors. However, it should be pointed out that, due to the common friction problem for the tendon-driven dexterous robot hand, both Shadow hand and KCL metamorphic hand can only provide limited force for grasping and manipulating meat. The friction between the meat and the finger tips is another problem for grasping robust because the meat is greasy material and thus appropriate glove needs to be identified for practical application. As for the metamorphic hand, although the reconfigurable palm helps increase workspace and dexterity of the robotic hand, it as well increases complexity

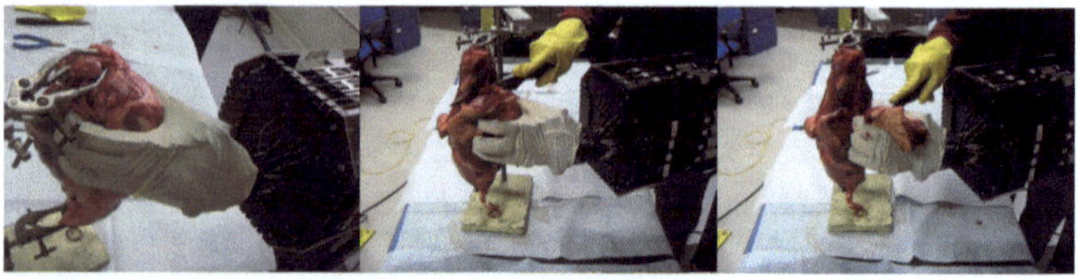

(a) Metamorphic hand integrated with a Fanuc robot arm at KCL in UK

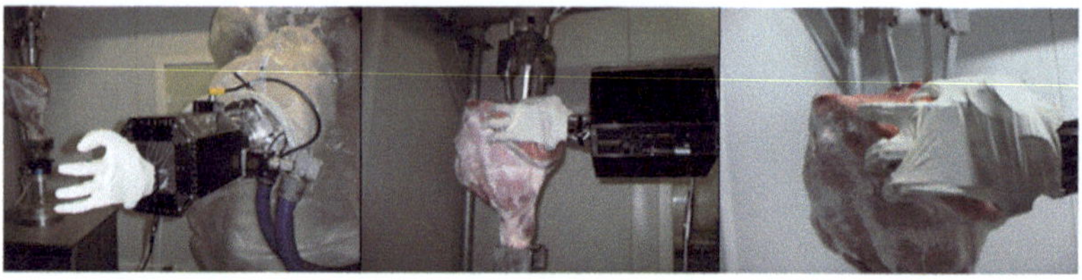

(b) Metamorphic hand integrated with an ABB robot arm at ADIV in France

Fig. 15. KCL Metamorphic hand applied in meat grasping for deboning operation

for trajectory planning and difficulty of precise control which is expected to be overcome with further introduction of extra sensors and distributed control.

5 Conclusions

This investigation aims to carry out the experimental application study of using dexterous robotic hands for deboning operation that is laborious and less safe for human operators as a pioneering study of establishing a hyper-flexible work cell for cutting, deboning and muscle extraction operation in meat industry. In this experiment, deboning operation performed by human operators was systematically investigated which provides sufficient background and information for the experiments of using dexterous robotic hand to replace the left-hand of human hand in deboning operation. In this investigation, two dexterous robotic hands, i.e. one Shadow hand and one KCL metamorphic hand were specifically designed and demonstrated in use of this application integrated with low-level control and sensor systems leading to the in-depth theoretical and experimental study for the robotization of deboning operation in meat industry. The experiment hence contributes to both industrialist in meat handling and grasping, and to academics in trajectory planning and cooperative manipulation using a multi-fingered robotic hand with the cross-disciplinary background of the consortium.

From the theoretical investigation and experimental exploration carried out in this research, it comes to the conclusion that the application of using dexterous robotic hand to replace human left-hand in deboning operation is to some extent feasible and practical meaningful, which not only provides an efficient way for solving labour lacking problem but also presents a human-robot co-working cell preventing human operator from serious injury. Steps for deboning operation and strategies for trajectory planning were proposed laying background for further research and application. The experiments thus proved some feasibility limits of applying dexterous robotic hands for deboning operation. However, some critical issues such as reducing tendon-driven friction of the hand and increasing contact point friction between the meat and the robotic hand still need further investigation and further research funding for such research is expected. The experiment, further, has helped to identify some improvements needed to be made for both Shadow hand and KCL metamorphic hand so as to carry out successfully the desired operation. The introduction of a reconfigurable palm for the KCL metamorphic hand on one hand help increase dexterity and workspace of the hand, but on the other hand increase complexity and difficulty for control of the whole hand. It is also found that a four-fingered robotic hand is sufficient for the operation. Linkage-based robotic hand can be used for the proposed application but the weight of the linkage-based robotic hand needs to be reduced. Further, simple robust grippers, and three-fingered robotic hands can also be used for the proposed experiment but they would not have sufficient flexibility for providing in-hand manipulation, changing grasp, rolling and twisting that are demanded by deboning operation.

Acknowledgments. The authors gratefully acknowledge the support from ECHORD project funded by the EC FP7 under grant agreement number 231143. Thanks is also due to ADIV in France for providing a flexible robotic platform for the experiments, under conditions close to those found in the slaughtering industry. The authors also thank the team members of Shadow Robot Company (Gavin Cassidy, Hugo Elias and Daniel Greenwald), King's College London (Ketao Zhang, Evangelos Emmanouil and Yun Qin), Institut Pascal (Laurent Sabourin, Chedli B. Bouzgarrou, Nabil Essahbi, Valentin Ciupe) and ADIV (Matthieu Alric) who have also participated to the DEXDEB project by providing technical support during the experiments.

References

1. Nof, S.Y.: Handbook of Industrial Robot. John Wiley & Sons Inc. (1985)
2. Guo, J., Lee, K.: Musculoskeletal Model for Analyzing Manipulation Deformation of an Automated Poultry-meat Deboning System. In: 2011 IEEE/ASME International Conference on Advanced Intelligent Mechatronics (AIM), pp. 1081–1086. IEEE Press, New York (2011)
3. Purnell, G.: Robotic Equipment in the Meat Industry. Meat Science 49(1), 297–307 (1998)
4. Daley, W.D.R., Wyvill, J.C., Thompson, J.C., Holcombe, W.D., McMurray, G.V.: Robotics and the poultry processing industry. In: Khodabandehloo, K. (ed.) Robotics in Meat, Fish and Poultry Processing, pp. 48–69 (1993)
5. Pettersson, A., Ohlsson, T., Davis, S., Gray, J.O., Dodd, T.J.: A hygienically designed force gripper for flexible handling of variable and easily damaged natural food products. Innovative Food Science & Emerging Technologies 12(3), 344–351 (2011)

6. Sorensen, S.E., Jensen, N.M., Jensen, W.K.: Automation in the production of pork meat. In: Khodabandehloo, K. (ed.) Robotics in Meat, Fish and Poultry Processing, pp. 115–147 (1993)
7. Kusuda, Y.: The use of robots in the Japanese food industry. Industrial Robot: An International Journal 37(6), 503–508 (2010)
8. Khodabandehloo, K.: Robotic handling and packaging of poultry products. Robotica 8(4), 285–297 (1990)
9. Gurie, G., Sabourin, L., Gogu, G., Lemoine, E.: Robotic Cell for Beef Carcass Primal Cutting and Pork Ham Boning in Meat Industry. Industrial Robot: An International Journal 37(6), 532–541 (2010)
10. Essahbi, N., Bouzgarrou, B., Gogu, G.: Soft Material Modeling for Robotic Manipulation. Applied Mechanics and Materials 162, 184–193 (2012)
11. Guire, G., Sabourin, L., Gogu, G., Lemoine, E.: Robotic Cell with Redundant Architecture and Force Control: Application to Cutting and Boning. In: 2010 IEEE 19th International Workshop on Robotics in Alpe-Adria-Danube Region (RAAD), pp. 99–104 (2010)
12. Zhou, D., Daley, W., McMurray, G.: Kinematics and Verification of a Deboning Device. In: 2009 International Conference on Mechatronics and Automation, ICMA 2009, pp. 2143–2148. IEEE Press, New York (2009)
13. Friedrich, W.E., Spooner, N.R., Lim, P.K.: Fish deboning requires knowledge of anatomy. Industrial Robot: An International Journal 22(5), 18–21 (1995)
14. Dai, J.S., Taylor, P.M., Liu, H., Lin, H.: Folding Algorithms and Mechanisms Synthesis for Robotic Ironing. International Journal of Clothing Science and Technology 16(1/2), 204–214 (2004)
15. Wurdemann, H., Aminzadeh, V., Dai, J.S., Reed, J., Purnell, G.: Category-based Food Ordering Processes. Trends in Food Science & Technology 22(1), 14–20 (2011)
16. Rodriguez-Leal, E., Dai, J.S.: From Origami to a New Class of Centralized 3-DOF Parallel Mechanisms. In: Proceedings of the ASME 2007 International Design Engineering Technical Conferences & Computers and Information in Engineering Conference, DETC2007-35516 (2007)
17. Brisson, M.: Method for Preparing a Piece of Meat, Element for Grasping a Piece of Meat. PCT patent, WO/1998/030107 (1998)
18. Jolliffe, I.T.: Principal Component Analysis, 2nd edn. Springer (2002)
19. Aminzadeh, V., Walker, R., Cupcic, U., Elias, H., Dai, J.S.: Friction Compensation and Control Strategy for the Dexterous Robotic Hands. In: Dai, J.S., et al. (eds.) Advances in Reconfigurable Mechanisms and Robots I, pp. 697–706 (2012)
20. Gan, D., Dai, J.S., Liao, Q.: Mobility Change in Two Types of Metamorphic Parallel Mechanisms. Transactions of the ASME: Journal of Mechanisms and Robotics 1(4), 041007 (2009)
21. Dai, J.S., Rees Jones, J.: Kinematics and Mobility Analysis of Carton Folds in Packing Manipulation Based on the Mechanism Equivalent. Journal of Mechanical Engineering Science, Proc. IMechE, Part C 216(10), 959–970 (2002)
22. Dai, J.S.: Robotic Hand with Palm Section Comprising Several Parts Able to Move Relative to Each Other, Patent WO/2005/105391, Priority Date: 10 November 2005, International Patent PCT/GB2005/001665, UK Patent GB04 095 48.5, 2004, Europe Patent: EP05740527.6, US Patent: US 11/587,766, China Patent: CN200580018189.6
23. Wei, G., Dai, J.S., Wang, S., Luo, H.: Kinematic Analysis and Prototype of a Metamorphic Anthropomorphic Hand with a Reconfigurable Palm. International Journal of Humanoid Robotics 8(3), 459–479 (2011)

24. Wei, G., Aminzadeh, V., Dai, J.S.: Prehension Analysis and Manipulability of an Anthropomorphic Metamorphic Hand with a Reconfigurable Palm. In: Proceedings of the 2011 IEEE International Conference on Robotics and Biomimetics, pp. 1116–1121. IEEE Press, New York (2011)
25. Dai, J.S., Wang, D.L., Cui, L.: Orientation and Workspace Analysis of the Multingered Metamorphic Hand Metahand. IEEE Transactions on Robotics 25(4), 942–947 (2009)
26. Cui, L., Dai, J.S.: Posture, Workspace, and Manipulability of the Metamorphic Multifingered Hand with an Articulated Palm. Transactions of the ASME: Journal of Mechanisms and Robotics 3(2), 021001 (2011)
27. Liu, T., Ting, K.: On the Rotatability of Spherical N-bar Chains. Transactions of the ASME: Journal of Mechanical Design 116(5), 920–923 (1994)
28. Wei, G., Aminzadeh, V., Emmanouil, E., Dai, J.S.: Structure Design, Kinematics and Grasp Constraint of a Metamorphic Robotic Hand for Deboning Operation. In: Proceeding of Design Engineering Technical Conferences & Computers and Information in Engineering Conference, DETC2013-13408 (2013)
29. Murray, R.M., Li, Z., Sastry, S.S.: A Mathematical Introduction to Robotic Manipulation. CRC Press (1994)

Part III
Human-Centered Robots

Part III which deals with human-centered robotics covers five ECHORD experiments that address different ways of interconnecting human and machine. This linkage is of a varying degree of closeness from wearable robots to ambient intelligence; it comprises haptics, multimodal interfaces and even empathic cognition. In the following a brief description of these contributions is provided.

The ASTROMOBILE experiment focused on the development of and experimentation with a robotic assistive platform integrated in smart environments, able to provide daily high quality services for "ageing well" using advanced and natural interaction with end-users. Design and implementation of the ASTRO robot was sustained by a multidisciplinary team in which technology developers, designers and end-user representatives collaborated using a user-centered design approach. The main aspects of ASTROMOBILE concerned (i) the improvement of ASTRO's behavior by means of a smart sensor network able to share information with the robot; (ii) the design of an advanced human robot interfaces based on natural language and able to enhance the usability and interaction of elderly users with the ASTRO robot; (iii) the design of ASTRO's appearance and functionalities by means of a substantial analysis of users' requirements and attitude to robotic technology to improve acceptability and usability.

The aim of the experiment Psychophysiological Interaction and Empathic Cognition for Human-robot Cooperative Work (PsyIntEC) was to investigate emotional states in human-robot interaction, and develop a feasibility demonstrator that addresses safe ergonomic and empathetic adaption by a robotic system to the needs and characteristics of a human co-worker during collaborative work in a joint human-robot work cell. The project consisted of three stages. Stage one was constructing a collaborative human-robot work cell where a human co-worker solves a reference puzzle task alone, collaborating with another human and collaborator with a human scale articulated robot arm. During the three tasks the human co-worker is equipped with biometric sensors measuring brain activity, facial muscle activity, skin conductance, heart rate and gaze. Stage two was developing a human cognition and affect model for human-robot interaction. The model uses real-time data from the biometric sensors to distinguish affective states such as relaxation, comfort, fear and anxiety in the human co-worker. Stage three was developing an adaptive robot decision model where the articulated robot arms adapt their behavior based on the affective states of the human co-worker.

In the project TUAV, the focus was on applications in which aerial robots interact actively with their surrounding environment and/or human operators. Teleoperated Unmanned Aerial Vehicle (TUAV) scenario involves a Vertical Take-Off and Landing (VTOL) vehicle which is remotely piloted by a human operator in order to approach a target infrastructure (i.e. obstacle) such as bridges, hydraulic dam walls, or building to inspect faults by means of one or several onboard cameras. Quite often, visual evaluation of the distance between the vehicle and the obstacle is difficult and often inaccurate. As a consequence, even experienced pilots may fail to safely pilot the vehicle without collision with the obstacle. Moreover, complex aerodynamic effects induced by strong and unpredictable wind gusts, and/or by interactions between the vehicle and the target complicate the matter even more. These difficulties are especially relevant when a UAV gets into close proximity of an obstacle which is a requirement for inspection tasks. The Flybox UAV developed by Skybotix technologies is used as a testbed, to develop control algorithms that can be used by general VTOL vehicles. It provides generous payload capability and processing power that allows the fitting of suitable sensors and actuators for realistic industrial applications. They accomplished the development and the implementation of a high-performance and robust flight control system for the Flybox UAV. Additionally, a first scheme that enables to safely operate a wide range of VTOL UAVs by an untrained user in a cluttered environment was also developed. In this scheme, the user operates the UAV through the use of a haptic joystick.

The MAAT project aimed at developing a new robotic system for the delivery of highly sophisticated therapy for stroke patients. The system is able to adaptively and dynamically change the complexity of the exercise, in accordance with patient requirements, but also to maximize patient motivation and continuously assess the progress of the recovery from the functional and neurological viewpoint, with special attention to the issue of safety in human-robot interaction. The main goal was to include the patient in the control loop and use multi-sensory data to dynamically change the complexity of the therapy and real-time display an immersive virtual reality environment, in accordance with specific patient requirements. To pursue this objective, a bio-cooperative control system has been developed; it is updated according to patient needs thanks to the information about patient global status, provided by a module for the evaluation of the patient's physiological state and biomechanical performance. Two prototypes of multimodal robotic platforms have been developed for validating the MAAT approach.

Body extenders are an emerging new breed of robots. These can be worn on the human body to amplify the force of the user while at the same time they allow the maximum freedom of limb movements. These devices facilitate a combination of the intelligence and perceptual capabilities of the human with the strength and force capability of the robots into a unique system. The overall objective of the TESBE experiment (Technologies for Efficient and Safe Body Extender) was to develop suitable core technologies able to make the use of these devices more efficient and safe. TESBE was completely successful, having achieved the development of all the three identified core technologies: a fast, accurate and robust force control has been developed to reduce the resistant forces perceived by the user during the body movements and, hence, to speed up the operations performed with the body extender. A collaborative control of the body extender's posture has been developed to prevent the overturning of the system under the action of the gravity and, hence, to guarantee the safety of the operator. A highly flexible, haptically enhanced gripper has been developed to allow the grasping of objects of different shapes, as well as making the operator aware of the possible slippage of the handled load, with the final aim of speeding up the operations performed with the body extender.

TESBE: Technologies for Efficient and Safe Body Extenders

Massimo Bergamasco[1], Fabio Salsedo[1], Simone Marcheschi[1], Giovanni Stellin[2], Gabriele Cingano[2], and Francesco Becchi[2]

[1] PERCRO Laboratory of Scuola Superiore Sant'Anna, Pisa, Italy
massimo.bergamasco@sssup.it
[2] TELEROBOT S.p.A., Genova, Italy
stellin@telerobot.it

Abstract. Body Extenders (BE) are an emerging class of wearable robots, aiming at physically supporting humans during the complex handling of materials in un-structured environments. In the framework of the TESBE experiment of the European RTD project ECHORD, PERCRO and Telerobot successfully developed three core technologies deemed as enabling to make safer and more efficient the use of BE in the envisaged application scenarios. In particular the newly developed force control has allowed to reduce by a factor of 5-7 times the resistance forces exerted by the device on the operator's body, during the tracking of its movements, the collaborative control of the BE posture has demonstrated its capability to prevent the overturning of the system under the action of gravity, when approaching its equilibrium boundary, and the new haptically enhanced gripper allowed to speed up the grasping of objects having different shapes and sizes, thanks to its original highly under-actuated mechanism that automatically adapts the orientations/positions of its multiple grasping surfaces.

Keywords: Body Extenders, Powered Exoskeleton, Wearable Robotics, Force Control, Under-actuated Gripper, Haptic Feedback.

1 Introduction

Body extenders (BE) are wearable powered exoskeletons able to track the natural movements of the human limbs and to amplify the force of the user during complex handlings of materials in unstructured environments. The TESBE experiment aimed at the development of core technologies, deemed as enabling for making safer and more efficient the use of BE in their envisaged application scenarios, like the removal of debris for the rescue of victims of natural disasters or the assembly of large products, like aircraft and shipyards. More specifically the following three core technologies have been developed in the framework of the TESBE experiment:

- Fast, accurate and robust force control: to allow faster and more intuitive load handlings, the tracking of the operator movements has to be performed exerting minimal resistant forces (i.e. forces not related to the handled load) in a large enough

F. Röhrbein et al. (eds.), *Gearing Up and Accelerating Cross-Fertilization between Academic and Industrial Robotics Research in Europe*, Springer Tracts in Advanced Robotics 94,
DOI: 10.1007/978-3-319-03838-4_12,

bandwidth. Furthermore the control of the system has to be sufficiently robust to ensure those performances for a relatively large variability of the mechanical impedances of the human limbs and of the load/environment.

- Collaborative control of the BE posture: due to the force amplification, the operator is not aware of the actual equilibrium conditions of the system (BE, operator and handled load). A collaborative control of the device posture is needed to prevent the overturning of the system under the action of gravity, distorting at minimum the intended motion of the operator.
- Development of a highly flexible haptically enhanced gripper: to widen the range of the objects that can be grasped by the BE, as well as to speed up the load grasping, a multi-fingered under-actuated gripper, providing haptic feedback to the operator to make him/her aware of both the grasping force and the slippage conditions, has been identified as core component to be developed.

The robotic platform used to carry out the research activity proposed in the TESBE experiment is the BE developed in 2009 by PERCRO, in the framework of a national RTD project funded by the Italian Ministry of Defence (see Fig.1).

Fig. 1. Picture of the BE worn on a human operator

The device is composed by a central body (the backpack) and four robotic limbs (2 legs and 2 arms), having a total of 22 degrees of freedom (DoFs), each actuated with a dedicated DC torque motor. Each of the two arms is equipped with a 1 DoF gripper, having two fixed jaws and one translating jaw. The grasping force exerted on the handled object is generated as an amplification of the operator's finger force acting on the commanding trigger of the device.

The motion intention of the user is inferred by suitable elaboration of the interaction forces and torques exerted by the operator on the BE in correspondence of 5 pre-defined interaction points, located at the level of the hand, the feet and the back of the operator. To achieve their measure, expressly conceived force/torque sensors have been developed, using commercial one-component axial load cells.

In the following of this paper the research activities and the achieved results relating to a specific core technology has been reported in a dedicated paragraph. General conclusions about the advancements allowed by the research carried out in TESBE and the technical challenges, that still have to be addressed to make BE really usable in real application scenarios, are reported in the last paragraph.

2 Fast, Accurate and Robust Force Control

2.1 Statement of the Problem

Even if the problem of the development of effective force controls suitable for robotic manipulators is not new and has been diffusely addressed in past researches, the case of BE has peculiarities that make it particularly challenging. To understand the terms of the problem, it is sufficient to analyze the mathematical model of the system relating to the case of a BE mechanics having only 1 DoF (see Fig. 2). In this case the BE mechanics consists of a link (LK), hinging about a fixed holding structure (HS), an actuator (A) generating a torque on the link, being composed by an electric motor and a mechanical transmission, and a force sensor (FS), located on the moving extremity of the link. The load (LD) to be handled is connected to the moving extremity of the link and the device is operated by grasping and moving the handle (H) connected to the load flange of the force sensor. The torque generated by the actuator is achieved by suitable elaboration of the measure of the interaction force (Fu) provided by the force sensor. The core component of this elaboration is a force controller that allows both the tracking of the operator movement and the amplification of his/her force, according to the general scheme depicted in Fig. 3.

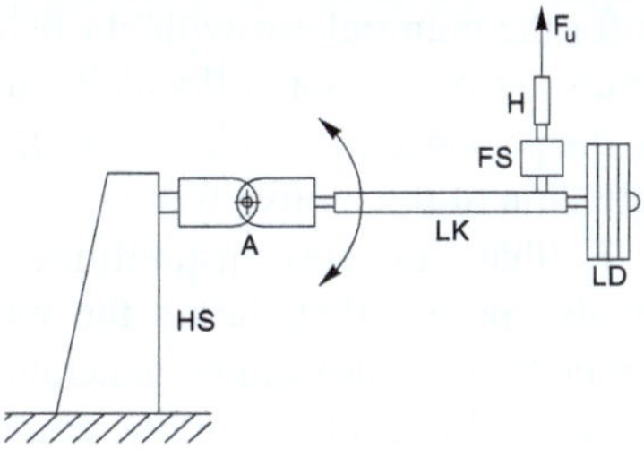

Fig. 2. Picture (left) and scheme (right) of the BE mechanics for the 1DoF case (the abbreviations are explained in the text)

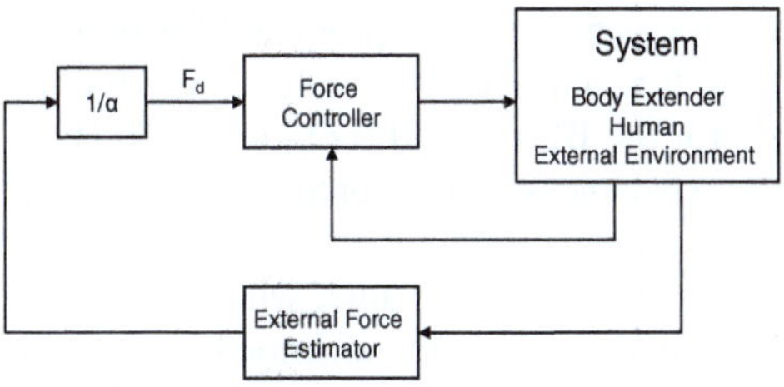

Fig. 3. General scheme of the BE controller

Taking into account the mechanical compliances of the mechanical transmission and of the link and assuming that the force sensor is sufficiently stiff, the system, consisting of the BE mechanics, the handled load and the human operator, can be modelled with constant lumped parameters according to scheme depicted in Fig. 4, that is sufficiently accurate for relatively low frequencies.

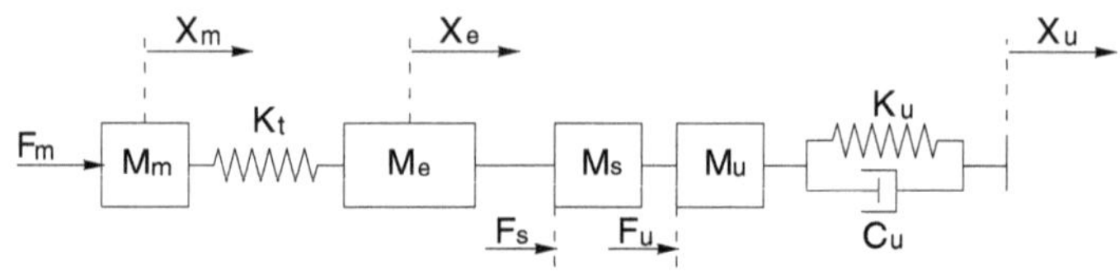

Fig. 4. Mathematical model of the plant for a 1 DoF BE mechanics

With reference to Fig. 4, the motor generates a magnetic force F_m, acting on the motor moving mass M_m, whose time-varying position is X_m. Through the mechanical transmission and the link, of equivalent serial stiffness K_t, a motive force is generated on the mass M_e, located at the moving extremity of the link, whose time-varying position is X_e. The force sensor, measuring the interaction force F_s, is located between the mass M_e and the mass M_s of the handle that, in turn, is connected to the equivalent mechanical impedance of the human operator limb. In its simplest form, that impedance can be modelled as a mass M_u connected to the intended limb position X_u through a spring of stiffness K_u and a damper of viscosity C_u.

From this simple model, it is possible to highlight the peculiarities of the BE that makes the development of a high performance force control particularly challenging.

As for the manipulator with elastic elements (see for example [1]), the actuator and the force sensor are not collocated, since a non negligible dynamic exists in between, due to the presence of the elasticity K_t and the mass M_e. This produces lags between the actuation of the motive force F_m and the measure of the corresponding interaction force F_s that can cause instabilities for relatively large gains of the force loop. It is worth also noting that, being the mechanical transmission low dampened, a large resonance (amplitude) can be generated in response to the motive force F_m.

However, differently from the common robotic manipulators, the mass M_e can be relatively large if compared with M_m and K_t. Indeed, the main contribution to the mass M_e is the mass of the handled load, that can be relatively large, since the BE is intended for handling heavy loads. This implies that the lag due to the dynamic existing between the motor and the sensor can be produced for relatively low frequencies that can fall in the range of frequency characterizing the movement of interest for the applications envisaged for the BE (from 0 Hz to about 10 Hz).

Moreover, from the robustness of the force controller point of view, the range of variability of the M_e parameter is relatively large, spanning from a minimal value, when no load is handled, to a maximal value, when the rated load is handled (50Kg). More in general, the equivalent mechanical impedance of the external environment, with which the BE interacts, can be very different not only for its value but also for its structure, ranging from a pure spring of relatively high stiffness, when the end effector of the BE interacts with a fixed rigid structure, to a pure mass of relatively large value, when the rated load is handled.

The other peculiarity of the BE case is due to the presence of the mechanical impedance of the human operator that directly influences the dynamics of the force F_s. This impedance is characterized by a large variability, not only for the relatively large ranges of its parameters, but also for its structure that depends on the attitude of the operator in holding and grasping the handle of the force sensor. Indeed, previous experimentations on the BE evidenced that the equivalent mechanical impedance could present an additional spring located between the handle and the equivalent mass of the limb, depending on how firmly the handle is grasped.

2.2 Background

Relevant researches relating to the development of efficient force control can be found in the fields of force amplifying wearable robots, haptic interfaces and robotic manipulators with elastic elements.

In the first field, the researches of Kazeroni are the most relevant for the control of the BE [2],[3],[4]. They demonstrate that an outer explicit force loop closed on an inner velocity/position loop can achieve a specified force amplification in a certain bandwidth, independently from the dynamics of the extender. Moreover, they also highlight that the stability does depend on the dynamics of the extender and that a trade off exists between robustness and performances. However, these researches do not directly tackle the problem of the elasticity of the mechanical transmission.

In the second field, the control of the energy flow to ensure the passivity of haptic interfaces, proposed by Hannaford [5], can effectively ensure the stability of the interaction force. However a method to evaluate the achievable performances is not provided. Moreover the non linear nature of the control can induce vibrations into the system. The method proposed by Hogan [6] to measure the performance of an impedance controlled device is useful to optimally dimension the parameters of a controller.

Among the control approaches having a strong theoretical ground and tackling directly the problem of the elasticity of the mechanical transmission, the feedback linearization ensure the achievement of given tracking performances in the position control of the end effector of a robotic device [7]. However its implementation requires the use of higher derivatives (3° order) and it relies on a good knowledge of the dynamic parameters of the manipulator. Other approaches, also having strong theoretical grounds, like the singular perturbation/integral manifold and energy shape method proposed by Spong [8] and Ortega [9], can effectively ensure the stability and accuracy of the position control of the end-effector of a robotic device with elastic joints, requiring lower computational complexity. However they also require a good knowledge of the dynamic parameters of the manipulator.

The position control for manipulator with elastic joint proposed by De Luca [10], use an accelerometer to estimate the states on the link side and generate a suitable damping action, without assuming the knowledge of the dynamics on the link side.

This approach has been deemed particularly suitable for the case of the BE, since solid state accelerometers can be easily integrated into the existing hardware of the device and the unrequired knowledge of the dynamics on the link side is well compatible with the variability of the mass of the handled load.

2.3 Formulation of the New Force Control

The new force control has been developed taking as reference the classical outer force loop-inner motion loop scheme as depicted in Fig. 5.

That scheme has been selected since it allows to implement a specified admittance of the BE and hence an intuitive behaviour of the device in response to the force exerted by the operator at the interaction points with the device. Indeed, it produces motions (e.g. velocity or acceleration vectors) that are always proportional and aligned with the force vectors exerted on them, making the BE behave like a common object (for example a mass), with which a human is used to interact. The simplest implementation of this general scheme consists in using a high gain velocity loop closed on the motor velocity as primary motion control and damper type desired admittance, i.e.

$$V_d = K_F \left(F_H - F_d \right) \tag{1}$$

From the tracking performances point of view, it is desirable having higher K_F to produce smaller resistance forces to the motion of the user, as well a large enough bandwidth to ensure that the BE behaves like the desired admittance in the frequency range of interest for the envisaged applications of the device (about 10 Hz).

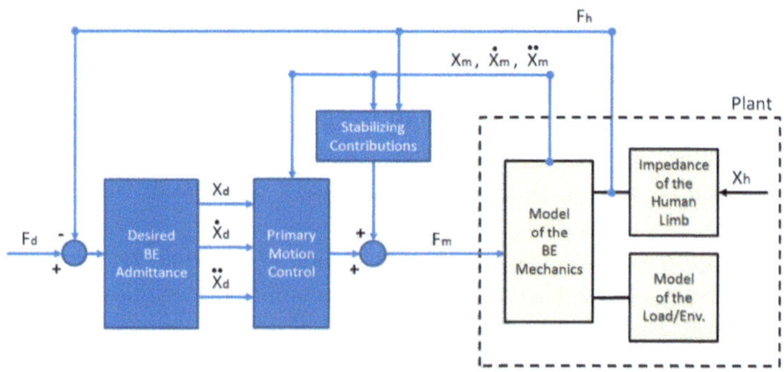

Fig. 5. General scheme taken as reference for the development of the BE force control

A quantitative evaluation of the maximum tracking performances achievable by this simple control, applied to the case of a 1DoF mechanics, shows that they are dramatically dependent on the relative positions of the couple of complex zeros of module S_h, characterizing the dynamics of the human impedance (natural frequency of the human), with respect to the couple of complex poles of module S_T, characterizing the dynamics of the BE and of the environment (natural frequency of the mechanics). This can be well noticed by Fig. 6 in which the Bode plots of the system open loop (OL) force transfer function are reported for the $S_h \leq S_T$ case (left), and for the $S_h \geq S_T$ case (right). Indeed, while the shape of the OL transfer function for the case $S_h \leq S_T$ guarantees a phase margin of the closed loop of at least 90°, and hence the gain K_F and the bandwidth can be theoretically infinite, to ensure a satisfactory phase margin of the closed loop for the case $S_h \geq S_T$ the gain K_F has to be drastically limited. Unfortunately, the second case is very likely occurring for the BE, due to the relatively large masses of the load that can be applied to the BE (that produce low value of S_T) and the relatively high stiffness of the human limb impedance that the operator can show at the force sensors (that produces high value of S_h).

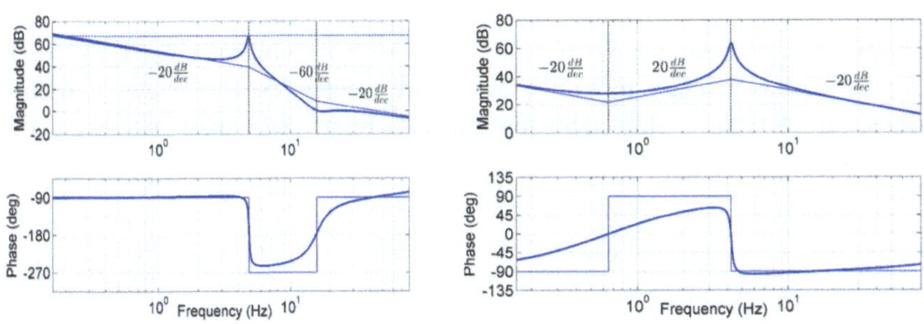

Fig. 6. Bode plots of the system OL force transfer function for $S_h \leq S_T$ case (left) and for $S_h \geq S_T$ case (right)

From the quantitative point of view, it is also worth highlighting that the maximum admissible value of the gain K_F depends on the amplitude of the resonance of the mechanics at S_T, having to be dimensioned in order to achieve a module of the OL transfer function less than unity in correspondence of the resonance. Being the mechanics low dampened, the amplitude at the resonance is very high and, hence, the gain K_F has to be kept very low. In order to overcome those limitations to the tracking performances, a more sophisticated motion controller is needed being able to raise the lowest characteristic frequency of the mechanics (S_T), possibly above the highest characteristic frequency of the human (S_h), and to reduce the amplitude of the resonance of the mechanics.

Even if, in principle, those conditions could be achieved using a suitable full state feedback, it is well known that, if the desired performances of the controlled mechanics are very different from those of the uncontrolled mechanics, the required gains of the full state feedback can be very high, with consequent risk of saturation of the actuators. Furthermore excessive gains can produce instabilities inside the primary motion control itself, if the estimations of the system states are slightly delayed with respect to the real ones.

These considerations lead us to find out a basic configuration of the motion control that could make the mechanics working on its highest possible natural frequency.

This can be better understood, thinking at two very different ways of driving a motor to make the link position track the target.

In the first way the motor is used as a position generator acting on the mechanical transmission. By controlling the position of the motor, it is possible to control the forces transmitted to the total mass M_T located at the BE end-effector and, hence, to control the position of the end-effector to make it tracking the target (see Fig. 7).

If the motor position X_M is directly driven by the target position X_d (i.e. $X_M = X_d$), the X_l/X_d transfer function results:

$$\frac{X_l}{X_d} = \frac{K_T}{K_T + K_h} \frac{1}{\left(\frac{s}{S_{Tp}}\right)^2 + 1} \quad \text{where} \quad S_{Tp} = \sqrt{\frac{K_T + K_h}{M_T}} \tag{2}$$

This is exactly the value of S_T if a high gain velocity control closed on the motor velocity is used as primary motion control. The mechanics also shows a significant resonance at $S = S_{Tp}$, that depends on the relatively small damping of the transmission.

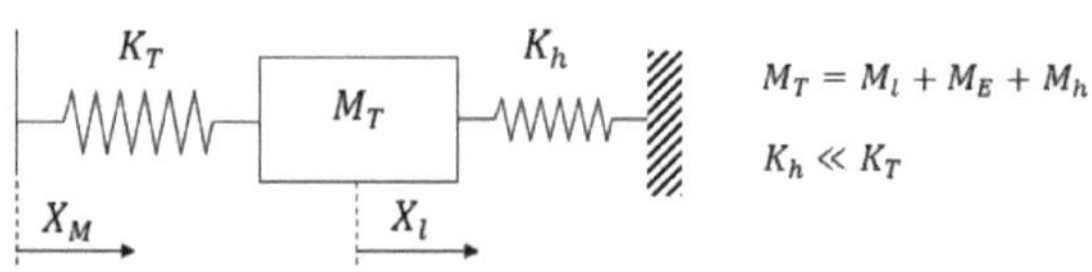

Fig. 7. Equivalent plant mechanics in case the motor is used as position generator

In the second way the motor is used as a force generator that acts on the mass of the motor and the mechanical transmission. By controlling the force of the motor, it is possible to control the force transmitted to the total mass M_T and, hence, to control its position to make it tracking the target (see Fig. 8).

If the target position X_d drives directly the force F_m generated by the motor (i.e. $F_m = X_d$), the X_l/X_d transfer function results:

$$\frac{X_l}{X_d} \cong \frac{1}{K_h} \cdot \frac{1}{\left(\frac{s}{S_L}\right)^2 + 1} \cdot \frac{1}{\left(\frac{s}{S_{Tf}}\right)^2 + 1} \quad \text{where} \quad \begin{aligned} S_L &= \sqrt{\frac{K_h}{M_M + M_T}} \\ S_{Tf} &= \sqrt{\frac{K_T}{M_{eq}}} \\ M_{eq} &= \frac{M_M \, M_T}{M_M + M_T} \end{aligned} \tag{3}$$

It is worth noting that now the natural frequency S_{Tf} of the mechanics is significantly higher than the one of the previous case. Indeed, for the BE mechanics, the minimum value of S_{Tf} would be about 7Hz, that is fairly close to the ideal desired value of 10Hz. Furthermore S_{Tf} is less affected by M_E, since M_{eq} is very close to the mass of the motor M_M, being M_M significantly lower than M_T. This means that S_{Tf} is quite constant in the range of variability of the plant.

On the other hand, the use of the motor as pure force generator introduces two other poles at a frequency S_L close to zero in the transfer function X_l/X_d, making it very different from the desired tracking (unity gain) up to the cut-off frequency. Moreover, also in this case there is a significant resonance of the mechanics, due to the small damping of the transmission.

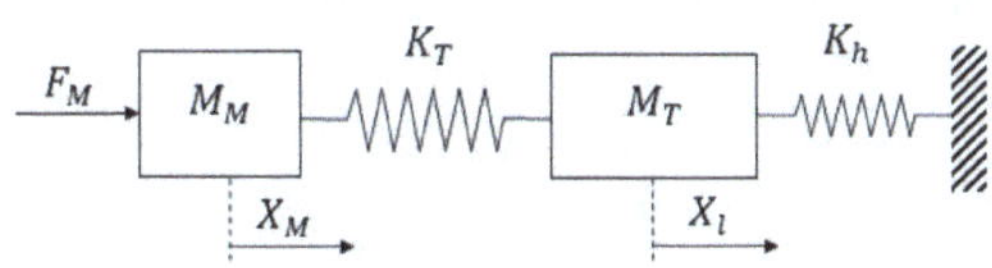

Fig. 8. Equivalent plant mechanics in case the motor is used as force generator

However these problems can be overcome, adding suitable feed-forward contributions and full-state feedback contributions the former being function of the desired motion of the end-effector (i.e. of its desired position, velocity and acceleration) and the second capable to slightly increase the natural frequency of the mechanics and to reduce the resonance at the natural frequency.

A suitable dimensioning of the parameters defining the two contributions allows X_l to track the target X_d with well damped behaviour, up to a cut-off frequency that is close to the natural frequency S_{Tf}, requiring, at the meantime relatively small values of the full state feedback gains (see Fig. 9).

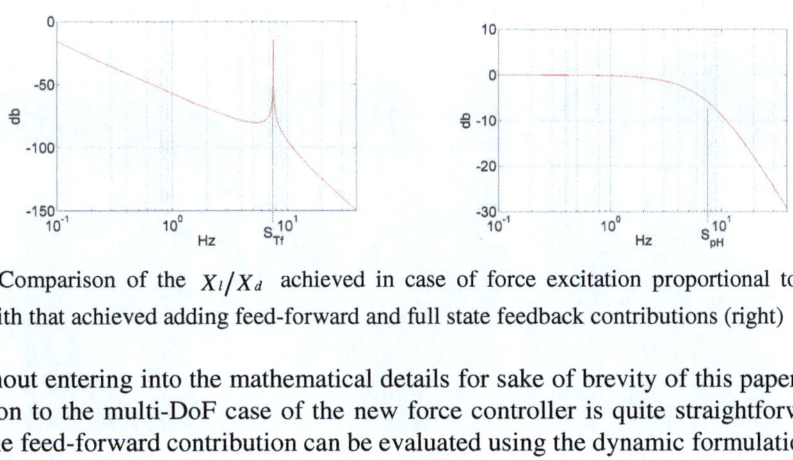

Fig. 9. Comparison of the X_l/X_d achieved in case of force excitation proportional to X_d (left), with that achieved adding feed-forward and full state feedback contributions (right)

Without entering into the mathematical details for sake of brevity of this paper, the extension to the multi-DoF case of the new force controller is quite straightforward, since the feed-forward contribution can be evaluated using the dynamic formulation of the rigid manipulator. Considering that for the BE applications the joint angular velocities are relatively small, the non linear contribution to the joint torque due to them have been neglected, to keep low the complexity of the run time calculations.

2.4 Experimental Validation of the Force Control

The new force controller has been validated firstly for the 1 DoF case and secondly for the multi-DoF case.

For the 1 DoF case, a test-bench has been purposely developed (see Fig. 2), using as actuator the standard actuation module implementing the BE flexion-extension joints (see [11] for a detailed description of the module), and a solid state accelerometer as additional sensor for the evaluation of the states on the end-effector side. A typical plot of the speed and resistance force occurring during the tracking of quasi-sinusoidal movement of the subject hand is reported in Fig. 10.

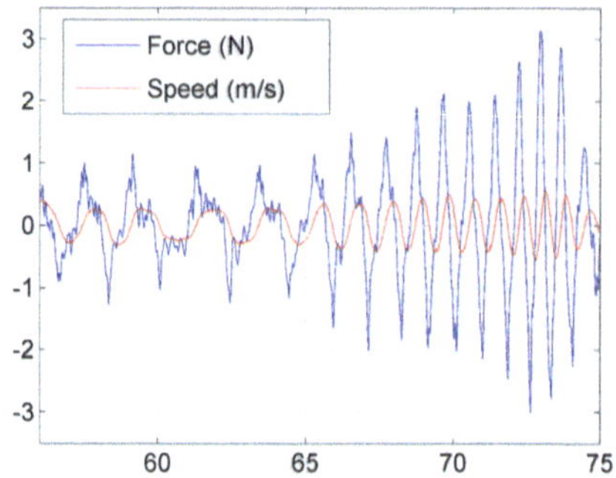

Fig. 10. Typical resistance force during the tracking of the hand movement (M_e = 30Kg)

The achieved experimental results are in good accordance with the theoretical prediction, being the stability of the system ensured for the whole specified range of variability of the handled load and of the limb mechanical impedance.

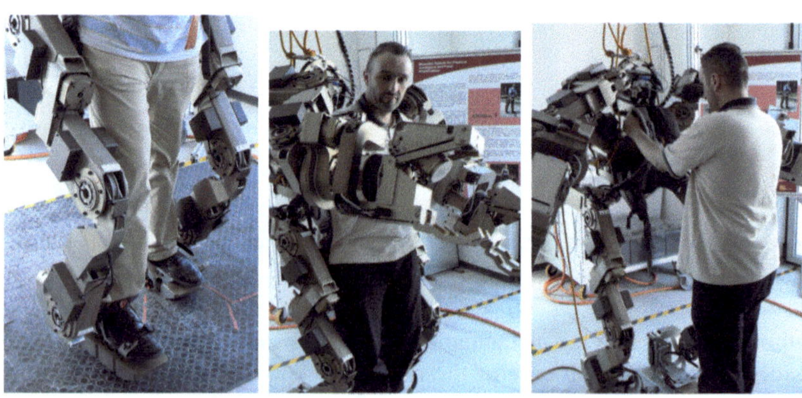

Fig. 11. Experimental setups for the multi-DoF tracking tests. Foot tracking (left), hand tracking (central), back tracking (right).

Furthermore the resistance forces are well under the threshold of 10N for hand movements with frequency contents in the specified bandwidth of the force controller (10Hz). For the multi-DoF case, the robotic limbs of the BE have been used as experimental setup to verify the soundness of the new controller (Fig. 11). Three component solid state accelerometers, located at the end-effector of the robotic limbs, have been used as additional sensors to evaluate the states of the end-effector motion.

Several sets of tests have been carried out for the different cases of tracking (foot, hand and back). Typical plots of the time histories of the speed and resistance force for the multi-DoF tracking tests achieved with the new force controller compared with those achieved with the original controller are reported in Fig. 12.

The comparison of the achieved performances highlights a decrease of the resistance forces of the order of 5-7 times, being their maximum values under the threshold of 10N, deemed small enough to not interfere with the normal motion habit of the user.

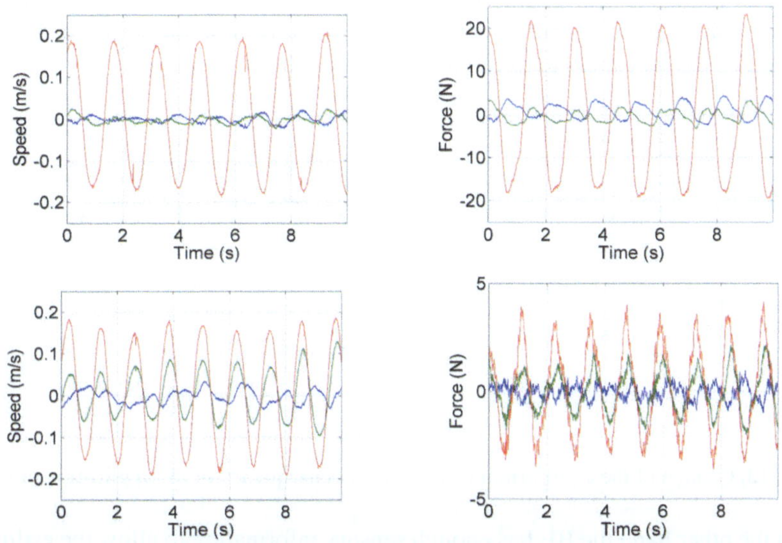

Fig. 12. Generated resistant force (right) in function of the hand speed (left), during longitudinal movements for the original force controller (top) and for the new force controller (bottom)

3 Collaborative Control

3.1 Statement of the Problem

According to the original formulation of the BE controller, the posture of the device, i.e. the set of the BE joints angular positions, is time by time uniquely defined by the posture of the limbs of the human operator. Even if this formulation allows the operator to control the BE movements with maximum freedom, it does not guarantee the equilibrium of the system under the action of gravity. This can be easily understood, considering that, due to the BE force amplification, the operator is not aware of the actual equilibrium conditions of the system, as evidenced in Fig.13.

When the BE is unloaded (on the left in Fig. 13), the operator is also unloaded and the posture of the BE corresponds to the one naturally reached by the operator to assure his/her own equilibrium under the action of the weight of his/her limbs. Since the mass distribution of the link BE is similar to that of the human limbs, the position of the center of gravity (CoG) of the BE projected on the support plane is substantially aligned with the one of the operator body. Hence, the operator, finding his/her own equilibrium, brings the BE to its own equilibrium.

When an external load is being applied to the BE (in the middle of Fig. 2), the CoGs of both the BE and the operator will move forward, but the corresponding displacements can be very different due to the BE force amplification. In particular if a large force amplification is set and a large load is applied, the position of the CoG of the operator will be only slightly modified, since only a fraction of the load is applied on his/her hands, while the displacement of the BE CoG is relatively large. Hence, the change of posture performed by the operator to find his/her own equilibrium when a load is applied can be not sufficient to ensure the equilibrium of the loaded BE (on the right in Fig. 2), and the system can overturn under the action of gravity.

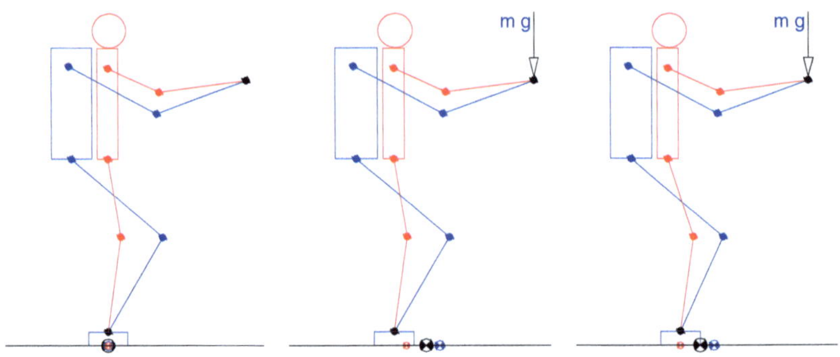

Fig. 13. Change of the equilibrium conditions in consequence of the application of the load

On the other hand the BE has enough sensory information to allow the estimation of its own equilibrium conditions and, hence, it can be envisaged a more sophisticated control of the BE posture, collaborating with the human and preventing the overturning of the system. More specifically the new controller of the BE posture has to verify the compatibility of the operator intention of motion with the system equilibrium and, if required, has to perform suitable distortions (i.e. modifications) of the operator intended motion, to bring it inside the system equilibrium boundary.

The main requirement for those distortions is that they have to be minimal, i.e. they have to assure the maximum possible freedom for the operator movements that is compatible with the system equilibrium.

3.2 Formulation of the Collaborative Control

The mathematical formulation of the collaborative control (CC) of the posture has been defined in the n-dimensional space of the n degrees of freedom of the BE (22), in which the operator intended motion is expressed.

The following simplifying hypotheses have been assumed for its definition:

1. at the beginning of the operations the system is in equilibrium, i.e. the projection of the CoG falls inside support polygon of the system;

2. the BE is loaded with a constant force applied at the level of the gripper, this force being produced by the weight of the handled object;
3. the only cause for the change of the BE CoG is the change of the BE posture (to track the operator's limb motion);
4. the velocity of the operator intended motion is sufficiently small to allow the assumption that the effect of the inertial forces in the determination of the BE zero moment point (ZMP) is negligible. Differently, the effects of the inertial forces have to be taken into account when a braking action is performed by the CC.

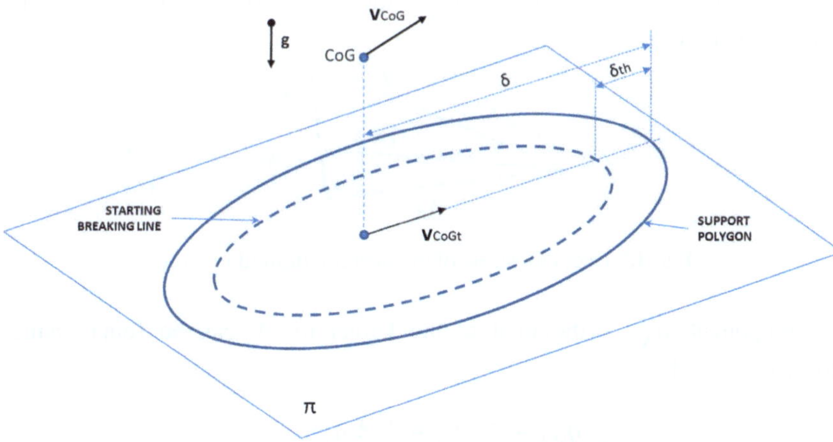

Fig. 14. Projection of the loaded BE CoG and its velocity on the support plane

Assuming the above hypotheses the following considerations hold (see Fig. 14):

- since the CoG position of the loaded BE is a weighted average of the CoG positions of the BE links and of the handled load, the velocity V_{CoG} of the CoG is also a weighted average of the CoG velocities of the BE links and of the handled load;
- since the velocities of the BE link CoGs and of the handled load can be expressed, for a given posture q of the BE, as a linear functions of the BE joint angular velocities $\dot{q}$, the velocity V_{CoG} can also be expressed as a linear function of the same variables;
- to ensure the fulfilling of the equilibrium conditions, only the projection V_{CoGt} of V_{CoG} on the support plane, has to be checked and, if required, reduced, through a suitable braking action, performed by the CC.

In conclusion, the verification of the fulfilling of the system equilibrium conditions implies the estimation of the projection V_{CoGt} of the CoG velocity V_{CoG} of the loaded BE on the support plane, that is a linear function of the BE joints angular velocities.

From a mathematical point of view, this estimation can be achieved through the definition of a suitable linear transformation $A(q)$ from the n dimensional space of the n degrees of freedom of the BE to the 2 dimensional space of the 2 components of the projected CoG velocity, that is $V_{cgt} = A(q)\dot{q}$.

Hence, being n relatively large, there is a large kernel of BE motions of dimension $m = n - 2$ that do not produce any change of the equilibrium conditions of the loaded BE. This large kernel can be exploited to define a minimal distortion of the operator desired motion that can keep the CoG of the loaded BE inside the support polygon.

Referring to Fig. 15, this can be done projecting the n-dimensional vector $\dot{q}_d$ of the BE joint angular velocities, representing the operator intended motion, into two components, the first $\dot{q}_{dk\perp}$ orthogonal to the kernel of A and the other $\dot{q}_{dk}$ belonging to its kernel.

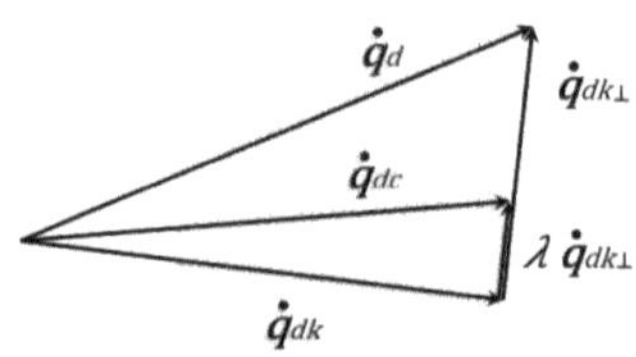

Fig. 15. Decomposition of the operator desired motion

The component $\dot{q}_{dk\perp}$ orthogonal to the kernel of A can be found using the pseudo-inverse of A:

$$\dot{q}_{dk\perp} = A^{\dagger}\, v_{cgt} = A^{\dagger} A\, \dot{q}_d \tag{4}$$

This component has the property to be the vector of the BE joint angular velocities having the minimal possible module that produces the given projected velocity V_{CoGt}.

The component $\dot{q}_{dk}$ belonging to the kernel of A can be found by subtracting $\dot{q}_{dk\perp}$ from $\dot{q}_d$:

$$\dot{q}_{dk} = \dot{q}_d - \dot{q}_{dk\perp} = \dot{q}_d - A^{\dagger} A\, \dot{q}_d = \left(I - A^{\dagger} A\right)\dot{q}_d \tag{5}$$

To keep the BE CoG inside the support polygon only $\dot{q}_{dk\perp}$ has to be checked and, if required, reduced, while $\dot{q}_{dk}$ can be sent to the motion controller of the BE unmodified, since it does not produce any change of the equilibrium conditions.

Hence, indicating with $\dot{q}_{dc}$ the target motion vector sent to the motion controller, the action of the CC can be formulated as:

$$\dot{q}_{dc} = \dot{q}_{dk} + \lambda\, \dot{q}_{dk\perp} \tag{6}$$

where λ is a scalar reduction factor that depends on the distance δ of the projected CoG from the support polygon (see Fig. 14), evaluated along the direction of V_{CoGt}.

In a simplified formulation of the braking action, to ensure that V_{CoGt} is zero when the projected CoG reaches the support polygon and to have a smooth intervention of the

CC, the reduction factor λ could be expressed using a predefined distance δ_{th}, after which the braking action is initiated (see Fig. 14):

$$\lambda = \begin{cases} 1 & \delta \geq \delta_{th} \\ \dfrac{\delta}{\delta_{th}} & \delta < \delta_{th} \end{cases} \tag{7}$$

However, the use of a predefined distance not depending on the value of V_{CoGt} when the braking is initiated, would in general produce inertial forces that can significantly displace the ZMP of the BE with respect to its CoG, giving rise to the possibility that the ZMP falls outside the support polygon.

To define an optimal braking action able to guarantee that the ZMP always falls inside the support polygon, let's consider the simple 1 DoF system depicted in Fig.16.

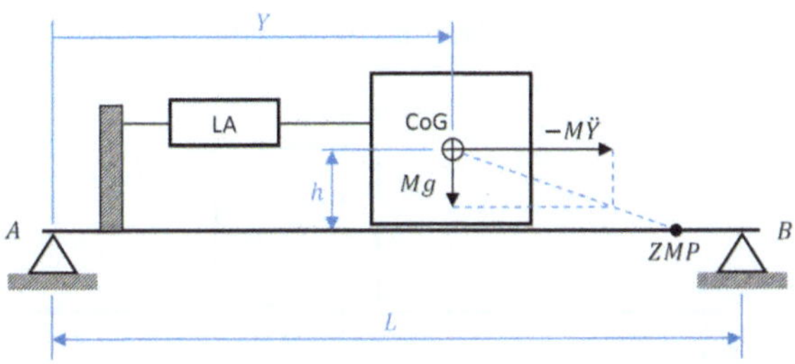

Fig. 16. Scheme of a 1 DoF system laying on two unilateral supports

With reference to the figure, a mass M, having its CoG identified by the time-varying position Y, can slide on a horizontal line laying on two unilateral supports located in points A and B, being moved by a controlled linear actuator LA. The external forces acting on the system are the vertical gravity force M_g and the horizontal inertial force $-M\ddot{Y}$ (in the figure depicted in case of deceleration) that produce a resultant force intersecting the supporting line in ZMP.

What it is shown by this simple model is that, during the deceleration of the mass moving towards the support point B, the position Y_{ZMP} of the ZMP is displaced forward with respect to the position of the CoG by a quantity that is proportional to the height h of CoG with respect to the supporting line and to the deceleration $-\ddot{Y}$, according to the following equation:

$$Y_{ZMP} = Y + \frac{h}{g}\left(-\ddot{Y}\right) \tag{8}$$

Indicating with δ_{SB} the distance of CoG from the support B when the braking action is initiated (braking starting point BSP in figure 17) and with δ_a the distance of the CoG from BSP when the braking action is completed (i.e. the CoG velocity is zero), the first requirement to ensure the equilibrium of the system is that $\delta_a \leq \delta_{SB}$.

Furthermore, indicating with δ the time-varying distance of the CoG from the BSP, during the braking, the second requirement is that:

$$\delta_{ZMP} = \delta + \frac{h}{g}\left(-\ddot{\delta}\right) \leq \delta_{SB} \tag{9}$$

The maximum admissible deceleration and, hence, the minimum braking length δ_a is achieved when $\delta_{ZMP} = \delta_{SB}$ for all the time during the deceleration. This brings to the following differential equation that defines the optimal braking action:

$$\frac{h}{g}\ddot{\delta} - \delta = -\delta_{SB} \tag{10}$$

As it is evident, higher δ_{SB} brings to smaller δ_a, since the admissible deceleration is also higher. At limit if $\delta_a = \delta_{SB}$ (limit condition still compatible with the system equilibrium), δ_{SB} is the smallest possible and, hence, also the limitation to the operator intended motion is the smallest possible.

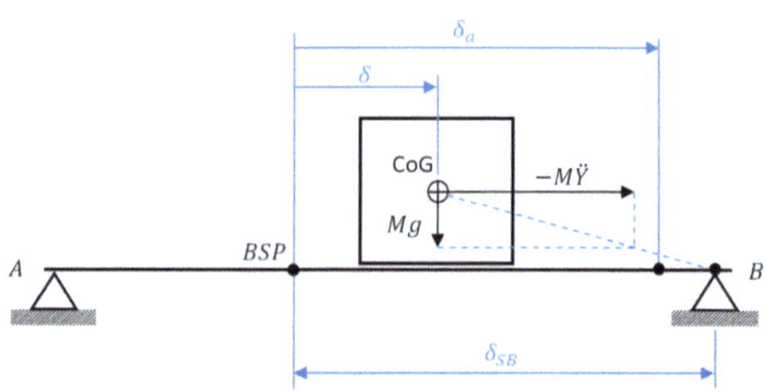

Fig. 17. Variables and parameters defining the 1 DoF System

In conclusion, the smallest distortion for the operator intended motion is achieved when the ZMP falls on the border of the support polygon at any time during the braking and the velocity of the projected CoG is zero when it reaches the border.

Solving the differential equation and finding the condition for which $\delta_a = \delta_{SB}$, it is possible to define the optimal velocity profile, that must be applied to the moving mass to guarantee the system equilibrium, and the distance δ_{SB} from the support point at which the braking action has to be initiated.

$$\dot{\delta} = V_0 \, e^{-t\sqrt{\frac{g}{h}}} \qquad \delta_{SB} = V_0 \sqrt{\frac{g}{h}} \tag{11}$$

It is worth noting that, differently from the simplified formulation of the braking action (see above), δ_{SB} is now dependent on the velocity of the moving mass at the time the braking is initiated and on the height h of the center of gravity of the moving mass from the soil. After the CC has been formulated for the 1 DoF case, it has been extended for the multi-DoF case. Without entering into the mathematical details for sake of brevity of this paper, it can be said that the extension to the multi-DoF case consists in finding an equivalent height of the CoG with respect to the support polygon, taking into account all the inertial forces acting on the BE during the braking action. The soundness of the CC has been verified using a numerical model of the BE, developed using SimMechanics of MatWorks.

3.3 Experimental Validation of the Collaborative Control

The experimental activities aiming at the verification of the efficacy of the CC to keep the ZMP of the real device inside its support polygon, evidenced the need to achieve a sufficiently accurate numerical model of the BE. This implied not only the definition and the implementation of suitable experimental procedures to find out accurate enough values of the model parameters, but also to increase the complexity of the model in order to take into account also the effects of the compliances of the transmissions and of the structural parts, as well as to identify suitable control strategy to optimally transfer to the soil the forces acting on the BE, assuring the perfect adherence of the BE feet with the soil, as hypothesized by the model.

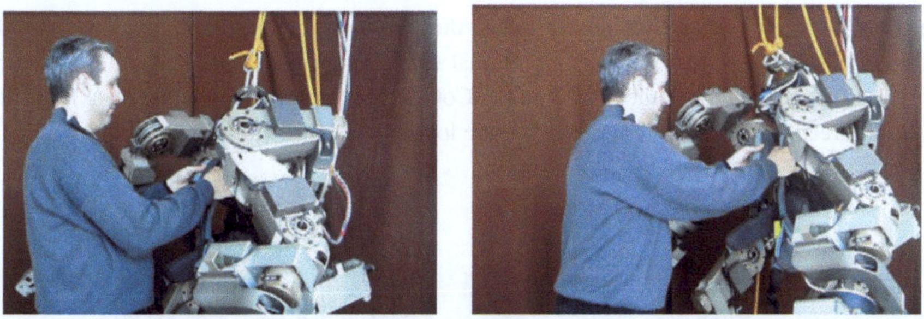

Fig. 18. Validation of the collaborative control with tests providing the physical interaction and the subject not wearing the BE

After a sufficiently accurate model of the BE has been achieved, the soundness of the CC has been verified with two distinct sets of tests, the first providing a physical interaction with the BE and the subject not wearing the device (Fig. 18) and the second providing a physical interaction and the subject wearing the device (Fig. 20).

This first set of tests consisted in activating the force control of the BE and in pushing and pulling with the hands the load flange of the force sensor located in the backpack, in order to achieve its lateral and longitudinal motion towards the equilibrium boundary of the device, being both the BE feet in contact with the floor and the arms kept in a fixed posture. These tests demonstrated the good capability of the CC to keep the BE ZMP inside its support polygon, as shown in the plot of the relevant variables, logged during a typical test (Fig. 19). The second tests consisted in wearing the BE and moving the trunk along the lateral and longitudinal directions in order to induce the motion of the device towards its equilibrium limits. Also in this case the tests showed a good capability of the collaborative control to keep the BE ZMP inside its equilibrium boundaries. Furthermore the subjects involved in the experiment declared that the resulting distortion of the desired movement realized by the CC (consisting in the specific case in a counter-rotation of the backpack when the CoG is approaching the equilibrium boundary) is very natural and well acceptable. More interestingly, after few trials, in the subject emerged the confidence in the capability of the device to assure the equilibrium up to the point that the subject is no longer concerned of the possible overturning of the system.

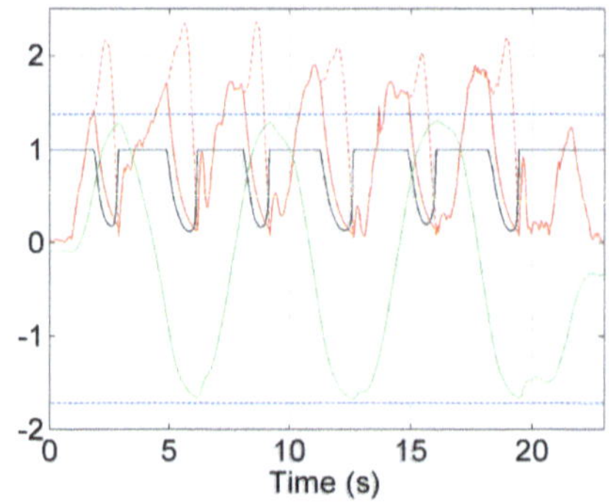

Fig. 19. Plot of the relevant variables, logged during the execution of the first set of validation tests of the CC (movement in the longitudinal direction): desired (dashed red) and corrected (continuous red) velocity of the projected BE CoG, velocity reduction factor (black), position of the CoG (green) and equilibrium limits in the longitudinal direction (dashed blue)

Fig. 20. Validation of the collaborative control with tests providing the physical interaction with a human wearing the BE during lateral motions (left) and longitudinal motions (right)

4 Highly Flexible Haptically Enhanced Gripper

4.1 Statement of the Problem

In the SoA there are few examples of wearable power exoskeletons provided with grippers; all of them are characterized by only one DoF and cannot provide any haptic feedback to user concerning the grasp stability. E.g., the upper-body exoskeleton "POWERLOADER" by Activelink Co. Ltd, from Panasonic group[1] declares a rated load of 1000 N for each arm (provided by a DC motor). The hydraulic arm extender developed by Kazerooni [12] is another good benchmark.

The original gripper developed by PERCRO (see Fig. 21), expressly conceived for the present BE, is another example of previous work in the field of wearable grippers. However, as the other existing devices, it has only DoF and provides haptic feedback only for the grasping force. It is well suited for grasping cylindrical and conical objects of different dimensions but it is inadequate for coping with different shapes.

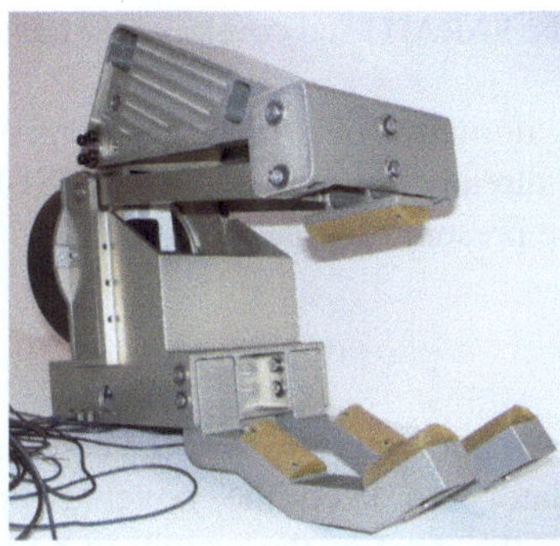

Fig. 21. The first BE gripper developed by PERCRO

Furthermore, due to the inherent relatively high mechanical stiffness of the gripper's jaws, high interaction forces can be produced by their possible collisions with the load, during the pre-grasping phase. This can produce the damage of the load, in case it is fragile, or its overturning, if its positioning is globally unstable. Finally, only a force feedback proportional to the grasping force is provided to operator that, hence, being not aware of the occurrence of the slippage, cannot effectively carry on the task. The consequences of these limitations are, also in this case, a slowing down of the operations and, hence, an increase of their completion time.

The design approach necessarily came from the SoA of advanced gripper for humanoid robots and prosthesis; more in detail from the exploitation of under-actuated mechanisms for grasping. Under-actuation means having fewer actuators than DoFs; the literature reports two main under-actuated approaches for grippers, the first based on tendon transmission and the second based on link transmission.

Tendon systems are generally adopted in order to minimize transmission dimensions but are limited to small grasping forces, while link systems are preferred for applications in which large grasping forces are required.

The work performed at University of Laval by Gosselin [13][14] represents without any doubt the state of the art for this typology of device and were assumed as starting point. The University of Laval group developed the gripper for space and industrial

[1] `http://psuf.panasonic.co.jp/alc/en/`

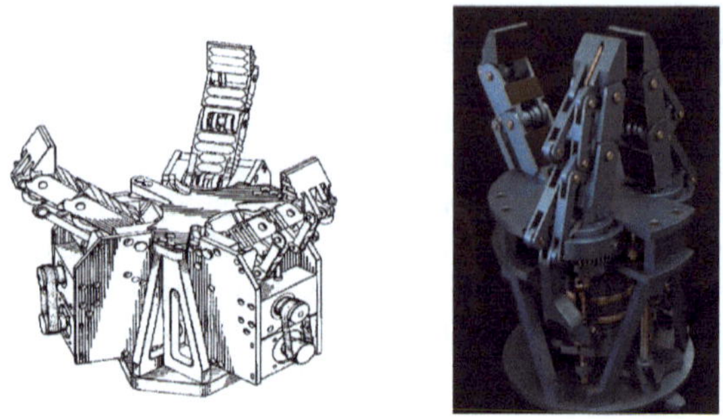

Fig. 22. The SARAH hand developed by Laval University

robotics, two application domains in which the controllability and reliability are mandatory. However, as already introduced, the TESBE requirements pushed the stated objective beyond the present state of the art.

4.2 Research Activity

The BE arm and previous gripper structures worked both as benchmark and first design input. It was thus mandatory to keep the novel gripper as much compact and lightweight as possible. As the most of the weight in a mechanism is imposed by the motors, only one motor was integrated in the system; nevertheless, the gripper was required to cope with a wide set of object shapes, materials and weights and a simple 1 DoF gripper cannot work properly.

As already introduced, an underactuated design approach was mandatory, since an underactuated gripper can adapt automatically the grasp around the object, furthermore with a proper pre-shaping, thus increased stability. The main specifications for the design were the tasks to be performed, the shape typologies, the range of dimensions, the max weight, the material typologies and other characteristics of the objects to be grasped (e.g. softness or fragility).

From the kinematics point of view, the movement of an underactuated mechanism depends on links length and on the elastic passive elements (which act as antagonistic actuators) at each joint. These rule the finger (flexion) movement until the finger touches an object (or a mechanical stop); once one (or more) contact points have been set, each finger automatically configures itself on the object. Considering a gripper with three or more fingers, this means that the grasping of an object is somehow automatic in a way that is closer to the human grasping than independent actuation.

From the point of view of dynamics, grasping is the ability of totally constraining a grasped object by fixing (locking) all the joints. Thus obtaining a stable grasp implies the finding of a configuration where all contact forces are in equilibrium.

The link transmission approach in under-actuation can be parameterized, according to the task to be performed and the objects shape. The single finger model can be observed in Fig. 23.

Given the only one motor, the motion is transmitted to two parallel ball-screws by means of a timing belt transmission. The ball-screw nuts are connected to a 2 DoF differential, which allows pitch and yaw free movements. The range of movements of the differential depends on each phalanx flexion range at any possible combination of finger configurations. The differential slides until contact occurs (e.g. object or mechanical stop) then it exploits pitch and/or yaw to close the kinematic chain.

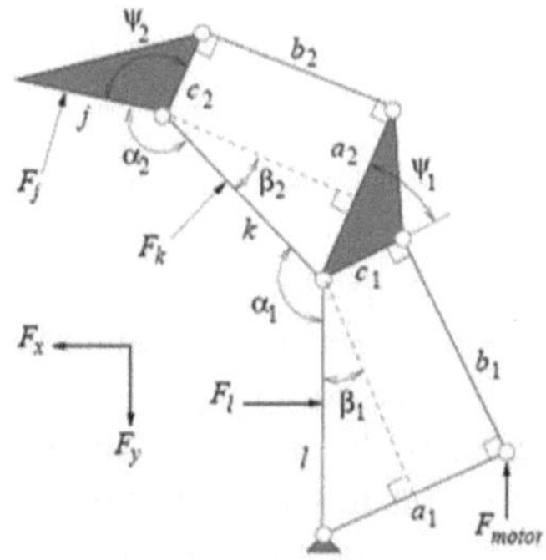

Fig. 23. Finger modelization

By the way, the underactuated approach was exploited not only in the finger design but also in the force distribution from the only actuator to the three fingers, as depicted in Fig. 24.

From a dynamic point of view, several grasp stability guidelines were considered:

- All the finger joints were provided with mechanical stops to avoid hyperextension
- The distal and middle phalanges were provided with torsion springs acting as passive actuators in extending the finger
- Once the object had been grasped, the equilibrium point had positioned on the last physical phalanx in all configurations
- The relation among the fingertip length, the distance between the joint and the contact point at the fingertip and the distal phalanx flexion were kept inside a critical stability range
- The reaction between the object grasped and any phalanx had always to be oriented towards the palm or another finger

Obviously, several efforts were dedicated to the definition of the commanding mechanism, in which the commanding force exerted by the user has been coupled with the force exerted by the BE gripper. The operator can control the fingers by means of a sensorized trigger, whose range is from +20° (open) to 5° (closed), according to ergonomics; it is provided with a load cell, which is connected to the differential linear travel. Its kinematics are connected to the differential movements; hence, the operator pushes on the trigger, the load cell gets both a position and force command and the gripper closes/opens – thus implementing a collaborative control.

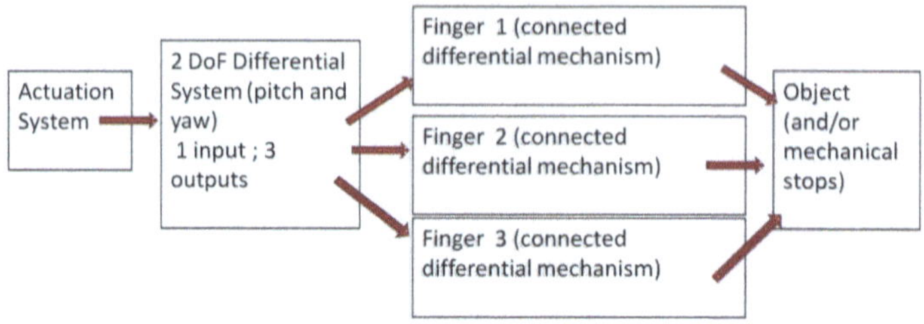

Fig. 24. Gripper transmission modelization

However, as already introduced, performing a grasp means solving a stability problem. When the gripper is about to grasp, each finger is in an unstable configuration and automatically searches for an equilibrium state. Given the underactuated kinematics, this means that even if the actuation motor has been stopped (but not powered off), each finger can go on moving and "palpates" the object until a local equilibrium state has been obtained. It has to be pointed out that this equilibrium cannot be defined "stable", even if the object has been steadily grasped. Stability problems may occur also in case the grasp has been obtained, because of this unstable equilibrium state. This means that, in case of external forces over a certain threshold applied to the object or gripper, or in case of inertial forces originated by a movement of the BE (e.g. arm flexion or during walking), the gripper may lose its equilibrium. From an external point of view, when at the contact point, small uncontrolled movements occur, it means that the system has entered an "incipient slippage" state. In other words, if not counterbalanced, each finger automatically starts to move and any distal phalanx oscillates over its contact point again. This behaviour may result in a grasp self-configuration or in a grasp loss. It is extremely important that the operator can understand that the gripper has entered this state or not. In the BE gripper, a novel approach was implemented. Given the collaborative control of the whole system, since a stopped trigger always corresponds to a stopped motor, the operator does not need any further position sensing to understand if the motor is moving or not. The BE gripper, therefore, was provided with 6 axis accelerometers placed in the distal phalanx dorsum (while the fingertips are made of ADIPOL, which is a resilient plastic material), which can discriminate the movements associated with incipient slippage, thus give a haptic feedback to the user by means of small vibrations generated by the gripper actuator. This is possible by working properly on vibration frequency and selecting one value felt by the user (e.g. 300 Hz) but that does not affect the under-actuated mechanism.

Thanks to the sensors and the trigger position, the operator can understand what is going on, identify the causes and react properly. E.g., in case the object is held stable but is about to break/deform, the operator can decrease the grasping force. On the other hand, in case of slippage, the object is about to move but the operator can exert more force to recover it. In both cases, the operator knows that the motor is not moving (because of the trigger) but cannot "understand" the finger behaviour without the provided haptic cue.

4.3 Results

The developed gripper can grasp different object shapes. It has been designed for grasping cylindrical and tapered objects with a diameter range from 120 mm to 220 mm, maximum weight of 50 Kg, while the gripper's weight is about 12 kg. The grasp typologies to be performed can be cylindrical and "power grasp" but the gripper can also lift objects using only two fingers as a forklift or pick up objects by means of only its distal phalanges (e.g. picking objects).

In any grasp task, the finger behaviour pushes the object towards the palm and can orient in an automatic search for stability. In the under-actuation scheme of the gripper, 3 independent fingers, having each 3 DOFs, are driven by means of only one motor (9 DOFs in total). The only actuation unit is equipped with a DC torque motor, able to deliver up to 5 Nm continuous torque. The hand closes in about 0,5 sec (no object) and exerts at any fingertip more than 500 N. Examples of grasps performed with the new gripper are pictured in Fig. 25.

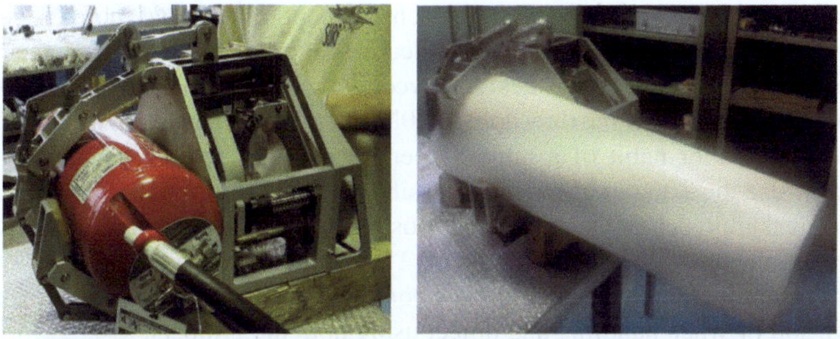

Fig. 25. The TESBE gripper: A cylindrical grasp; A two-fingers lateral grasp

The gripper is provided with a sensorized trigger connected to the differential, which enables the operator to control the gripper closure/opening and the force exerted in a natural way. The accelerometers embedded in the distal phalanxes are able to detect several sources of accelerations (mechanism vibrations, phalanx accelerations, collisions etc.) as shown by the acquired signals reported in Fig. 26, 27. However, in order to use them to generate a haptic feedback on the device commanding trigger, a suitable filter remains to be defined to discriminate the slippage of the grasped object.

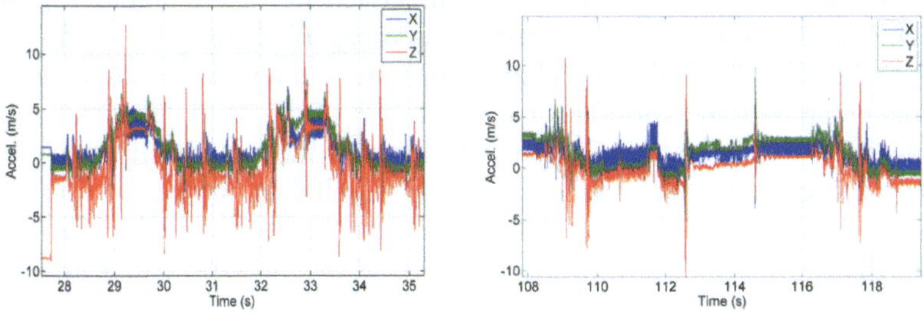

Fig. 26. Raw data from accelerometers when the gripper closes and returns to an open configuration (with no object - left; with a 220 mm diameter cylinder - right)

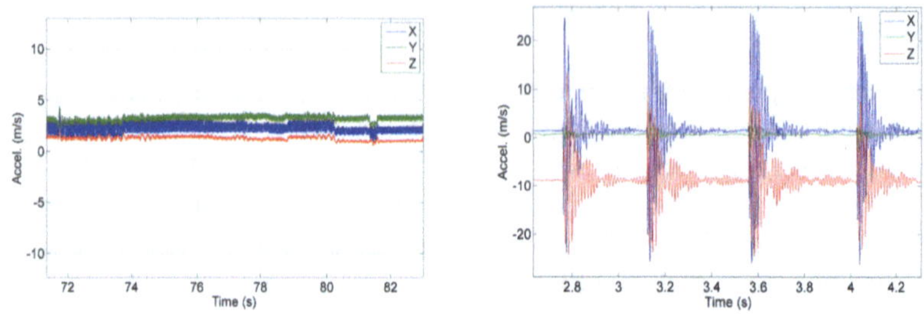

Fig. 27. Raw data from accelerometers during stable grasp (on the left) and during stable grasp affected by external bumps and forces applied to the grasped object (on the right)

5 Conclusions

The three core technologies developed in the framework of the TESBE experiment represent a significant step forward in the achievement of safer and more efficient BE.

The new force controller has allowed to decrease the resistance forces in the range of 5-7 times with respect to the ones achieved with the original controller, being in absolute terms well below the threshold of 20N, deemed sufficiently low to not produce change of the motor habit of a human operator. Furthermore, the underlying basic approach to use the actuator as much as possible as "pure torque generator", in contrast to the original approach of using them as "position generators", makes the device more compliant to the unstructured environment. This is of central importance, for example, to assure an optimal load distribution between the BE two feet in case of irregular and unknown soil or when handling and object using both the grippers.

The numerical verification of the CC and the validation tests have demonstrated its efficacy to keep the BE ZMP inside the support polygon. This has been achieved through a well formulated "minimal intervention" on the BE posture that is function of the distance of the BE projected CoG with respect to the boundary. However, significant work still remains to carry out in order to prevent the overturning of the system in all the cases that can occur in real application scenarios, like time-varying applied loads, variable support polygon and uneven and unstable soils.

The newly developed gripper can be claimed as the first developed highly under-actuated grasping device (9 DoFs and only one motor) providing haptic feedback to the operator, expressly conceived to be integrated into force amplifying wearable robots. The validation test demonstrated its good capability to stably grasp objects of different shapes and dimensions, giving the operator suitable kinaesthetic feedback to control the grasping force. However the generation of the haptic feedback relating to the slippage of the object, that is achieved through the superposition of vibrations generated on the commanding trigger, requires significant work to define a suitable filter able to discriminate slippage conditions in the signals acquired by the embedded accelerometers.

Acknowledgements. This work was supported by the EU Project EC FP7-ICT-231143 ECHORD.

References

1. Volpe, R., Khosla, P.: A theoretical and experimental investigation of explicit force control strategies for manipulators. IEEE Transactions on Automatic Control 38(11), 1634–1650 (1993)
2. Kazerooni, H., Mahoney, S.L.: Dynamics and Control of Robotic Systems Worn By Humans. ASME Journal of Dynamic Systems, Measurements, and Control 113(3), 379–387 (1991)
3. Kazerooni, H., Guo, J.: Human Extenders. ASME Journal of Dynamic Systems, Measurements, and Control, 281–289 (1993)
4. Kazerooni, H., Racine, J.-L., Lihua, H., Steger, R.: On the Control of the Berkeley Lower Extremity Exoskeleton (BLEEX). In: Proceedings of the 2005 IEEE International Conference on Robotics and Automation, pp. 4353–4360 (2005)
5. Hannaford, B., Ryu, J.-H.: Time domain passivity control of haptic interfaces. In: IEEE International Conference on Robotics and Automation, vol. 2, pp. 1863–1869 (2001)
6. Buerger, S.P., Hogan, N.: Complementary Stability and Loop Shaping for Improved Human–Robot Interaction. IEEE Transactions on Robotics 23(2), 232–244 (2007)
7. Spong, M.: Modelling and control of elastic joint robots. ASME Journal of Dynamic Systems, Measurement, and Control 109, 310–319 (1987)
8. Spong, M.W.: On the force control problem for flexible joint manipulators. IEEE Transactions on Automatic Control 34(1), 107–111 (1989)
9. Ortega, R., Kelly, R., Loria, A.: A class of output feedback globally stabilizing controllers for flexible joint robots. IEEE International Journal of Robotics and Automation 11(5), 766–770 (1995)
10. De Luca, A., Schröder, D., Thümmel, M.: An acceleration-based state observer for robot manipulators with elastic joints. IEEE International Conference on Robotics and Automation 1, 3817–3823 (2007)
11. Bergamasco, M., Salsedo, F., Marcheschi, S., Lucchesi, N., Fontana, M.: A novel compact and lightweight actuator for wearable robots. In: IEEE International Conference on Robotics and Automation, pp. 4197–4203 (2010)
12. Kazeroni, H.: Human Power Extender: An example of Human-Machine Interaction via the Transfer of Power and Information Signals. In: 5th International Workshop on Advanced Motion Control, pp. 565–572 (1998)
13. Lalibertè, T., et al.: Underactuation in robotic grasping hands. Machine Intelligence & Robotic Control 4(3), 1–11 (2002)
14. Lalibertè, T., Gosselin, M.C.: Underactuation in space robotics hands. In: Proceeding of the 6th International Symposium on Artificial Intelligence and Robotics & Automation in Space (2001)

Improving Domiciliary Robotic Services by Integrating the ASTRO Robot in an AmI Infrastructure

Filippo Cavallo*, Michela Aquilano, Manuele Bonaccorsi, Raffaele Limosani, Alessandro Manzi, Maria Chiara Carrozza, and Paolo Dario

The BioRobotics Institute of Scuola Superiore Sant'Anna, Pisa, Italy
f.cavallo@sssup.it

Abstract. This work describes the ECHORD Experiment ASTROMOBILE, a project aimed to design, develop and test a system for favourable independent living, improved quality of life and efficiency of care for senior citizens in domestic environments. The system, composed of a mobile robotic platform (called ASTRO) and an Ambient Intelligent Infrastructure that actively cooperated between them and with the end-user, was designed and implemented with a user-centred design approach, involving different stakeholders. The system was designed to deliver services to users, like drug delivery, stand support, reminding, info-entertainment. The design took advantages of the integration of robotic platforms with smart environments, to provide to users higher quality and localization based services. Senior end-users were involved in the experimentation of the system in the DomoCasa Living Lab and feedbacks were gathered for the technology assessment. Particularly, this paper demonstrates the general feasibility of the ASTROMOBILE system and thanks to users feedbacks its acceptability and usability.

Keywords: Companion robot, ambient assisted living, autonomous robot, smart environment, ambient intelligence, user localization.

1 Introduction

Recent studies highlight that the global demographic trend of ageing people is rapidly growing. In 2007, it was underlined that the number of senior citizens was increasing by 8 million per year and by 2030 is expected to reach 24 million per year [1]. In order to provide assistance to senior citizens, the Society has to sustain high costs, that will increase in terms of medical cures, socio-medial services and social activities. In addition, the difficulty to manage this situation is also aggravated by the current financial crisis [2]. To induce important social and economical benefits, it is crucial to preserve the independence of the elderly, delay and reduce institutionalization and favour the participation to community life. Robotics and ICT technologies represent a profitable solution for more efficient

* Corresponding author.

F. Röhrbein et al. (eds.), *Gearing Up and Accelerating Cross-Fertilization between Academic and Industrial Robotics Research in Europe*, Springer Tracts in Advanced Robotics 94,
DOI: 10.1007/978-3-319-03838-4_13,

management and delivery of health and social care, as well as improvement of independent living, quality of life and efficient cares and assist people to live longer and more comfortably in their own homes [3].

The idea of using robotic systems for domiciliary assistance has been investigated in different works that highlighted the potentiality to enhance the independent living of elderly people and provide health care assistance and support in their daily activities and to improve their well-being and efficiency in care through Ambient Assisted Living (AAL) services [4–10]. In such services, mobile robotic systems become a sort of companion (co-worker) that operates with end-users [11] and provides dedicated services anywhere and any time, thus requiring high interaction capabilities and dependability and acceptability features. As a matter of fact, a service based on mobile robotic systems requires some properties that allow to enhance the Quality of Service (QoS), among which:

1. technical efficacy in providing service with high robustness and safety;
2. efficiency and readiness in providing correct services to the users;
3. easy usability in using the service;
4. appropriate aesthetics and functionalities.

Recently some works demonstrated that the integration of mobile robotic systems in smart domestic environments could reduce the necessity of robots with a large set of complex on-board sensors [26] and consequently provide high quality, cost effective and safety services. One of the most important requirements for robotic systems is the ability to reason about the places and times when and where user is in the domestic environment in order to navigate toward him/her when the service is activated and the user is not in the proximity of robot, i.e. in another room. This task could be very difficult for a single robot because requires several complex ability and sensors to seek the user in the home; however a sensor network based on simple and affordable sensors can more appropriately detect the user and share the relative position with the robot that, as a consequence, can in real time know the position of the user and provide the service immediately and in a reasonable time.

The capability of a robot to efficiently seek the user by exploiting an Ambient Intelligent (AmI) infrastructure is still a present scientific challenge. In [12], a robot combined its laser range finder for legs recognition with environmental Presence Infrared Sensors (PIR) able to estimate the position of the user. In [13], a companion robot and a domestic sensor network was simulated to evaluate the advantages of using an ubiquitous robotic framework approach with respect to the traditional robot-centred approach. It was demonstrated through extensive tests that the companion robot, integrated in a sensor network for real-time user localization and home status monitoring, resulted feasible, efficient and time effective.

In this context, this work presents the ASTROMOBILE Experiment, which aimed to design, develop and test the ASTROMOBILE System for favourable independent living, improved quality of life and efficiency of care and to demonstrate its general feasibility, scientific and technological effectiveness and acceptability by end-users. Particularly this paper focuses on the description of the ASTROMOBILE system, composed of the ASTRO robot and an AmI Infrastructure, that used a top-down technique to localize user inside the domestic environment. Starting from a *User Centered Design* approach and after the analysis of the user requirements, specific design activities on the appearance and functionalities of the robotic platform were carried out. The use of an advanced human robot interface, based on natural language, enhanced the interaction and the usability of the ASTRO robot. In addition, the ability to exploit the information coming from the smart sensor network improved the behaviour of the robot in terms of quality and performance of the services.

Section 2 describes the *User Centered Design* approach. Section 3 shows the components of the ASTROMOBILE system and Section 4 highlights the deep interaction between the robot and smart environment. At the end, Section 5 gives the details of the experimental phase.

2 Methodology

The development of the ASTROMOBILE system and its relative services were performed using a User Centered Design approach in order to address the criteria of acceptability and usability, that have been demonstrated to be crucial for plausible deployment and exploitation in real socio-medical contexts [14]. For this purpose, 21 senior citizens of the Municipality of Peccioli, an Italian small town where the ASTROMOBILE Experiment was tested, were recruited and involved in the design, development and experimental process of the system. Particularly, the development was scheduled in the following phases (summarized in Fig. 1):

1. identification of end-users' needs and definition of the ASTROMOBILE system's requirements, functionalities and relative application scenarios (11 elderly);
2. identification of technological aspects for the prototyping of the ASTROMOBILE system;
3. development of ASTROMOBILE prototype (21 elderly);
4. experimentation and evaluation of the ASTROMOBILE system with senior citizens in the DomoCasa Living Lab (15 elderly).

The key point of this approach was the continuous involvement of elderly in almost all phases to make this research really effective in terms of acceptability and dependability. During the first phase, a focus group with 11 elderly was organized to create the first concept of the system; multiple- choice and free- response questionnaires were used to collect ideas and motivations on the functionalities of the ASTROMOBILE system (Table 1), on the different modalities to interact with it (natural language, touch screen, colored lights, remote controller) and on the appearance characteristics (shape, dimensions, colors and materials). During the second phase, the technicians identified the technological opportunities and challenges to implement such ASTROMOBILE system according to the end-users' requests and requirements. During the third phase, the ASTROMOBILE system was completely developed and, interestingly, 21 elderly participated in

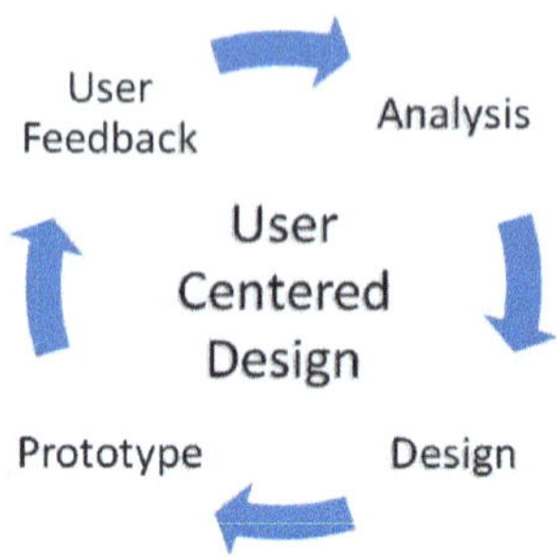

Fig. 1. Main developmental phases of ASTROMOBILE with an User Centred Design Approach

training the speech recognition software on the most common 145 words of the scenarios. Finally 15 elderly were asked to concretely test the ASTROMOBILE system in the seven realistic scenarios of the DomoCasa Living Laboratory and answer to evaluation questionnaires.

3 Description of the System

The ASTROMOBILE system was composed of two main coexisting agents that were able to cooperate between them and with end-users:

- the ASTRO robot;
- an Ambient Intelligent (*AmI*) Infrastructure, implemented by means of two Wireless Sensor Networks (WSN), one for environmental monitoring, called Sensor Network (*SN*), and another for localization, called Localization Network (*LN*).

3.1 ASTROMOBILE Architecture

The ASTROMOBILE System was conceptually implemented as a modular architecture structured in four different layers (see Fig. 2):

- the physical layer, composed of the hardware components, such as the mobile platform, Wireless Sensor Networks, touch screen, speaker and microphone, smartphone, Inertial Measurement Unit and PC;
- the low-level processing and communication layer, composed of six functional modules (Locator, Navigator, Sensor, GUI, Speech and Inertial);
- the high-level planning layer, composed of the Astrologic module, that was conceived to implement easy algorithms of context awareness and decision making and manage high-level tasks, such as "agenda and reminder", "web interface control" and "Skype-like applications";

Table 1. Scenarios used in ASTROMOBILE project

Service	Id	Scenarios
	S1	- The user needs ASTRO because he/she needs a physical help to stand up from the sofa or the bed
User needs ASTRO to carry out some daily activities	S2	- The user calls ASTRO because he/she needs one of objects set on the ASTRO's pocket (TV remote controller glass of water, etc.);
	S3	- The user calls ASTRO to call caregiver because he/she feels sick (Skype from bed);
	S4	- The user calls ASTRO because he/she needs to access to entertainment tools (music, pictures, videos, etc.);
ASTRO autonomously helps user in appropriate situations	S5	- ASTRO moves to user for reminding him/her to take drugs or appointments;
	S6	- ASTRO moves to user because there is a critical alert in the house (i.e. the door is open while he/she is in bedroom);
Caregiver remotely drives ASTRO to assist the user	S7	- The caregiver is outside the home and connects remotely to ASTRO asking it to move to the user and activate the webcam to assess and support the user

– the service layer, represented by the different services that the system was able to provide (see Table 1).

The six modules of the low-level processing and communication layer were assigned to a specific function and strictly interconnected to each other.

Locator. The Locator module was represented by a ZigBee Localization Network that was able to estimate the positions of the end-user in the domestic environment by using both a range based and area based localization procedure. The end-user's position was estimated measuring the Received Signal Strength Indicator (RSSI) from the exchanged messages between the mobile wearable nodes, placed on the user, and the fixed anchor nodes in the environment. Localization was addressed at two levels, i.e. infra-room localization and in-room localization. Firstly a simple area based localization algorithm compared RSSI values from all anchors to identify the occupied room by using a threshold algorithm [15]. Then, a range based localization method was used for a more precise "in-room" user localization. The positions of the anchor nodes in the workspace were defined according to the most probable user trajectories in the activity of daily life (e.g in front of the bed or near the household appliance) and a Hata propagation model was used for user-anchor relative distance estimation.

Navigator. The Navigator module was implemented to control the autonomous movement of robot and to acquire information from sensors installed on the robot, such as odometry and laser. Navigator communicated with the robot actuators and sensors by means of the Player framework [16]. Navigation was based on two algorithms: Wavefront for the global navigation, which was able to generate a sequence of waypoints that had to be reached by the robot [17], and VFH+ for the local navigation and obstacle avoidance [18].

Sensor. The Sensor module was represented by a ZigBee Sensor Network, maintained by a PC server placed in the environment, which was able to collect information coming from environmental and wearable sensors and provide some communication actions for remote users, robot and end-user.

Graphical User Interface (GUI). The GUI module consisted of a graphical user interface including multi level submenus for easily providing functionalities to end-users, such as multimedia functionalities (news, music and video), Skype calls with relatives or caregivers and monitoring of a desired area of the home. The GUI was also used as interface during dialogues between robot and user (described in the scenarios), as drug reminder or alerts. User could navigate inside different screens (or menus) using a touch screen or by vocal commands processed by the speech module.

Speech. The Speech module represented the natural language interface between the end-user and the robot by means of appropriate tools of speech recognition and vocal synthesis. Speech recognition was based on the Simon open-source speech recognition program, produced by Simon Listens [19]. The speech synthesis was a Text-To-Speech based on MaryTTS System [20].

Inertial. The Inertial module was a data logger for inertial data (3-axis accelerometers, gyroscopes and magnetometers) coming from the (Inertial Measuring Unit (IMU) placed inside the head of the robot.

A client-server communication between modules was developed using a shared communication software bus, using the D-Bus system [21].

3.2 The ASTRO Robot

The ASTRO robot was built on the commercial mobile platform SCITOS G5 (Metralabs, Germany) that, based on a differential drive and a weight and payload of respectively 60 Kg and 50 Kg, was able to move up to 1.1 m/s. ASTRO was equipped with a laser range sensor SICK S300 and a 9-axis IMU that allowed it to safely navigate in the domestic environment, avoiding obstacles and planning trajectories. The aesthetics, accessories and interfaces were appropriately designed on the basis of the end-users' requirements analysis, preliminarily identified in the first developmental phase. As a matter of fact the ASTRO (Fig. 3) was designed with a human-like shape and stylized eyes and mouth, made of a grey rigid thermoplastic ABS material and some lateral interchangeable spongy textiles on the chest and head. Based on the end-users' needs, ASTRO included a handle, as support for standing up or walking, and a glove box for

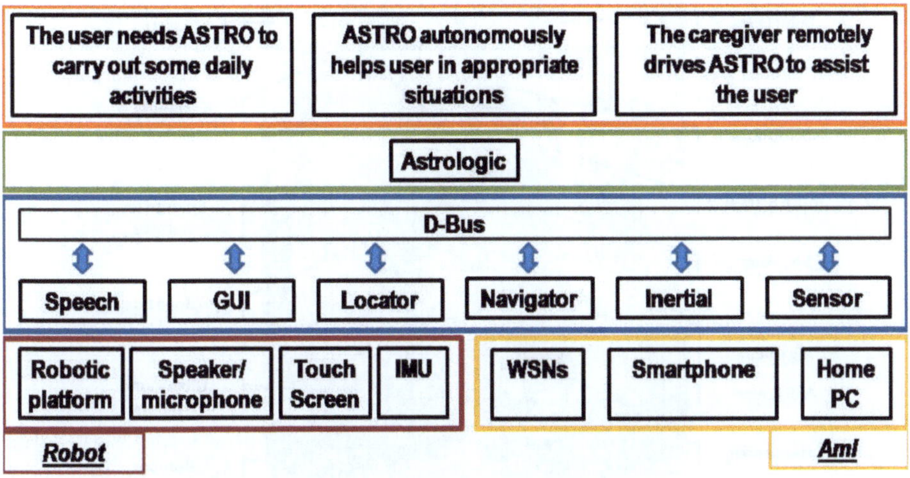

Fig. 2. ASTROMOBILE architecture

transporting small objects on the back. It was also equipped with three bumpers, implemented with two red push-buttons on the chest at the level of the handle and with a circular bumper on the base of robotic platform. In order to favor the interaction with end-users, ASTRO included an adjustable touch screen, red/blue/green-colored Light Emitted Diodes (LEDs) inside the eyes and a set of flexible microphone and speakers for the speech recognition implemented with the Simon open-source software (Simon Listens, Austria) [19].

3.3 AmI Infrastructure

The AmI Infrastructure was implemented by means of the two wireless sensor networks, SN and LN, suited to respectively monitor the environment and localize the users. SN and LN were designed as two different modular mesh networks respectively on the ZigBee channel 26 and 20 (Ember ZigBee-Pro stack) in order to improve the robustness and reliability of the networks. This choice was due to avoid medium access conflicts and interferences between ZigBee nodes, ensuring a better dependability in case of communication fault on a channel and an improved communication reliability by means of the mesh topology and multi-hop message delivery strategy. LN was composed of a ZB Coordinator (ZC), a Data Logger (DL), fixed Anchor Nodes (AN) and a wearable Mobile Node (MN) on user. ANs were equipped with HG2409P directive horizontal linear polarized antennas (Hyper Gain, USA), that were installed to spot the more frequently accessed areas of the environment with the main radiation lobe. The MN, with an omnidirectional horizontal polarized antenna, was worn by the user as a necklace. ZC was used to hold and maintain the entire LN, while the DL was used to store the measured data for localization. The localization cycle (5Hz) consisted of the measurement of the Received Signal Strength Indication (RSSI) from all

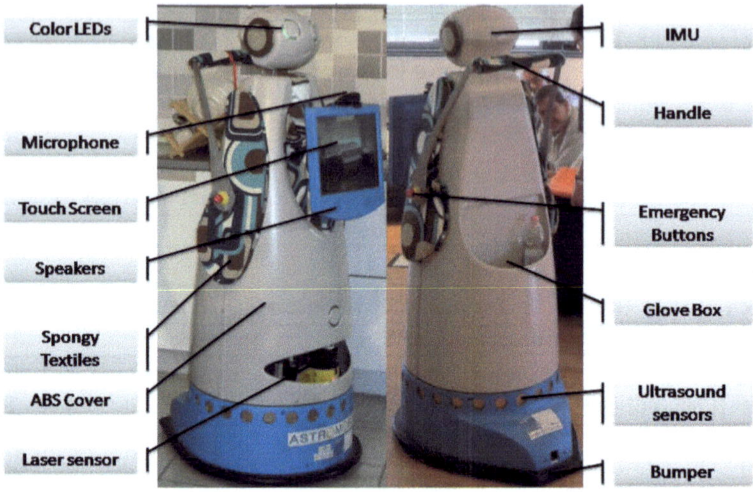

Fig. 3. Picture of ASTRO robot

ANs, in occasion of the reception of a message from the MN, and next in the transmission of such RSSIs to the DL, where the position of the MN was computed. SN monitored the environment by means of appropriate sensor nodes on doors, beds and sofas. Particularly, the information from these sensors were used to identify the presence of end-users on beds or sofas and then also combined with data from LN.

4 Implementation

The main idea of the ASTROMOBILE Experiment was far from a single, complex and autonomous robot in the environment; on the contrary the ASTRO robot was conceived to deeply cooperate with the AmI to perform complex tasks. The high-level planning layer, composed of the Astrologic module, represented the core of the AmI Infrastructure and it was implemented to coordinate and plan the execution of services, such as "agenda and reminder", "web interface control" and "communication applications" (see Table 1). For instance, considering one of the scenarios, i.e. the end-user calls ASTRO to request help in a certain situation, the Astrologic module used easy algorithms of context awareness and planning to manage the service and activate the agents to execute some consequential actions in relation to the end-users interaction (Table 2).

The role of Astrologic module was crucial because allowed to exploit the benefits of the integration of the AmI Infrastructure and the ASTRO robot, thus favouring the delivery of effective and high quality services. This integration expanded the sensing and planning capabilities of the ASTROMOBILE system in delivering services, allowing the ASTRO robot to know, for example, the user position even if the user was out of its sensing range. The capability of the AmI

Table 2. Description of the scenario in which end-user calls the ASTRO Robot to request help

Actors	Actions
End-User	- User sits on the sofa
ASTROMOBILE System	- ZigBee presence sensor on sofa transmits information to the Sensor module - Sensor module pushes this information to the Locator - Locator estimates and refreshes user position
End-User	- User calls ASTRO robot using the smartphone to ask help in standing up from sofa
ASTROMOBILE System	- Vocal commands are sent from the smartphone to the Speech module - Speech module recognizes the command and sends it to Astrologic - Astrologic checks robot availability and sends the command Go to user to the Navigator - Navigator asks to Locator the user position - Navigator moves the robot to the given position and informs Astrologic when the position is reached
End-User	- User cooperates with the system giving simple movement voice commands (as move forward, turn right) to accomplish the task as the user really needs
ASTROMOBILE System	- Speech module recognizes user commands and sends them directly to the Navigator - Navigator moves the robot following user vocal commands
End-User	- The user stands up from sofa using the robot handle

to inform the robot about the user's position or events in the environment allowed a continuous robot planning and made the ASTRO robot more effective in service delivery, avoiding long user seek procedures. Indeed the AmI Infrastructure continuously tracked the user position and sent it to ASTRO Robot, that was immediately able to know the target to reach and plan how to reach it (Fig. 4).

Table 3. Average time over 15 trials to reach the target point

User Position/Target Point	ASTRO starting position	Distance $[m]$	Time $[s]$	Notes
sofa in living room	middle of corridor	~ 4	37.14 ± 0.47	
bed in bedroom	middle of corridor	~ 6	75.22 ± 10.54	narrow passages

4.1 Top-Down Localization Approach

A Top-Down localization approach using sensor fusion between *LN* and *SN* nodes was attempted in order to localize the user in a reliable manner (Fig. 5).

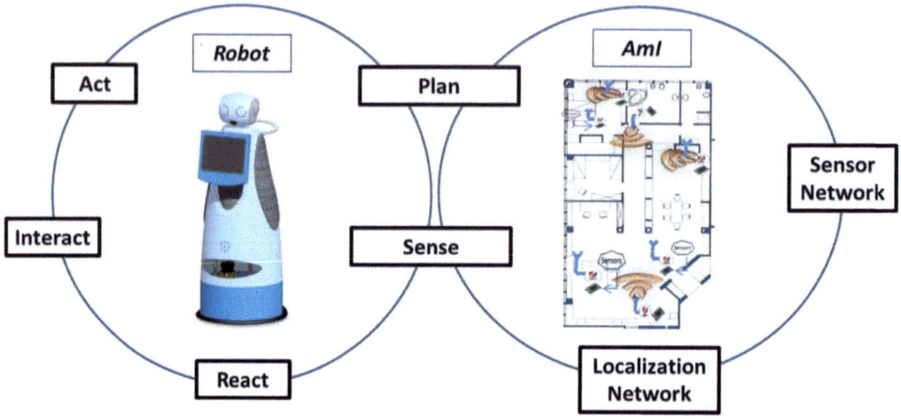

Fig. 4. Robot and AmI cooperation: AmI improves robot sensing and planning capabilities

This localization approach consisted of two main phases: firstly the user was roughly located in the room by means of the Area Based (AB) localization; then he was more finely located using the Range Based (RB) localization.

The AB localization was performed measuring the *RSSI* from all ANs and computing the user's position as the barycentre of the area where the anchor with higher *RSSI* was placed. As demonstrated by [15], if two adjacent ANs were placed at a distance greater than $4\,m$, then the accuracy could reach 98% within 15 seconds of continuous *RSSI* measurement. Therefore in ASTROMOBILE each anchor was associated to an area of $16\,m^2$. In [22], the authors demonstrated that the higher was the distance between transmitter and receiver, the higher was the probability to observe higher variance of the *RSSI*. Based on these results, the *LN* anchors were cast at a distance greater than $4\,m$, and RSSI was trusted to have an acceptable signal to noise ratio if MN was located within $2\,m$ from an anchor. For each AN a calibration procedure was performed to identify the threshold corresponding to a value of *RSSI* at a distance of $2\,m$. If the measured *RSSI* from an anchor was higher than the threshold, the MN was trusted to be located in the same room of the anchor. Once the user was located in a room, the anchors in that room were used to perform a RB localization, using observed *RSSI* and a linear Path Loss (PL) model as suggested in [24] and in [25]. If the estimated distance using the PL model was comparable with the room dimensions, the user was located at that distance on the anchor's antenna boresight, otherwise the user position was set on the center of the room. The use of sectorial antennas allowed to improve the localization accuracy over specific areas of interest, minimizing the possibility to have comparable *RSSI* observations outside the anchors workspace.

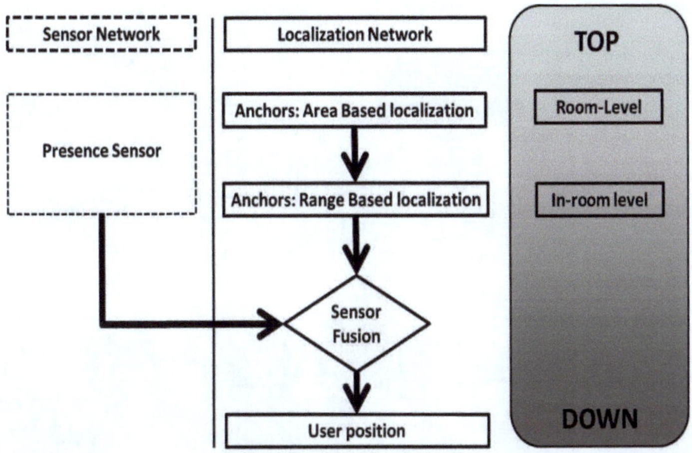

Fig. 5. Top-Down localization approach: user position was estimated in sequential steps, from a room-level location estimation, to a more sharp distance estimation between user and anchors in a room

Calibration. The user's localization was preliminarily anticipated by a two-levels calibration procedure. Firstly the *RSSI* Path Loss (PL) was modelled as a linear model using linear regression as described in equation 1,

$$RSSI_{(d)} = m * d + RSSI_{(d=0)} \tag{1}$$

where $RSSI_{(d)}$ was the *RSSI* at distance d, m was the slope and $RSSI_{(d=0)}$ was the *RSSI* value at the distance $d = 0$ (line/ordinate-axes interception). Slope m and d parameters were estimated for each anchor using MATLAB linear fitting toolbox. The second step consisted of associating a specific area to each anchor, depending on the localization accuracy to be reached and the number of installed anchors. Particularly, four areas were chosen, i.e. the kitchen, the bedroom, the corridor and the living-room. The *RSSI* observations collected during the calibration of the range based localization method were plotted respect to the relative distance between anchor and MN. The *RSSI* recorded at $2\,m$ from each anchor was used as threshold and MN was associated to the area belonging to the anchor that observed an *RSSI* greater than its threshold.

5 Experiments and Validation

The ASTROMOBILE System was tested in the DomoCasa Living Lab, that reproduces a fully furnished apartment of about $120m^2$ appropriately equipped with the above described ASTRO Robot and the wireless LN and SN (Fig. 6).

According to the identified users' needs, the seven scenarios, described in Table 1, were evaluated with 15 senior volunteers (male: 3, female: 12, mean age: 74, min age: 66, max age: 84, primary school: 9, secondary school: 5, university:

Fig. 6. Experimental sessions with elderly people in the DomoCasa Living Lab

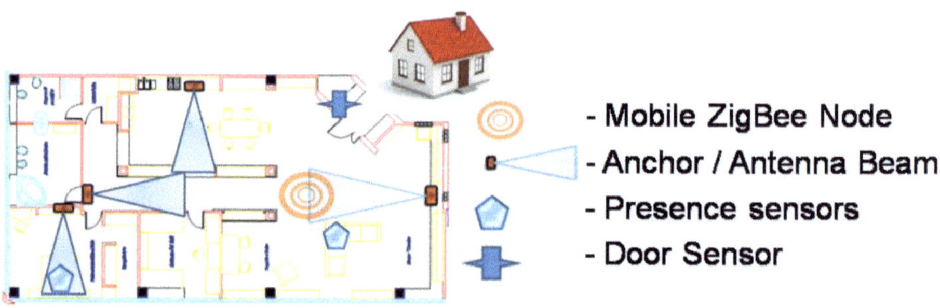

Fig. 7. Representation of the DomoCasa map, WSNs agents and position of the wireless nodes

1, living alone: 6), that accepted to participate in the experimentation after informed consensus.

During the experimentation, the LN was implemented with four anchors and a wearable mobile node, while the SN was composed of a presence sensor on bed and sofa and a door switch sensor on the main door; the ASTRO Robot was programmed to autonomously navigate in the environment, to cooperate with the AmI and to interact with users by means of an appropriate Graphical User Interface and a Speech Recognition module (Fig. 7).

An ad-hoc questionnaire was conceived in order to collect elderly volunteers' opinions. In particular, the questionnaire investigated the following aspects: functions and scenarios, aesthetics of ASTRO robot, control interfaces, interaction between user and ASTRO and general judgement about system utility and adaptability in personal habits and life style. The experiments were carried

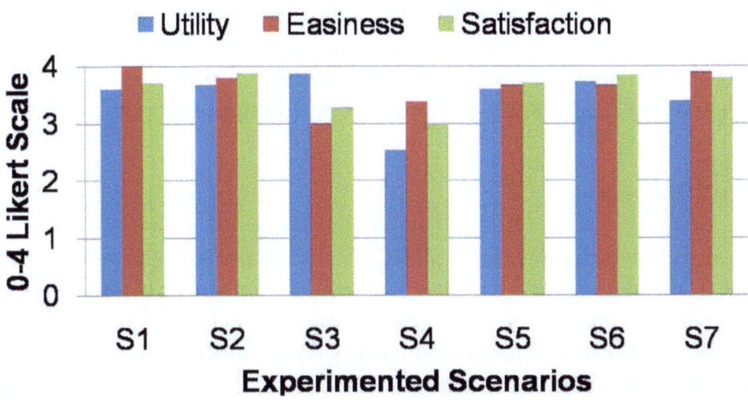

Fig. 8. Perceived Utility, Easiness and Satisfaction from volunteers, expressed in a 0-4 Likert Scale for each scenario (S1: Support in standing up, S2: Object transportation, S3: Help request /Communication, S4: Entertainment, S5: Reminder, S6: Environmental alert, S7: Remote control for caregiver)

out in different sessions, in which the 15 elderly volunteers concretely tested the ASTROMOBILE system.

Concerning the functions and services provided by the ASTROMOBILE system, volunteers involved in the experimentation had a positive perception in terms of utility, easiness and satisfaction (Fig. 8). Most of them were considered very useful with the exception of the entertainment scenario (S4), because they believed that television and radio were already adequate for this necessity and the robot could be used in more important tasks. About the easiness of using the ASTROMOBILE system, interviewees did not find any particular difficulty, with the exception of the communication and entertainment scenarios (S3, S4), because the relative Graphical User Interface was composed of a functional menu with many options (more than four options for the list of contacts and of communication types) and therefore it was perceived not so clear and readable. Asking elderly experimenters if they were satisfied about how ASTROMOBILE system worked in the scenarios, they replied positively for most of scenarios; they only remarked the low quality of the ASTRO's speakers that were too noisy in the communication and entertainment scenarios (S3, S4).

By the aesthetic point of view, all involved users (100%) agreed that the ASTRO Robot was nice, friendly, appeal and safe; on the other hand the 64% of them thought that ASTRO had a big size to be used in their apartments.

The three human robot interfaces, developed on the ASTRO Robot, i.e. GUI on touch screen, natural language and blue/red/green LED based eyes, were respectively evaluated in terms of usability with 3,60 / 3,20 / 3,73 points in the 0-4 Likert scale. Particularly, the speech recognition module did not work reliably, because sometimes they had to repeat the vocal commands, and the vocal synthesis was expected to be more quick and direct toward users. However asking subjects about their favourite interface, most of them pointed out the vocal one, since this was perceived more natural (93% of them).

Most of elderly subjects generally thought that the ASTROMOBILE system was useful and also that it could be enough integrated in their life style. A very interesting aspect emerged from the question about a possible ASTRO purchase: only two persons (13%) excluded completely this opportunity.

6 Conclusions

The work developed during the ASTROMOBILE project showed how a robotic technology for "ageing well" applications could be used in a real domestic environment with real end-users. This result could be achieved by the cooperation of the ASTRO robotic platform with smart environments and users.

The multidisciplinary approach was crucial in all developmental phases, i.e. the design phase of the system to identify the user needs, the technical development and the experimentation, according to a *User Center Design* approach. Cooperation with end-users became very important also during a task, where users and robot worked together to accomplish a service, in a co-working behaviour (i.e. during navigation tasks to decrease the error on the final position with respect to the goal). This was essential to perform tasks in the environment in an effective and efficient way, improving the quality of services provided to end-users. Sensors in the environment were used to monitor the state of the home and to inform the platform about current position of the user, in order to provide services where robot had to reach him/her.

Future work will focus on studying and developing the dependability of the system as well as the improvement of the whole ASTROMOBILE system. In particular, speech recognition and vocal synthesis will be investigated to enhance the capability of recognition (considering also commercial products) and to make interaction more natural, introducing instruments like dialogue manager. Future studies on AmI and WSNs will concentrate on optimization to enhance performances of networks, but also to use the information provided to improve the acceptability in terms of Human Robot Interaction.

In conclusion, this work demonstrated that the robotic technology for "ageing well" applications is nowadays promising and feasible.

Acknowledgements. The ASTROMOBILE Experiment was supported by the ECHORD project, funded by the EU-FP7-ICT-2007.2.2, (Agreement N.231143).

References

1. Plowman, T., Prendergast, D., Roberts, S., Sherry, J.: The Global Aging Experience Project. Ethnographic Research, Intel Research (2010)
2. European Commission - Economic Policy Committee, The 2012 Ageing Report: Economic and budgetary projections for the 27 EU Member States (2010-2060) (2012)

3. van der Broek, G., Cavallo, F., Wehrmann, C.: AALIANCE Ambient Assisted Living Roadmap, vol. 6. IOS Press (2010)
4. The MOBISERV Project (An Integrated Intelligent Home Environment for the Provision of Health, Nutrition and Mobility Services to the Elderly), `http://www.mobiserv.eu/index.php?lang=en`
5. The Florence Project (Multi Purpose Mobile Robot for Ambient Assisted Living), `http://www.florence-project.eu`
6. The CompanionAble Project (Integrated Cognitive Assistive & Domotic Companion Robotic Systems for Ability & Security), `http://www.companionable.net`
7. The KSERA Project (Knowledgeable SErvice Robots for Aging), `http://ksera.ieis.tue.nl`
8. The SRS Project(Shadow Robotic System for Independent Living), `http://srs-project.eu`
9. Care-O-Bot 3, `http://www.care-o-bot.de/english/index.php`
10. Robot Maid, `http://www.lunegate.com`
11. European Robotics Technology Platform (EUROP), Robotic Visions To 2020 And Beyond - The Strategic Research Agenda For Robotics In Europe (2009)
12. Volkhardt, M., Mueller, S., Schroeter, C., Gross, H.M.: Playing hide and seek with a mobile companion robot. In: 2011 11th IEEE-RAS International Conference on Humanoid Robots (Humanoids), pp. 40–46. IEEE (October 2011)
13. Yachir, A., Tari, K., Amirat, Y., Chibani, A., Badache, N.: QoS based framework for ubiquitous robotic services composition. In: IEEE/RSJ International Conference on Intelligent Robots and Systems, IROS 2009, pp. 2019–2026. IEEE (October 2009)
14. Heerink, M., Krse, B., Evers, V., Wielinga, B.: Measuring acceptance of an assistive social robot: a suggested toolkit. In: The 18th IEEE International Symposium on Robot and Human Interactive Communication, ROMAN 2009, pp. 528–533 (2009)
15. Yan, R.H., Yu, C.H., Ding, I.J., Tsai, K.C.: Wireless Sensor network Based Smart Community Security Service. In: The 2nd Workshop on Wireless, Ad Hoc, and Sensor Networks, p. 13 (2006)
16. Player web page, `http://playerstage.sourceforge.net/`
17. LaValle, S.M.: Planning algorithms. Cambridge University Press, Cambridge (2006)
18. Ulrich, I., Borenstein, J.: VFH+: reliable obstacle avoidance for fast mobile robots. In: Proc. 1998 IEEE International Conference on Robotics and Automation, ICRA 1998, Leuven, Belgium, pp. 1572–1577 (1998)
19. Simon Listens official web page, `http://simon-listens.org`
20. Modular Architecture for Research on speech sYnthesis (MARY) official web page, `http://mary.dfki.de`
21. D-Bus web page, `http://www.freedesktop.org/wiki/Software/dbus`
22. Graefenstein, J., Bouzouraa, M.E.: Robust method for outdoor localization of a mobile robot using received signal strength in low power wireless networks. In: IEEE International Conference on Robotics and Automation, ICRA 2008, pp. 33–38. IEEE (May 2008)
23. Dabin, J.A., Ni, N., Haimovich, A.M., Niver, E., Grebel, H.: The effects of antenna directivity on path loss and multipath propagation in UWB indoor wireless channels. In: 2003 IEEE Conference on Ultra Wideband Systems and Technologies, pp. 305–309. IEEE (November 2003)

24. Whitehouse, K., Karlof, C., Culler, D.: A practical evaluation of radio signal strength for ranging-based localization. ACM SIGMOBILE Mobile Computing and Communications Review 11(1), 41–52 (2007)
25. Ding, X., Zhao, H., Zhu, J., Zhang, K., Li, D.: A Novel Localization Algorithm Based on RSSI for Wireless Sensor Networks. In: 2011 7th International Conference on Wireless Communications, Networking and Mobile Computing (WiCOM), pp. 1–4. IEEE (September 2011)
26. Arndt, M., Berns, K.: Mobile Robots in Smart Environments: The Current Situation. In: Autonomous Mobile Systems 2012, pp. 39–47. Springer (2012)

Psychophysiological Interaction and Empathic Cognition for Human-Robot Cooperative Work (PsyIntEC)

Johan Hagelbäck[1], Olle Hilborn[1], Petar Jerčić[1], Stefan J. Johansson[1], Craig A. Lindley[2], Johan Svensson[1], and Wei Wen[1]

[1] Blekinge Institute of Technology, Karlskrona, Sweden
{johan.hagelback,olle.hilborn,petar.jercic, stefan.johansson,johan.svensson,wei.wen}@bth.se
http://www.bth.se/com/cogneuro
[2] Intelligent Sensing Laboratory, CSIRO, Australia
craig.lindley@csiro.au

Abstract. The aim of the PsyIntEC project is to explore affective and cognitive modeling of humans in human-robot interaction (HRI) as a basis for behavioral adaptation. To achieve this we have explored human affective perception of relevant modalities in human-human and human-robot interaction on a collaborative problem-solving task using psychophysiological measurements. The experiments conducted have given us valuable insight into the communicational and affective queues interplaying in such interactions from the human perspective. The results indicate that there is an increase in both positive and negative emotions when interacting with robots compared to interacting with another human or solving the task alone, but detailed analysis on shorter time segments is required for the results from all sensors to be conclusive and significant.

Keywords: human-robot interaction, psychophysiology, affective modeling, robotics.

1 Introduction

The target industrial scenario of the PsyIntEC project is that of a small to medium enterprise engaged in rapid prototyping of novel devices. Production is typified by one-off assemblies or small batch runs of prototypes that often go through several iterations, customer specifications vary widely, and engineering and technical staff are few and have both diverse tasks and highly developed expertise. The one-off or small batch prototyping workshop application is one in which robot systems are not typically used, due to the excessive time and cost of programming and hardware configuration in relation to the small numbers of produced units.

Prototype production frequently requires much more cognitive problem solving on the fly during production than highly pre-planned, larger scale production runs, often with many prototype iterations investigating new components and configurations. Unlike mass production lines, workstations need to be flexible and support human engineers in performing diverse tasks for which they may have highly variable skills.

F. Röhrbein et al. (eds.), *Gearing Up and Accelerating Cross-Fertilization between Academic and Industrial Robotics Research in Europe*, Springer Tracts in Advanced Robotics 94,
DOI: 10.1007/978-3-319-03838-4_14,

Uncertainties are much greater during prototype development, and this creates greater emotional and attentional demands upon human workers. The PsyIntEC project focuses upon demonstrating the feasibility of robotic guidance, support and facilitation of collaborative human-robot prototype production. It emphasizes support for human emotion and attention regulation, modulation and assessment (e.g. maintaining optimum levels of human attentional engagement in the task at hand) during cooperative human-robot task performance, based upon the use of psychophysiology data to measure human affective, emotional and cognitive states.

The first step in generating behavioral adaption in robots based on affective states in human co-workers is to understand the affective queues present in the interaction. By doing so, we can possibly use that knowledge to detect, and let the robot act according to the presumed state in the interaction.

1.1 Tower of Hanoi

The Tower of Hanoi (ToH) puzzle was selected as the reference task for the study. The reasons are that it is an easy task for a robot arm to handle, there exists an optimal solution, and solving the puzzle is a reasonable challenge for most humans. ToH is originally a single player game, and in collaborative gameplay the human-human or human-robot take turns to complete the game.

ToH is a mathematical puzzle game consisting of three rods, and a number of disks of different sizes that can slide onto any rod. The goal of the puzzle is to start from a given configuration of the disks on the leftmost peg and to arrive in a minimal number of moves at the same configuration on the rightmost peg. The puzzle has the following rules [1]:

- Only one disk may be moved at a time.
- Each move consists of taking the upper disk from one of the rods and sliding it onto another rod, on top of the other disks that may already be present there. A disk can also be placed on an empty rod.
- No disk may be placed on top of a smaller disk - i.e., disks are only allowed to be moved to empty rods or be placed on top of larger disks.

1.2 Hardware System

The hardware system contains two Adept Viper S650[1] 6 DOF robot arms with Robotiq Adaptive 2-finger Grippers[2] as end effectors. The robot arms are controlled using an Adept SmartController CX control box and two Adept Motionblox-40-60R power adapters. The end effectors are controlled using two Robotiq K-1035 control boxes. An overview of the hardware platform used in the human-robot work cell is shown in Fig. 1.

[1] `http://www.adept.com/products/robots/6-axis/viper-s650/general` [Accessed 27/05/2013 09:09].

[2] `http://robotiq.com/en/products/industrial-robot-gripper` [Accessed 27/05/2013 09:11]

For psychophysiological measurements a Biosemi ActiveTwo[3] system with sensors measuring electrocardiography (ECG), electromyography (EMG), galvanic skin response (GSR) and electroencephalography (EEG) is used. A Microsoft Kinect camera is used to track the moves made by a human or robot during a ToH game. A single PC running Windows 7 is controlling the system.

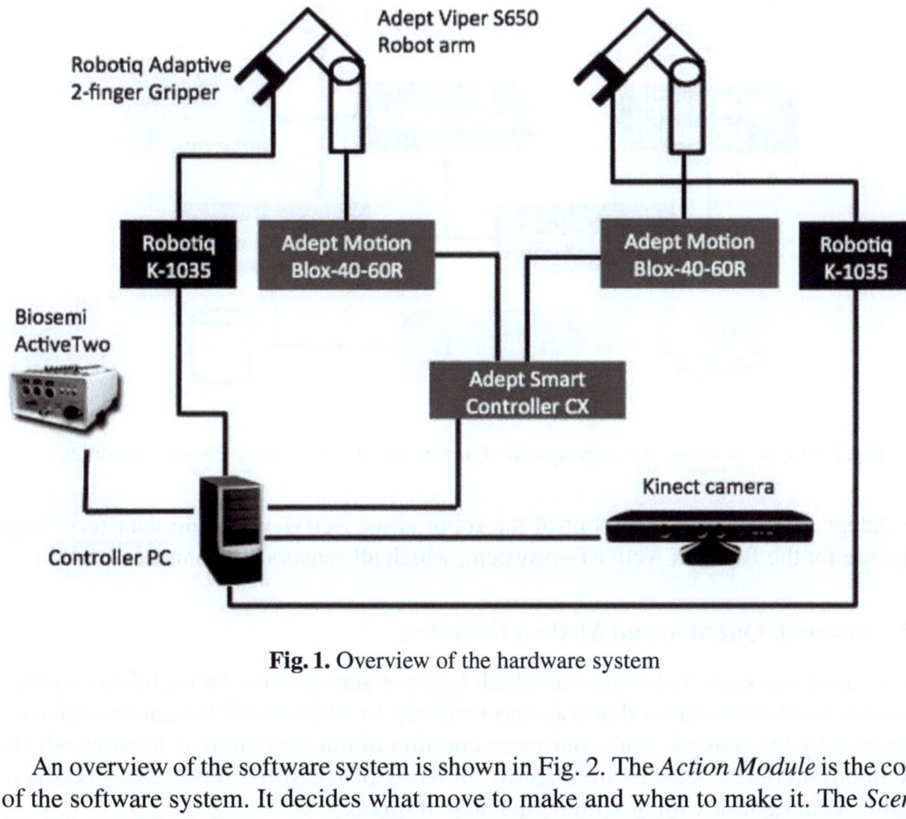

Fig. 1. Overview of the hardware system

An overview of the software system is shown in Fig. 2. The *Action Module* is the core of the software system. It decides what move to make and when to make it. The *Scene Module* provides information about the moves made and which player is next to make a move. All move events (move made and timestamp for the move) are stored to disk using the *MoveLogging Module*. The *RobotControl Module* is responsible for executing a move. It involves how to move the robot arms to pick up and drop the specified game disk, and when to close/release the grippers. The *Scene Module* uses Microsoft Kinect SDK[4] to connect to the Kinect camera, and the EmguCV[5] (a C# version of OpenCV) library to construct a scene of the current game state. The *RobotControl Module* uses

[3] `http://www.biosemi.com/products.htm` [Accessed 27/05/2013 09:12].

[4] `http://www.microsoft.com/en-us/kinectforwindows/develop/overview.aspx` [Accessed 27/05/2013 09:14].

[5] `http://www.emgu.com` [Accessed 27/05/2013 09:14].

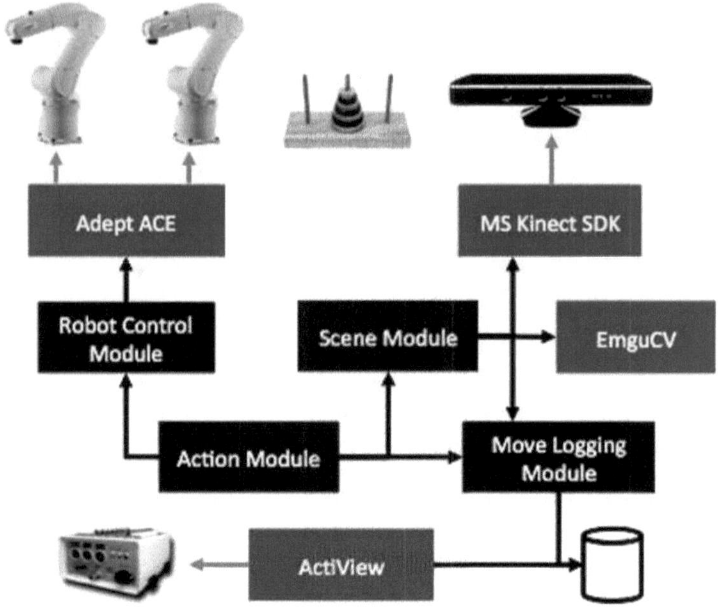

Fig. 2. Overview of the software system. Gray boxes are third-party libraries/modules.

the Adept ACE software[6] to control the robot arms. *ActiView*[7] is the data recording software for the Biosemi ActiveTwo system, which all sensors are connected to.

1.3 Research Question and Method Overview

The goal of our study is to find out which types of sensors may be useful in creating a human-robot workstation that uses bio-feedback to adapt to the human emotions as measured by the sensors. Thus, our main question in this first study is to show which psychophysiological sensors are effective in terms of measuring differences between solving a puzzle task with or without the help of robots.

We limit this question by testing a selected set of sensors on the task of cooperatively solving ToH. This is done by conducting experiments with four different conditions as illustrated in Fig. 3. The first condition is a human completing the reference task alone, the second human-human collaboration and the third and fourth human-robot collaboration. In the fourth condition the robot is unpredictable in terms of speed and path taken when moving the disks. In the experiments continuous measures of ECG, EMG, GSR and EEG were recorded during the completion of the reference task. The dependent variables involved quantitative measurements for evaluating task performance and

[6] `http://www.adept.com/products/software/pc/adept-ace/general` [Accessed 27/05/2013 09:15].

[7] `http://www.biosemi.com/software_biosemi_acquisition.htm` [Accessed 27/05/2013 09:16].

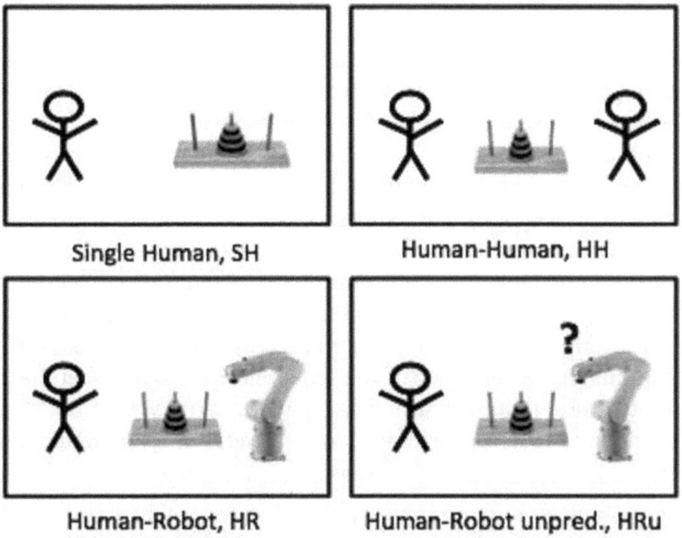

Fig. 3. Illustration of the four conducted experiment conditions. The order varies between participants.

psychophysiological emotional and qualitative measurements for checking perceived emotions.

2 Psychophysiological Measurements

In psychophysiological measurements, a number of different sensors can be used to measure physiological response to emotional states. In the PsyIntEC project ECG, EMG, GSR and EEG have been used. Below is an overview of each sensor type and how it can be used in emotion detection.

2.1 Electromyography

Facial EMG measures voltage levels on the surface of the skin on the locations of facial muscles involved in the generation of affective facial expressions, as an indication of emotional valence (pleasure to displeasure). Increased activity over the cheek (zygomaticus major) and periocular (orbicularis oculi) muscle regions increase with positive valence, while EMG activity over the brow (corrugators supercilii) muscle region increases with negative valence and decreases with positive [2]. A study by Janke showed substantially reduced EMG activity associated with inwardly-directed anger, compared with a group expressing induced anger [3]. This was interpreted as providing support for the view of Fridlund that facial expressions of emotion express information rather than necessarily reflect all felt emotion [4].

2.2 Galvanic Skin Response

Galvanic skin response (GSR) measures skin conductance as an indication of emotional arousal, attention, alertness and effort. Wu et al. showed an average detection accuracy of 45-73% for negative emotions and 62-78% for positive emotions using 30 extracted features [5]. There is also some evidence of GSR being a relatively strong indicator of negative emotions but fewer correlations have been found for positive emotions [6]. GSR measurements are taken from the surface of the skin, where electrical conductivity is affected by sweat released by eccrine sweat glands in response to physiological and emotional arousal. GSR includes short-term phasic responses to specific stimuli, and relatively stable, longer-term tonic levels. GSR may be measured in terms of resistance, conductance, or absolute voltage [6]. Measurements are typically taken from the palm of the hand, the fingers, or the soles of the feet, where eccrine sweat glands are most densely distributed, providing the strongest signal variations. One drawback with the use of GSR in the field is its susceptibility to influences of temperature, humidity and mechanical pressure at the points where sensors are attached.

2.3 Electroencephalography

Electroencephalography (EEG) measures electric activity of the brain that reaches the scalp, indicating cognitive/conscious states of attention, alertness and drowsiness. The interpretation of EEG data has largely been framed in terms of specific frequency bands associated with different cognitive and emotional states [6]: alpha waves (8–13 Hz) with relaxation, reflection and inhibition; beta waves (14–30 Hz) with alertness, anxiety, concentration, mental and physical activity; delta waves (0.5–3.5 Hz) with deep sleep; theta waves (4–7 Hz) with drowsiness, pleasure, displeasure, idling and inhibition; kappa waves (10 Hz) with thinking; and gamma waves (resting frequency around 40 Hz, modulated by 3–5 Hz) with cross-modal perception and perceptual recognition. Evidence suggests that stronger alpha waves in the right frontal hemisphere are associated with withdrawal (negative emotions), while those in the right frontal hemisphere are associated with approach (positive emotions) [2].

2.4 Electrocardiography

Electrocardiography (ECG) measures the rate and regularity of heartbeats by detecting the peaks of the highly positive R-waves in the signal [7]. It also measures other things like the size and position of the heart chambers, but that is irrelevant for our purposes here. A study by Malmstrom et al. used heart rate and skin conductance (GSR) to measure arousal when watching a stressor motion picture film [8]. The study showed a correlation with arousal, closely paralleling skin conductance variations. A study by Drachen et al. showed a correlation between heart rate and arousal (excitement) in video game players [9]. Increased heart rate is a good indicator of arousal, especially for negative emotions [6]. Ya et al. showed over 90% detection accuracy of both positive (joy) and negative (sadness) emotions for multiple subjects based on ECG recordings [10].

3 Experimental Setup

A crossover study with controlled experiments was conducted in a laboratory setting. Lighting and temperature was controlled and the laboratory room had minimal distraction from the outside. Subjects were seated in a chair with fixed height and a predefined position. The height and position were constant during all experiment sessions in order to be able to compare the data between different sessions. Two experimenters were always present in the laboratory room to monitor the experiments, but they were completely hidden behind a screen and were instructed to be as quiet as possible. The presence of the experimenters was for safety reasons when using the robot arms and to monitor that all data were recorded correctly. Surveillance of the robot arms and subject was done using live feed from a video camera. In addition the robot control software included an emergency stopping sequence that would be triggered if defects in the program execution were detected.

The GSR, EMG and ECG signals were recorded using the Biosemi ActiveTwo system using active electrodes. The electrodes were prepared with conductive gel and attached to the participant as follows:

Table 1. The placement of the electrodes used for the GSR, EMG and ECG signals

Signal	Placement
EMGC	Corrugator supercilli (eyebrow)
EMGZ	Zygomaticus major muscles (edge of the mouth)
ECG	Sternum top (chest)
	Sternum bottom (chest)
GSR	Medial phalanx (middle joint on left index and ring finger)
Ground /	Earlobe
Reference	Zygomatic bone (cheek bone)

The EEG signals were recorded using Biosemi ActiveTwo system and Biosemi Headcap with active electrodes[8]. In the experiments EEG signals from left and right frontal, central, anterior temporal and parietal regions (F3, F4, C3, C4, T3, T4, P3, P4 positions according to the 10–20 system and referenced to Cz) were used, in total eight channels. Anderson and Sijercic used 6 channels in [11], Ying et al. also used 6 channels [12] and Petrantonakis and Hadjileontiadis only used 3 channels [13]. A Microsoft Kinect camera was used to monitor the moves made by the humans and, in human-robot collaboration, the robot arms. The camera software stored a log file with a timestamp for each move, which move was made, and if the move was optimal (ToH has a mathematically optimal solution) or not.

The system used the ActiView recording software to record and store data from the Biosemi ActiveTwo system. ActiView, the robot control software and the Kinect camera software were running on the same computer in order to sync timestamps between the different data files.

[8] `http://www.biosemi.com/products.htm` [Accessed 27/5/2013 09:17].

All experiments were recorded using a digital camcorder placed in front and to the right of the participant and elevated to get a better overview of the experiment. The camcorder was equipped with a microphone that provided the opportunity to record utterances from the subject during the experiments in the attempt to model the oral communication between the subject and experimenter or vocalizations when the subject was alone. The same camera was also used for surveillance of the subjects and robot arms in case of any emergency.

The experiments were carried out at the Robotics and Cognition laboratory at Blekinge Institute of Technology, Karlskrona, Sweden. The Ethical Review Board in Lund, Sweden, approved all experiments (reference number 2012/737).

3.1 Experiment Procedure

A demonstration of the setup is shown in Fig. 4. It shows one of the authors sitting at the human-robot work cell with all the sensors attached.

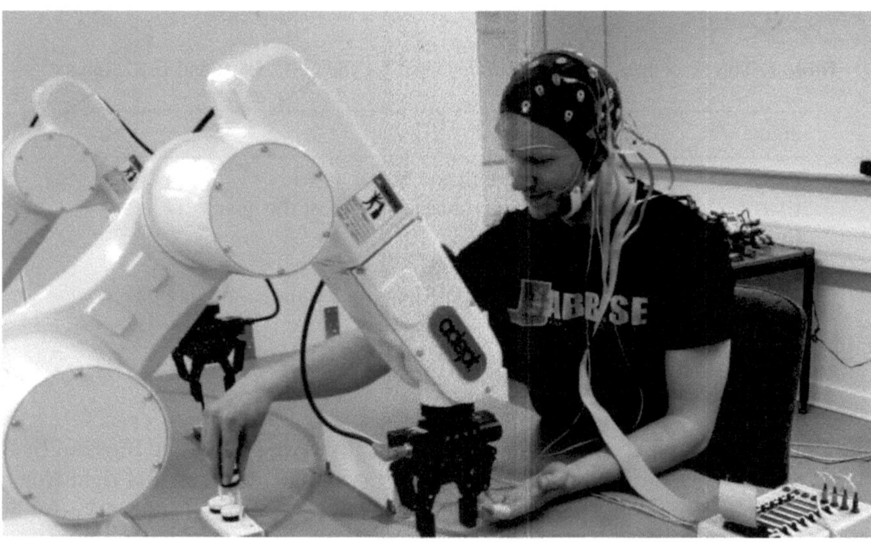

Fig. 4. One of the authors demonstrating the experimental setup

Upon arrival, the participant followed a documented procedure in order to give all participants the same treatment as far as possible:

1. After entering the lab room each participant was seated in a fixed chair at the table and faced the game task at 50–60 cm distance.
2. The participants were given written information about the experiment and a description explaining the Tower of Hanoi puzzle. Before starting the experiment session, the participants played a practice ToH game with three disks in order to get

them acquainted with the task. They were also given written information explaining psychophysiological measurements and that the data was stored anonymously. When the participants agreed to take part in the experiments they signed an informed consent form.

3. Before the experiment started participants filled in a questionnaire that included questions about age, their familiarity with the Tower of Hanoi task, board games in general, and solving mathematical problems.
4. The psychophysiological sensors measuring ECG, EMG, GSR and EEG were attached. Participants were asked to relax for four minutes in order to acquire a baseline for the psychophysiological data.
5. Each participant performed an experiment with four conditions: SH, HH, HR and HRu. The order of the conditions was varied between participants in order to minimize ordering effects. Each experiment condition was conducted as follows:
   (a) The participant played a ToH game while continuous measures of psychophysiological data and video were recorded.
   (b) After a game was finished, the participant was asked to mark his/her emotional state on the Geneva Emotion Wheel (GEW) [14]. The purpose of GEW is to log the subjective feelings of the subjects during the experiment period.
   (c) The operator instructed the participant about what task to perform next by showing information signs on a laptop placed next to the subject. The operator controlled the laptop using remote desktop.

For each of the four conditions, the participant played the ToH game three times (thus in all, a total of 12 games were played for each participant). Each experiment took around 90 minutes to complete. Note that the analysis of the GEW is out of scope for this part of the study.

In total 70 subjects participated in the study. 58 were male and 12 female. Ages varied between 19 and 31 with a mean of 23.6. The participants were students at Blekinge Institute of Technology and were recruited by advertisement in the university corridors and by recruiting during lectures. Each participant received a cinema ticket (worth approximately €9) for participating in the study.

4 Data Processing

The psychophysiological data was used to estimate the emotional states of each participant according to the two-dimensional valence and arousal scale. *Valence*, describing if the emotion is negative, positive, or neutral, and *arousal*, describing the physiological activation state of the body ranging from low to high.

Each participant data set was checked for errors. Errors were usually caused by problems like participants that did not completely understand the rules of the ToH game, or failure of one or more sensors. This caused whole or parts of the data to be invalid, for some participants. More specifically, 5% of the EMG currogator, 13.2% of the EMG zygomatic, 17.7% of the EEG, 6.1% of GSR, and 21.4% of ECG.

The signals were recorded continuously from when the sensors were attached to the subject and until the experiment ended. To analyze the relevant data, a segment for each game (three games per condition x and participant i) and a segment for the baseline period were extracted from the signals and the rest of the data (from waiting periods between games) were discarded. The first minute of the baseline period was skipped to remove initial non-relaxed activation in subjects. For the rest of the baseline period, a baseline value $b_{i,s}$ was calculated for each sensor s. The ECG data was an exception and was further processed to extract the heart rate before the baseline was calculated. To compare psychophysiological data between participants several signal features were extracted [13,15]. The features used are mean value, minimum value, maximum value and standard deviation.

For a condition x and a sensor s, the mean values $\overline{v_{x,s}}$ are defined as follows:

$$\overline{v_{x,s}} = \frac{1}{N} \sum_{i \in N} \frac{1}{|G_x|} \sum_{g \in G_x} (\overline{v_{i,g,s}} - b_{i,s}) \quad (1)$$

where N is the total number of individuals participating in the experiments, G_x is the set of games using condition x, $\overline{v_{i,g,s}}$ is the average of the values of sensor s in game g with participant i, and $b_{i,s}$ is the baseline value for i and s. The maximum value $v_{x,s}^{max}$ is calculated in the following way:

$$v_{x,s}^{max} = \max_{i \in N} \max_{g \in G_x} (\max(v_{i,g,s}) - b_{i,s}) \quad (2)$$

The minimal value is calculated in a similar way:

$$v_{x,s}^{min} = \min_{i \in N} \min_{g \in G_x} (\min(v_{i,g,s}) - b_{i,s}) \quad (3)$$

We will focus on analyzing the overall mean, minimum, maximum, and standard deviation values for each condition. Note that the minimum value is the smallest observable value in a condition, and maximum is the largest observable value. Analysis of the first and second difference is out of scope for this part of the study. Averages can be used to detect systematic changes over time, although it might miss specific spikes in otherwise relative low activity since these are averaged out. To compensate for this, minimum and maximum values were included, since the magnitudes of the largest spikes (positive and negative) are given by these. To show an example of an EMG zygomatic signal, see Fig 5.

5 Data Analysis

For each data segment the mean, minimum and maximum for each sensor were calculated and the baseline (average activation of the baseline period) was subtracted. A data segment is the sensor signal data measured during one ToH game. Each subject therefore has three segments for each condition SH, HH, HR and HRu (three games were played in each condition). The mean, min and max values for all eight EEG sensors were averaged to get single values for EEG activation.

The values for each condition were then averaged over all 70 subjects. Each condition has in total 210 data segments (70 participants and three data segments per participant). The average values over all data segments were then compared between the four different experiment conditions. The results are presented in Table 2.

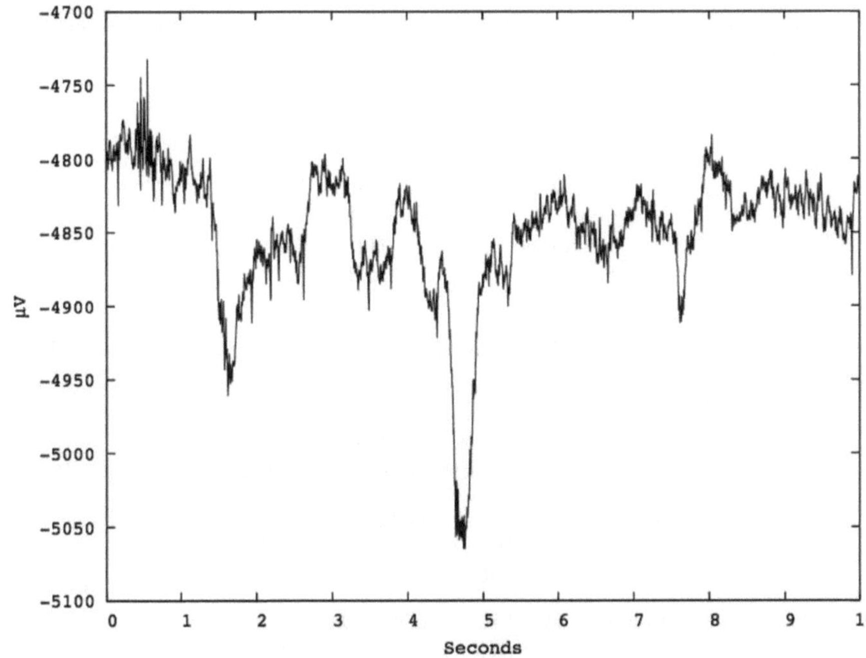

Fig. 5. A EMG zygomatic recording from one subject

Table 2. The averages of all participants' sensor data, as well as the overall minimum and maximum for each sensor and each condition

	SH				HH			
Sensor	Avg	StdDev	Min	Max	Avg	StdDev	Min	Max
EMGC	1894	4311	-9055	11060	1938	4277	-9251	11947
EMGZ	1583	3540	-5308	9665	1105	3111	-6421	8502
GSR	846.8	811.8	-985.5	3447.1	1130	774.3	-525.2	3227.6
Heart rate	2.816	5.911	-11.46	21.74	2.300	5.384	-12.44	15.26
EEG	919	2431	-3937	6600	1166	2523	-3948	7687
	HR				HRu			
	Avg	StdDev	Min	Max	Avg	StdDev	Min	Max
EMGC	1997	4051	-6443	11740	2439	4623	-7989	11966
EMGZ	1092	3555	-6931	9759	1498	3898	-6818	9853
GSR	767.7	896.6	-1281.2	3302.6	688.1	874.3	-1433.5	2791.4
Heart rate	1.427	5.783	-12.97	19.86	0.4977	4.638	-11.08	13.10
EEG	1146	2586	-3287	7425	1212	2576	-3899	7790

The averages that differed between the conditions were calculated using one-way analysis of variance (ANOVA), and are presented in Table 3. A problem with this is that ANOVA assumes that the observations should be independent of each other, which is not

Table 3. One-way analysis of variance (ANOVA) for the normalized sensor values, significant values at $\alpha = 0.01$ marked with *

Sensor	Significance
EMG corrugator	0.564
EMG zygomatic	0.405
GSR	0.000*
Heart Rate	0.001*
EEG	0.716

Table 4. Post-hoc analysis using Fisher's least significant difference for Heart Rate and GSR, significant values at $\alpha = 0.01$ marked with *

		Experiments		
Sensor	Experiment	HH	HR	HRu
GSR	SH	0.001*	0.348	0.062
	HH		0.000*	0.000*
	HR			0.345
Heart Rate	SH	0.393	0.022	0.000*
	HH		0.146	0.003*
	HR			0.122

the case here since the same people played all four conditions. The ordering effects are averted (or at least minimized) by having different orders of the various conditions, but they are still not independent of each other. To lower the risk of committing a type I error (i.e. to detect a difference where there actually is none), the more stringent $\alpha = 0.01$ was used in the analysis.

5.1 EMG Corrugator

The activity levels for the EMG sensor placed on the corrugator muscle are shown in Fig. 6. Increased activity is a good indicator for negative valence (see Section 2.1). The results indicate almost equal levels for SH, HH and HR with increased activity for HRu. This may suggest that the unpredictable robot behavior can produce negative emotions. The differences are however not significant, see Table 3.

5.2 EMG Zygomatic

The activity levels for EMG over the zygomatic muscle are depicted in Fig. 7. Increased activity is a good indicator for positive emotions (see Section 2.1). The results indicate equal activation levels for HH and HR and increased activity for SH and HRu. This may suggest that playing alone compared to playing with an experimenter produces more positive feelings, maybe due to fear of performing badly when collaborating with an experimenter. The EMG corrugator results also indicated increased activity for HRu,

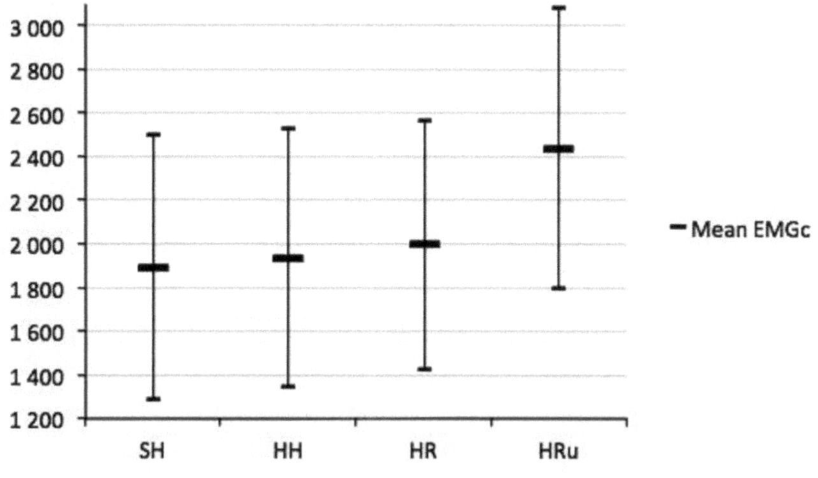

Fig. 6. Mean value (with confidence interval) for the EMG corrugator for the four different experiment conditions

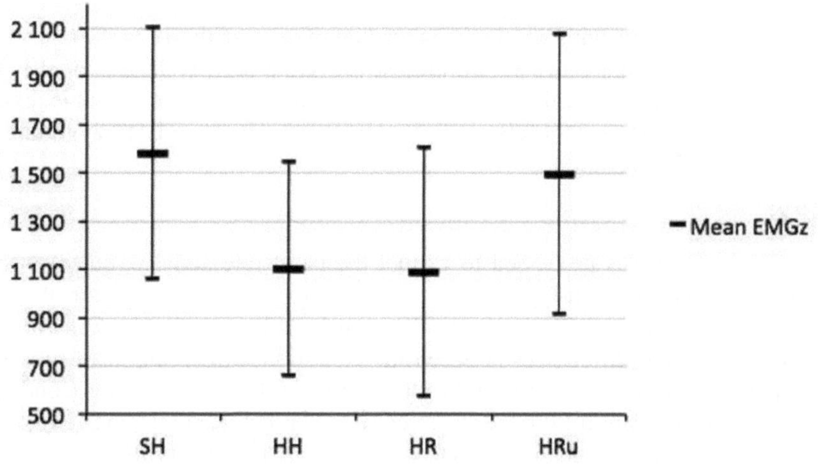

Fig. 7. Mean value (with confidence interval) for the EMG zygomatic for the four different experiment conditions

suggesting that playing with the unpredictable robot can produce both positive and negative emotions. As in the case of EMG corrugator, the differences are not statistically significant (see Table 3).

5.3 GSR

The activity levels for the GSR sensor are shown in Fig. 8. GSR is a strong indicator of arousal, especially for negative emotions (see Section 2.2). The results show high

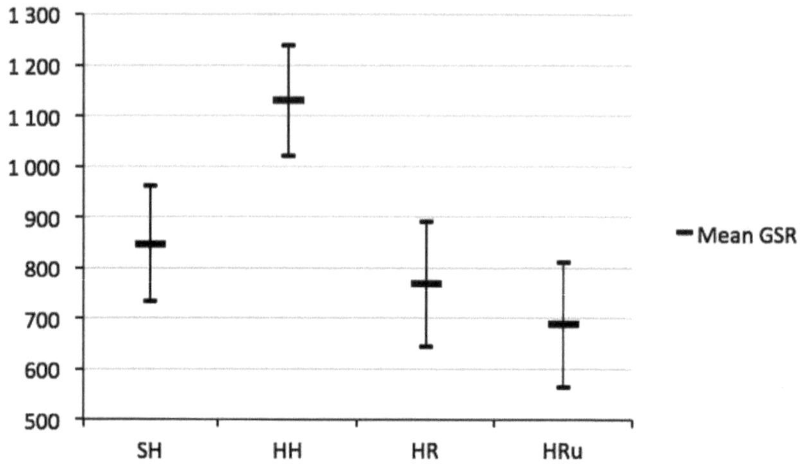

Fig. 8. Mean value (with confidence interval) for the GSR for the four different experiment conditions

activation when playing with another human (significant at the $\alpha = 0.01$ level, see Table 4), suggesting high arousal, and almost the same activity levels for the other game types. This might be an indication of increased negative stress for subjects when playing with an experimenter, as also indicated by the results from EMG zygomatic.

5.4 ECG

The raw ECG signal was processed to extract the heart rate activity (pulse) in beats per minute. This is shown in Fig. 9. Heart rate activity is a good indicator of arousal for both positive and negative valence (see Section 2.4). The results show small, but in some cases significant differences between the conditions where the heart rate in the single human condition is slightly higher than in the robot collaboration conditions. There is also a significant difference in the comparison between HH and HRu, where the subjects show a higher mean heart rate when playing together with another human, compared to playing with the robot (see Table 4 for an overview). This is an indication of increased arousal when playing alone or with another human, compared to playing with robots. This can be an indication for subjects feeling calmer when playing with robots, relying on the robots to "do the work".

5.5 EEG

The EEG activity levels averaged over all eight sensors are shown in Fig. 10. The results indicate quite similar activity levels for HH, HR and HRu and a slightly decreased activity for SH. This suggests that all tasks where the subject has to collaborate with either a human or robots require increased attention and cognitive activity (see Section 2.3). However, there are no statistically significant differences, as can be seen in Table 3.

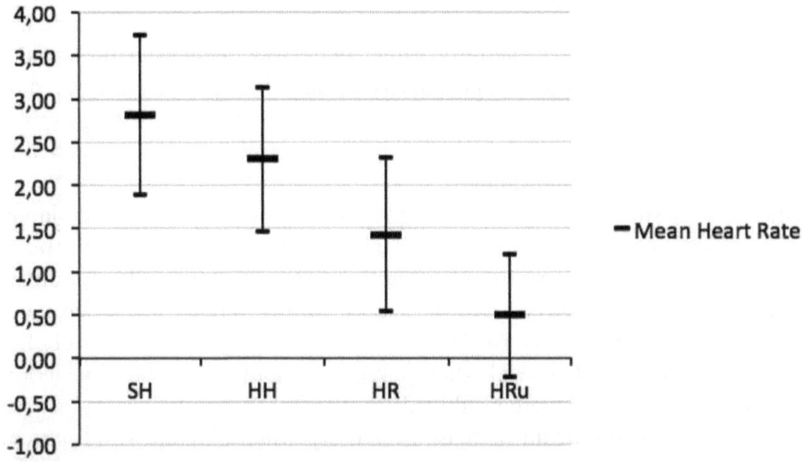

Fig. 9. Mean value (with confidence interval) for the Heart rate activity for the four different experiment conditions. Note that the values are normalized w.r.t. the baseline.

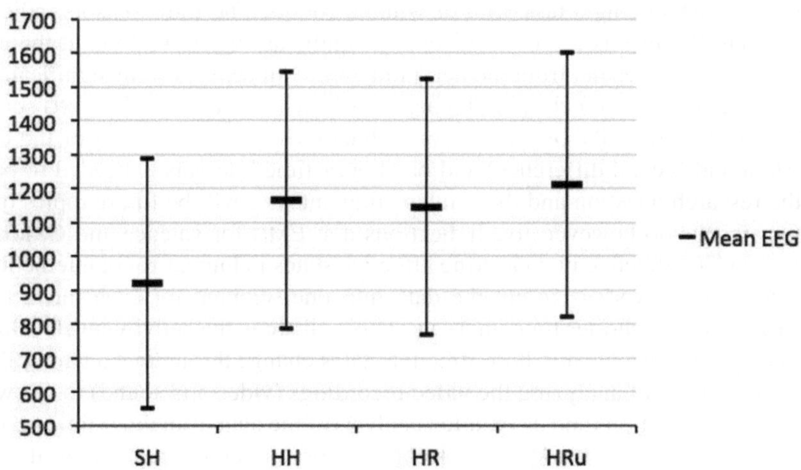

Fig. 10. Mean value (with confidence interval) for the EEG for the four different experiment conditions

6 Discussion, Conclusions and Future Work

The results presented in this part of the study are from playing a complete Towers of Hanoi game. A game often lasts several minutes and averaging the activity of a sensor over such long time periods can disguise short-term activity variations.

The two EMG sensors showed clear differences between playing alone or with another human compared to the unpredictable robot. This suggest that playing with the unpredictable robot can produce both positive and negative feelings in subjects, the differences are however not statistically significant. It is reasonable to assume that the differences are through spikes of activity, rather than slow changes. This would mean that if the EMG sensors registered high activity during short periods of time, much of that would be lost when averaging over the whole condition due to resting periods between the spikes. A more detailed analysis on shorter time intervals is needed to see short-term changes in emotional states.

The EEG sensors showed increased attention and cognitive load for all collaborative tasks, but no differences were shown between collaborating with a human or robots. As for the EMG sensors, the differences in EEG activity are not statistically significant.

The heart rate activity showed small but significant differences between the game types, with a decrease in activity for HRu compared to playing without robots. This is an indication of increased arousal when playing without robots, but does not say if positive or negative emotions were produced.

The GSR showed a significant increase in activity when playing with a human experimenter. This is a sign of increased arousal, maybe due to fear of performing badly. It does however not say if the feelings experienced were positive or negative.

The data analysis performed cannot clearly answer the research question of what psychophysiological sensors are effective in terms of measuring differences in affective states between solving a task with or without robots. The EMG sensors indicated a difference but the results are not statistically significant when we look at the average values alone. The results from the heart rate and GSR sensors were significant but showed no clear difference between playing with or without robots. The EEG sensors might be indicators of collaborative or non-collaborative tasks. Analysis of the other features (first and second difference) and on shorter time intervals is needed to better answer the research question and these interesting analysis will be future topics of research. Our results do however give indications that EMG for valence and GSR/heart rate for arousal are effective in measuring affective states in human-robot interaction.

A possible future work is to cut the data into one segment for each move to be able to do analysis for shorter time intervals. This will show the activity levels in more detail and may give insights into how affective states change throughout a game. There is also the possibility of analyzing the video recordings (video and sound) to see when participants show emotions and do detailed analysis of the data from those time periods.

Another possible future work is to correlate the psychophysiology data to the subjective feelings of participants as stated in the GEW forms. This could give valuable insights into whether participants that subjectively felt stressed in the human-robot interaction also experienced increased psychophysiological activity and therefore possibly show stronger correlations in the data between playing with or without robots.

The next step in the PsyIntEC project is to develop a human cognition and affect model (a form of knowledge base) based on the experiment results. The model will have real-time automated state updates based upon data inputs from biometric interfaces. The purpose is to detect changes in affective states in the human co-worker. An analysis of the data on shorter time intervals is needed to develop this model.

The cognition and affect model will then be used to develop an adaptive robot decision model to complete reference tasks while optimizing human safety, job satisfaction/engagement and task performance. Robot performance parameters will be modified in real time based upon the human affect model.

A future experiment is planned to compare the human affective perception of the adaptive robot decision model compared to the human-robot and human-robot unpredictable conditions used in the experiments described in this part of the study.

Acknowledgments. This work was supported by the EU Project EC FP7-ICT-231143 ECHORD.

References

1. Lucas, E.: Récreations Mathématiques, vol. 3. Gauthier Villars, Paris (1893)
2. Larsen, J., Berntson, L.G., Poehlmann, K.M., Ito, T., Cacioppo, J.: The Psychophysiology of Emotion. In: Lewis, R., Haviland-Jones, J.M., Barrett, L.F. (eds.) The Handbook of Emotions, 3rd edn. Guilford Press (2008)
3. Janke, L.: Facial EMG in an anger-provoking situation: individual differences in directing anger outwards or inwards. International Journal of Psychophysiology 23(3), 207–214 (1996)
4. Fridlund, A.J.: Evolution and facial action in reflex, social motive, and paralanguage. Biological Psychology 32(1), 3–100 (1991)
5. Wu, G., Liu, G., Hao, M.: The analysis of emotion recognition from GSR based on PSO. In: 2010 International Symposium on Intelligence Information Processing and Trusted Computing (2010)
6. Andreassi, J.: Psychophysiology: Human behavior and physiological response, 5th edn. Lawrence Erlbaum Associates Inc., US (2007)
7. Houghton, A.R., Gray, D.: Making sense of the ECG, 3rd edn. Hodder Education (2012)
8. Malmstrom, E., Opton, E., Lazarus, R.: Heart Rate Measurement and the Correlation of Indices of Arousal. Psychosomatic Medicine (1965)
9. Drachen, A., Nacke, L., Yannakakis, G., Pedersen, A.: Correlation Between Heart Rate, Electrodermal Activity and Player Experience in First-Person Shooter games. In: Proceedings of the 5th ACM SIGGRAPH Symposium on Video Games (2010)
10. Ya, X., Guangyuan, L., Min, H., Wanhui, W., Xiting, H.: Analysis of affective ECG signals toward emotion recognition. Journal of Electronics 27(1) (2010)
11. Anderson, C., Sijercic, Z.: Classification of EEG Signals from Four Subjects During Five Mental Tasks. Psychophysiology 16, 202–203 (1979)
12. Ying, N.Y., Saratchandran, P., Huang, G.B., Sundarajan, N.: Classification of Mental Tasks from EEG Signals using Extreme Learning Machine. International Journal of Neural Systems 16, 29–39 (2006)
13. Petrantonakis, P.C., Hadjileontiadis, L.J.: Emotion Recognition from Brain Signals Using Hybrid Adaptive Filtering and Higher Order Crossings Analysis. IEEE Transactions on Affective Computing 1, 81–97 (2010)
14. Scherer, K.: What are emotions? And how can they be measured? Social Science Information 44(4), 695–729 (2005)
15. Picard, R.W., Vyzas, E., Healey, J.: Toward machine emotional intelligence: analysis of affective physiological state. IEEE Transactions on Pattern Analysis and Machine Intelligence 23, 1175–1191 (2001)

Bilateral Haptic Teleoperation of an Industrial Multirotor UAV

Sammy Omari[1], Minh-Duc Hua[2], Guillaume Ducard[3], and Tarek Hamel[3]

[1] Autonomous Systems Lab (ASL), ETH Zurich, Switzerland & Skybotix AG, Luegislandstrasse 105, Zurich, Switzerland
omaris@ethz.ch

[2] Institut des Systèmes Intelligents et de Robotique (ISIR), CNRS-UPMC, France
hua@isir.upmc.fr

[3] Laboratoire d'Informatique, Signaux et Systèmes (I3S), CNRS-UNS, France
ducard(thamel)@i3s.unice.fr

Abstract. This chapter presents an intuitive laser-based teleoperation scheme to enable the safe operation of a multirotor UAV by an untrained user in a cluttered environment using a haptic joystick. An obstacle avoidance strategy is designed and implemented to autonomously modify the position setpoint of the UAV if necessary. This scheme includes a novel force-feedback algorithm to enable the user to feel surrounding environment of the UAV as well as the disturbances acting on it. The stability analysis of the whole teleoperation loop, including the nonlinear dynamics of both UAV and joystick, is provided. The implementation of the teleoperation scheme on the Flybox hexacopter platform by the company Skybotix is described. Finally, experimental results and videos are reported to demonstrate the successful implementation and the performance of the overall system.

Keywords: Aerial Robotics, Teleoperation, Haptics and Haptic Interfaces.

1 Introduction

The next generation of unmanned aerial vehicles (UAVs) will be capable of executing missions that are too dangerous, too difficult or simply impossible for humans. This is the case, for example, when it is necessary to inspect power lines, buildings, bridges, tunnels, hydraulic dam walls, pipelines, etc. Also, after a natural or industrial disaster, the inspection of the inside of a collapsed house, building or plant, by a small UAV would represent a considerable advantage in terms of time-saving, cost and human risk.

There are different modes of inspection with a UAV. 1) Fully autonomous mode: the UAV takes off, reaches a target, performs the task at target (delivers payload or inspects target), and finally returns to base. 2) Teleoperation mode: the aerial robot is remotely operated by a human pilot while the teleoperation loop is ensuring the stability and safety of the robot.

F. Röhrbein et al. (eds.), *Gearing Up and Accelerating Cross-Fertilization between Academic and Industrial Robotics Research in Europe*, Springer Tracts in Advanced Robotics 94,
DOI: 10.1007/978-3-319-03838-4_15,

The teleoperation mode is particularly relevant when the visual inspection requires a human to make the assessment. In the teleoperation scheme, a human pilot controls the UAV with an input device, such as a joystick, using visual cues. However, visual evaluation of the distance between the vehicle and the obstacles is usually difficult. Therefore, even experienced pilots may fail to safely pilot the flying robot without collisions with obstacles. Moreover, in the vicinity of structures, complex aerodynamic effects induced by strong and unpredictable wind gusts may complicate the task of the human pilot.

To promote the use of UAVs as a tool for professional inspectors, the operation of an aerial robot should be straight-forward without the need of extensive training. Therefore, the goal of an efficient teleoperation scheme is to enable an inexperienced human pilot to perform complex and accurate inspection manoeuvres without touching obstacles despite aerological disturbances. To this end, at the core of every teleoperation system, there is 1) a controller that stabilizes the UAVs (unstable) rotational and translational dynamics, 2) an obstacle avoidance algorithm to ensure safe operation in presence of obstacles 3) feedback to the user, in order to provide the user accurate information on the surroudings of the UAV.

Several teleoperation schemes have been proposed to solve these requirements [1–7]. For example, in [8] awareness of obstacles in the UAV environment is rendered to the pilot through changing the stiffness of the joystick. In [3] an artificial force field translates the environmental constraints into force-feedback in the user's joystick. In [1], haptic force feedback is generated based on optical flow measurement data. Vision from onboard cameras is used to close the teleoperation feedback loop in [9] and [10], for example. The concept of a virtual slave UAV is introduced in [5] and extended in [7] to multidimensional and underactuated case. In [7], the concept of multi-state energy tank is introduced to ensure the passivity property of the teleoperation loop, by associating every action of the slave UAV with an energy expense, made available by the multi-state energy tank. A hierarchical control strategy is employed in which the high-level controller handles the teleoperation loop, whereas the low-level controller regulates the dynamics of the vehicle. The stability of each control level ensures the stability of the complete system.

The key contribution of this chapter is the design of an intuitive laser-based teleoperation scheme to enable the safe operation of a multirotor UAV by an untrained user in a cluttered environment. This scheme includes 1) a novel, laser-based force-feedback algorithm that enables the user to feel the texture of the environment, 2) a novel mapping function that allows to teleoperate the UAV in an unlimited workspace in position control mode with a joystick which has a limited workspace, and 3) a laser-based obstacle avoidance strategy which autonomously modifies the position setpoint of the UAV, independently of the pilot's commands. Moreover, a stability analysis proves the stability of the complete teleoperation loop made of the subsystems 1) master joystick and 2) slave UAV. Additionally, the description and implementation of the hardware and software architecture used aboard the Flybox hexacopter platform by the company Skybotix is discussed. Finally, experiments demonstrate the performance of the teleoperation loop using an haptic joystick and a hexacopter UAV equipped with a 2D laser-range scanner as input for both obstacle avoidance and haptic rendering.

2 Bilateral Teleoperation Scheme

2.1 Architecture of the Teleoperation Loop

The considered bilateral haptic teleoperation scheme, as depicted in Fig. 1, consists of a fully-actuated 3 degrees of freedom (dof) haptic joystick (master) and an underactuated VTOL UAV (slave). The user interacts with the UAV and its surrounding environment using the haptic joystick by imposing a force $\mathbf{F}_h$ on it.

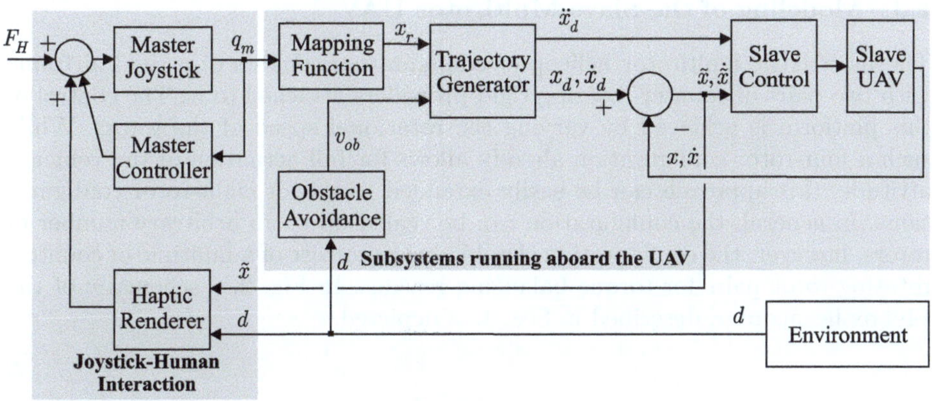

Fig. 1. Scheme of the bilateral teleoperation loop

A mapping function translates the position of the end effector of the haptic joystick $\mathbf{q}_m \in \mathbb{R}^3$ into a reference position for the UAV $\boldsymbol{x}_r \in \mathbb{R}^3$. To avoid obstacles, the mapping function also incorporates the output of the obstacle avoidance algorithm into the computation of the reference position $\boldsymbol{x}_r$. This reference position is then translated into a dynamically feasible reference trajectory that the position controller of the UAV tracks. In order to give the user some feedback about the operation of the UAV, a force vector $\mathbf{F}_e$ is haptically generated on the joystick. The force vector is a function of the surrounding obstacles, the control error, etc.

2.2 Modeling of the Master Haptic Joystick

Let $\mathbf{q}_m \in \mathbb{R}^3$, $\dot{\mathbf{q}}_m \in \mathbb{R}^3$, and $\ddot{\mathbf{q}}_m \in \mathbb{R}^3$ respectively denote the position, velocity, and acceleration of the end effector of the haptic joystick, expressed in the joystick frame $\mathcal{F}^m$. The joystick under consideration is a fully-actuated system. Thus, it can be described by the following Euler-Lagrange equation:

$$\mathbf{M}_m(\mathbf{q}_m)\ddot{\mathbf{q}}_m + \mathbf{C}_m(\mathbf{q}_m, \dot{\mathbf{q}}_m)\dot{\mathbf{q}}_m + \mathbf{g}_m(\mathbf{q}_m) = \mathbf{F}_m + \mathbf{F}_h + \mathbf{F}_e \qquad (1)$$

where $\mathbf{M}_m(\mathbf{q}_m)$ the joystick inertia matrix, $\mathbf{C}_m(\mathbf{q}_m, \dot{\mathbf{q}}_m)$ representing the Coriolis and centrifugal effects, $\mathbf{g}_m(\mathbf{q}_m)$ the vector of gravitational forces, $\mathbf{F}_m$ the local control force, $\mathbf{F}_h$ the human force acting on the joystick, and $\mathbf{F}_e$ denoting a haptic force generated on the joystick. For the sake of simplicity, the enclosed parameter(s) of $\mathbf{M}_m(\mathbf{q}_m)$, $\mathbf{C}_m(\mathbf{q}_m, \dot{\mathbf{q}}_m)$ and $\mathbf{g}_m(\mathbf{q}_m)$ are omitted subsequently. For later use, let us recall the following well-known property:

Property 1. (see [11]) The resulting matrix $\dot{\mathbf{M}}_m - 2\mathbf{C}_m$ is skew symmetric, i.e. $\forall \mathbf{x} \in \mathbb{R}^3$, $\mathbf{x}^\top(\dot{\mathbf{M}}_m - 2\mathbf{C}_m)\mathbf{x} = 0$.

2.3 Modeling of the Slave Multirotor UAV

The most basic multirotor helicopter configuration consists of a rigid airframe with two pairs of counter-rotating rigid propellers attached to it. The control of this platform is achieved by varying the rotational speed of the rotors. While such a four-rotor configuration already allows for full actuation of the vehicle's attitude, this approach can be easily extended to six- or eight-rotor configurations. In general, the configuration can be scaled up to an arbitrary number of rotors, however, the configuration should always consist of a multiple of counter-rotating rotor pairs for torque balancing reasons. In Fig. 2, a schematic of the Flybox hexacopter, described in Sec. 4, is depicted.

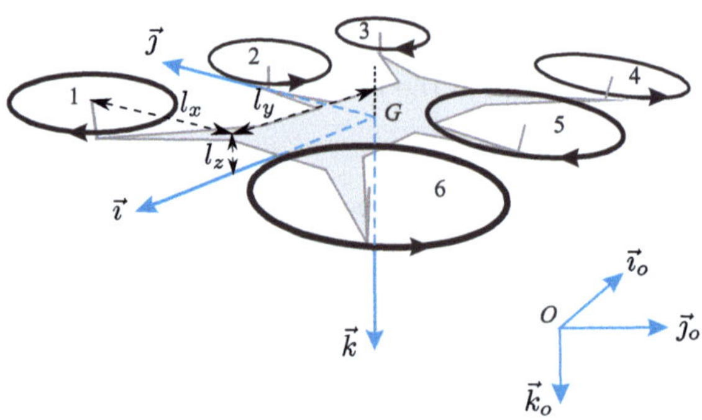

Fig. 2. Schematic representation of a six-rotor UAV platform. Inertial frame $\mathcal{F}^0$ with origin O and the body-fixed frame $\mathcal{B}$ with origin G.

The following notation is introduced. The vehicle's center of mass (CoM) is denoted as G, its mass m, and its inertia matrix $\mathbf{J}$. Let $\mathcal{F}^0 = \{O; \vec{\imath}_o, \vec{\jmath}_o, \vec{k}_o\}$ and $\mathcal{F}^s = \{G; \vec{\imath}, \vec{\jmath}, \vec{k}\}$ denote the inertial frame and the body frame attached to the vehicle, respectively. Let $\boldsymbol{x} \in \mathbb{R}^3$ denote the position of the vehicle's CoM expressed in $\mathcal{F}^0$. The rotation matrix representing the orientation of the frame $\mathcal{F}^s$ relative to the frame $\mathcal{F}^0$ is $\mathbf{R} \in SO(3)$. The vehicle's velocity and the wind velocity are both expressed in the frame $\mathcal{F}^0$ are denoted as $\dot{\boldsymbol{x}} \in \mathbb{R}^3$ and $\dot{\boldsymbol{x}}_w \in \mathbb{R}^3$, respectively. Let $\boldsymbol{\omega} \in \mathbb{R}^3$ be the angular velocity of the frame $\mathcal{F}^s$ expressed in $\mathcal{F}^s$. The canonical basis of $\mathbb{R}^3$ is denoted $\{\mathbf{e}_1, \mathbf{e}_2, \mathbf{e}_3\}$. Let $\mathbf{d}_i = [d_{1,i}\, d_{2,i}\, d_{3,i}]^\top \in \mathbb{R}^3$ be the position of the i-th rotor expressed in the body-fixed frame $\mathcal{F}^s$. We define $\mathbf{d}_i^\perp = [d_{1,i}\, d_{2,i}\, 0]^\top \in \mathbb{R}^3$ as the component of $\mathbf{d}_i$ perpendicular to $\mathbf{e}_3$. Let the thrust direction of all rotors be parallel to $\mathbf{e}_3$ in $\mathcal{F}^s$. The notation $_\times$ represents the skew-symmetric matrix associated with the cross product, i.e. $\mathbf{u}_\times \mathbf{v} = \mathbf{u} \times \mathbf{v}$, $\forall \mathbf{u}, \mathbf{v} \in \mathbb{R}^3$. The Euclidean norm in $\mathbb{R}^n$ is denoted as $|\cdot|$.

Dynamic Model of Vehicle

Following the model proposed in [26], the i-th rotor, turning at velocity ϖ_i, generates a thrust force $\mathbf{F}_{t,i} = c_T \varpi_i^2 \mathbf{e}_3$ and an aerodynamic torque $\mathbf{Q}_i = \lambda_i c_Q \varpi_i^2 \mathbf{e}_3$ with the aerodynamic constants c_T, c_Q and $\lambda_i = \{-1, 1\}$, depending on the direction of rotation of the rotor (cw: $\lambda_i = 1$, ccw: $\lambda_i = -1$). The remaining aerodynamic forces and torques (mostly due to drag by the fuselage) are summed up in a vector $\mathbf{F}_{aero} \in \mathbb{R}^3$ and $\boldsymbol{\Gamma}_{aero} \in \mathbb{R}^3$ respectively. The vehicle is subject to gravity $mg\mathbf{e}_3$.

Applying the Newton-Euler formalism, one obtains the following equations of motion of the vehicle [21]:

$$\begin{cases} m\ddot{\boldsymbol{x}} &= \mathbf{R}\sum_i \mathbf{F}_{t,i} + mg\mathbf{e}_3 + \mathbf{F}_{aero} & \text{(2a)} \\ \dot{\mathbf{R}} &= \mathbf{R}\boldsymbol{\omega}_\times & \text{(2b)} \\ \mathbf{J}\dot{\boldsymbol{\omega}} &= -\boldsymbol{\omega}_\times \mathbf{J}\boldsymbol{\omega} + \sum_i (\mathbf{Q}_i + \mathbf{d}_i \times \mathbf{F}_{t,i}) + \boldsymbol{\Gamma}_{aero} & \text{(2c)} \end{cases}$$

2.4 Mapping of the Joystick Workspace to the UAV Workspace

Because of the limited joystick workspace and the unlimited UAV workspace, recent teleoperation schemes directly map the joystick position to the velocity setpoint of the UAV [6], [7]. This comes at the cost of not being able to perform precise, position controlled flights as, for example, needed in inspection tasks. We propose a novel mapping function between the joystick workspace and the UAV workspace that overcomes this limitation and enables position controlled UAV flight in an arbitrarily large UAV workspace.

In this scheme, as shown in Fig. 3, the position of the joystick's end effector $\mathbf{q}_m$ is mapped to a reference position $\boldsymbol{x}_r$ of the slave UAV using the mapping

$$\begin{cases} \boldsymbol{x}_r(t) &= \mathbf{K}_m \mathbf{q}_m(t) \min\left(1, \frac{r_\star}{|\mathbf{q}_m(t)|}\right) + \boldsymbol{x}_c(t) \\ \boldsymbol{x}_c(t) &= \int_0^t \boldsymbol{\Psi}_m(\mathbf{q}_m(s))\, ds \\ \boldsymbol{\Psi}_m(\mathbf{q}_m(t)) &= \mathbf{K}_v \mathbf{q}_m(t) \max\left(0, 1 - \frac{r_\star}{|\mathbf{q}_m(t)|}\right) \end{cases} \quad (3)$$

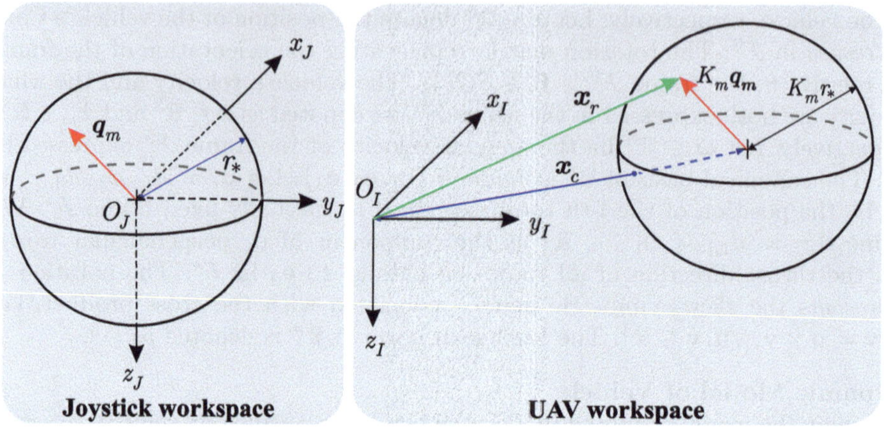

Fig. 3. Mapping of the joystick position $\mathbf{q}_m$ in the joystick workspace (left) into the position reference $\boldsymbol{x}_r$ of the UAV in the UAV workspace (right). In this figure, the diagonal scaling matrix $\mathbf{K}_m$ has identical elements.

with $\mathbf{K}_m$, $\mathbf{K}_v$ being two diagonal, positive definite scaling matrices. The term $\boldsymbol{x}_c$ is the center of operation in the UAV workspace, around which the user can operate the UAV. While the user is operating the joystick end effector closer than the distance $r_\star$ to the origin of the joystick workspace, the joystick is within the "position sphere". In this state, the center of operation $\boldsymbol{x}_c$ remains constant since $\boldsymbol{\Psi}_m(\mathbf{q}_m) = 0$. The mapping directly relates the joystick postion q_m to the UAV position setpoint around the center of operation by $\boldsymbol{x}_r = \mathbf{K}_m \mathbf{q}_m \frac{r_\star}{|\mathbf{q}_m|} + \boldsymbol{x}_c$.

When the position of the joystick $\mathbf{q}_m$ is moved outside of the "position sphere", the center of operation starts moving in the direction of $\mathbf{q}_m$ since $\boldsymbol{x}_c$ is now evolving as the integral of $\boldsymbol{\Psi}_m(\mathbf{q}_m) = \mathbf{K}_v \mathbf{q}_m (1 - \frac{r_\star}{|\mathbf{q}_m|})$. The speed at which the center of operation is moving is determined by the matrix $\mathbf{K}_v$. When the end effector is moved back inside the sphere, the term $\boldsymbol{x}_c$ remains constant again since $\boldsymbol{\Psi}_m(\mathbf{q}_m) = 0$, thus resulting in a new center of operation around which the UAV can be operated.

2.5 Trajectory Generator

The implementation of the position controller of the UAV requires a position setpoint up to its second derivative[1]. Since commercially available haptic joysticks only provide position and velocity of the end effector, the acceleration setpoint cannot be computed directly. Instead, we propose a low-pass filter to generate the setpoints

[1] For the computation of the term $\dot{\boldsymbol{\eta}}_d$ of Eq. 20, the trajectory is required up to its third derivative (see e.g. [12] for explanations). Since we do not fly acrobatic maneuvers, we can neglect this term and therefore only estimate the setpoint up to its second derivative.

$$\begin{cases} \dot{\hat{\boldsymbol{x}}}_r := \mathrm{sat}_{v_{\max}} \dot{\hat{\boldsymbol{x}}}_r \\ \ddot{\hat{\boldsymbol{x}}}_r := \mathrm{sat}_{a_{\max}} (-k_d \dot{\hat{\boldsymbol{x}}}_r - k_p(\hat{\boldsymbol{x}}_r - \boldsymbol{x}_r)). \end{cases} \tag{4}$$

with the classical saturation function $\mathrm{sat}_{\Delta}(\mathbf{x}) := \mathbf{x} \min(1, \Delta/|\mathbf{x}|)$. Using the saturation functions, we define maximal desired accelerations and velocities. Additionally, the aggressiveness of the trajectory generation can be tuned by adjusting the gains k_p and k_d.

Similarly, for the obstacle avoidance, we augment the obstacle velocity using a low-pass filter (see [4])

$$\dot{\hat{\mathbf{v}}}_{\mathrm{ob}} = -k_p \hat{\mathbf{v}}_{\mathrm{ob}} + \mathbf{v}_{\mathrm{ob}}, \tag{5}$$

where $\mathbf{v}_{\mathrm{ob}}$ is specified in the next subsection. Finally, the UAV's reference trajectory is expressed as

$$\begin{cases} \boldsymbol{x}_d = \hat{\boldsymbol{x}}_r + \int \hat{\mathbf{v}}_{\mathrm{ob}}\, dt \\ \dot{\boldsymbol{x}}_d = \dot{\hat{\boldsymbol{x}}}_r + \hat{\mathbf{v}}_{\mathrm{ob}} \\ \ddot{\boldsymbol{x}}_d = \ddot{\hat{\boldsymbol{x}}}_r + \dot{\hat{\mathbf{v}}}_{\mathrm{ob}} \end{cases} \tag{6}$$

2.6 Obstacle Avoidance

By assuming that the position error of the UAV remains small, we can work directly on the UAV setpoint as input to the obstacle avoidance scheme. The underlying idea of the obstacle avoidance scheme is to reshape the velocity setpoint of the UAV in a way that it avoids the surrounding obstacles [4].

For the operation of the obstacle avoidance algorithm, we assume to have a knowledge of the metric distance of the UAV setpoint to the surrounding obstacles. In this approach, the obstacles are represented using a sparse cloud of obstacle features. The obstacle features can come from a variety of sensors, e.g. measurement data from an onboard laser scanner or the point correspondences of a sparse monocular or stereo vision SLAM algorithm [16].

As depicted in Fig. 4, for each obstacle feature that is closer than some distance $d_i < d_\star$ from the reference position $\mathbf{x}_d$, a repelling velocity is computed as

$$\mathbf{v}_{rep,i} = -\chi(d_i)\boldsymbol{\eta}_i, \tag{7}$$

where $\boldsymbol{\eta}_i$ is the unit vector pointing from the reference position $\mathbf{x}_d$ to the obstacle feature. The function $\chi(d_i)$ is a smooth, non-increasing function that approaches infinity for d_i approaching the radius r_u of the UAV. An example of such a function is provided in Section 4.3. The resulting reference obstacle avoidance velocity v_{ob} is computed as the average of all repelling velocities

$$\mathbf{v}_{ob} = \frac{1}{N} \Sigma_{i=1}^{N} \mathbf{v}_{rep,i}. \tag{8}$$

Remark 1. In a particular case where only one planar obstacle (wall) is detected, the average repelling velocity can be approximated by $\mathbf{v}_{ob} = -\chi(d)\boldsymbol{\eta}$, where d is the distance between the reference position to the wall, and $\boldsymbol{\eta}$ is the unit normal

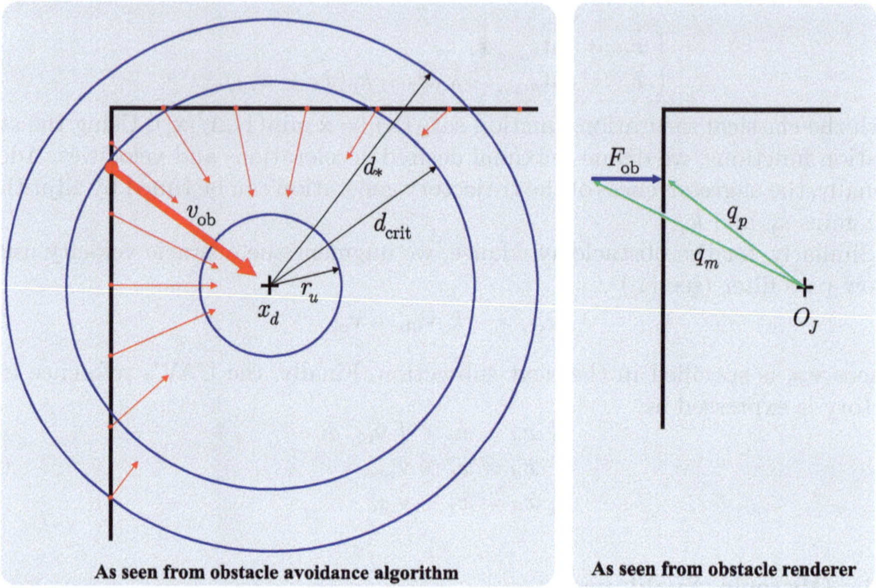

Fig. 4. Top-Down view of scenario where UAV is operated close to a corner. The left plot displays the obstacle avoidance algorithm. Each laser measurement (red dot) inside the circle with radius $d_\star$ generates a repulsive velocity $\mathbf{v}_{\text{rep}}$ (red vector). The resulting obstacle velocity $\mathbf{v}_{ob}$ (bold red vector) is the average of all $\mathbf{v}_{\text{rep}}$. On the right, the same scenario as seen from the obstacle renderer in the joystick workpsace. The user tries to penetrate the virtual obstacle by setting $\mathbf{q}_m$ inside the obstacle. The resulting haptic feedback $\mathbf{F}_{ob}$ corresponds to a stiff spring between $\mathbf{q}_m$ and its projected position on the obstacle surface $\mathbf{q}_p$.

vector pointing towards the wall and expressed in the frame $\mathcal{F}^i$. Noting that $\dot{d} = -\dot{\mathbf{x}}_d^\top \boldsymbol{\eta}$ and denoting $\sigma := -\hat{\mathbf{v}}_{ob}^\top \boldsymbol{\eta}$, one deduces from Eqs. (5) and (6) that

$$\begin{cases} \dot{d} = -\dot{\hat{\mathbf{x}}}_r^\top \boldsymbol{\eta} + \sigma \\ \dot{\sigma} = -k_p \sigma + \chi(d) \end{cases} \tag{9}$$

From these equations, using the boundedness of $-\dot{\hat{\mathbf{x}}}_r^\top \boldsymbol{\eta}$ and fact the non-increasing function $\chi(d)$ tends to infinity when d tends to r_u, one can prove the existence of a positive number $d_{min} > r_u$ such that $d(t) > d_{min}$, $\forall t > 0$, provided that $d(0) > d_{min}$. This means that the collision of the reference UAV and the wall is avoided. The proof of the property is based on a Lyapunov argument and can be found in [4]. On the other hand, one ensures that $\chi(d)$ and, consequently, v_{ob} remain bounded. Besides, one can also easily deduce from Eq. (9) the boundedness of σ, $\dot{d}$ and, consequently, of $\dot{\mathbf{v}}_{ob}$. Then, from Eqs. (5) and (6), one also ensures the boundedness of $\dot{\mathbf{v}}_{ob}$, $\dot{\mathbf{x}}_d$, $\ddot{\mathbf{x}}_d$, and $\mathbf{x}_d^{(3)}$.

2.7 Haptic Rendering

In earlier works on bilateral haptic teleoperation (e.g. [3], [5], [6]), the environment was haptically rendered to the user using a potential wall function. The result was a sluggish and soft sensation of the environment since the user could only feel a gradually increasing force when approaching an obstacle. Therefore, we design a different strategy to give the user the sensation of feeling the rigid environment.

In a first step, a polygonal 3D model is generated from the obstacle features. The obstacle model is then mapped into the joystick workspace using the linear mapping

$$\mathbf{v}_j = \mathbf{K}_m^{-1}(\mathbf{v}_w - \mathbf{x}_d), \tag{10}$$

with $\mathbf{v}_w$ being a vertex of the polygon in the world frame and $\mathbf{v}_j$ the corresponding vertex in the joystick frame. When the user now penetrates this virtual object with the end effector, a stiff spring pulls the end effector back to the surface. This gives the user the sensation of touching the real environment as if he was located directly on top of the position reference of the UAV. This method is closely related to the god-object rendering method that was proposed in [17] to give the user of virtual reality simulation a haptic sensation of the virtual rigid objects. We define the obstacle force $\mathbf{F}_{ob}$ as

$$\mathbf{F}_{ob} = -\text{sat}_{\Delta_{ob}} k(\mathbf{q}_m - \mathbf{q}_p), \tag{11}$$

with $k >> 1$ and $\mathbf{q}_p$ the projected position of $\mathbf{q}_m$ on the surface. A graphical representation of the rendering process is displayed in Fig. 4. For the computation of $\mathbf{q}_p$, the reader is referred to [17].

On top of the obstacle force, we propose to haptically display the position control error as a spring force:

$$\mathbf{F}_{err} = -\text{sat}_{\Delta_1} k_1(\mathbf{x} - \mathbf{x}_d). \tag{12}$$

Using this spring force, the user can feel the inertia of the UAV when changing the setpoint. In general, all effects that cause a momentary position control error, such as external disturbances, are displayed using this force.

When doing a transition of the joystick's end effector from inside to outside the "position sphere", the user should feel a sensation that resembles the penetration of a membrane using a needle. Therefore, we construct a membrane force $\mathbf{F}_{\text{mem}}$, as depicted in Fig. 5, with which the user only feels the resistance of the virtual membrane when going outside the sphere but not when entering it back.

In the end, the environment force $\mathbf{F}_e$ is constructed as the sum of all forces:

$$\mathbf{F}_e = \mathbf{F}_{err} + \mathbf{F}_{ob} + \mathbf{F}_{mem}. \tag{13}$$

At this point, it is to be noted that no component of the teleoperation scheme requires the use of a haptic joystick for proper operation. Since we use a velocity setpoint shaping approach for obstacle avoidance, the position reference of the UAV will be pushed away from the obstacle irrespective of the force feedback.

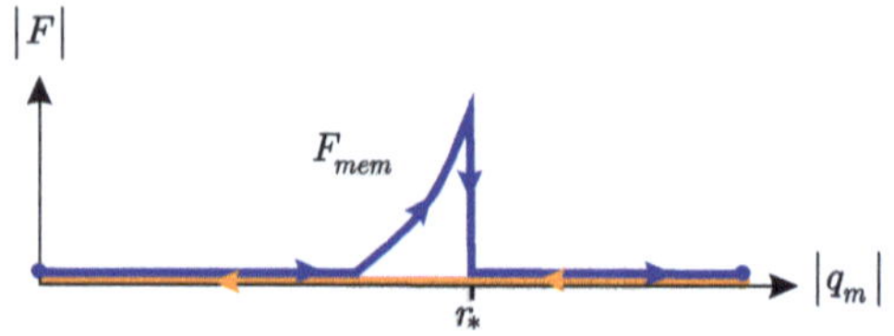

Fig. 5. Force profile to render the virtual membrane sensation

Therefore, the proposed teleoperation scheme can be interfaced and controlled using any input device such as a remote control or a tablet pc, in those cases without the need of generating haptic feedback.

3 Control Design

3.1 Control of the Master Joystick

The local control force $\mathbf{F}_m$ of the master joystick is designed in order to ensure that the joystick's end effector is pushed back into the "position sphere" when the human and environmental forces are null (i.e. $\mathbf{F}_h = \mathbf{F}_e = 0$). Consequently, if the user releases the joystick, the UAV is stabilized in position. The following control expression of $\mathbf{F}_m$ is proposed:

$$\begin{aligned}\mathbf{F}_m =& -(\mathbf{C}_m\boldsymbol{\Lambda}_1 + \lambda_2\mathbf{I}_3)\boldsymbol{\Psi}_m(\mathbf{q}_m) \\ &- \left(\mathbf{M}_m\boldsymbol{\Lambda}_1\frac{\partial\boldsymbol{\Psi}_m(\mathbf{q}_m)}{\partial\mathbf{q}_m} + \lambda_2\boldsymbol{\Lambda}_1^{-1}\right)\dot{\mathbf{q}}_m + \mathbf{g}_m,\end{aligned} \tag{14}$$

with $\boldsymbol{\Lambda}_1 \in \mathbb{R}^{3\times3}$ a positive diagonal gain matrix and $\lambda_2 \in \mathbb{R}$ a positive gain.

3.2 Control of the Slave UAV

Model for Control Design and Rate Control

The UAV model (2) can be rewritten as

$$\begin{cases}\Sigma_1 : \begin{bmatrix} m\ddot{\boldsymbol{x}} \\ \dot{\mathbf{R}}\end{bmatrix} = \begin{bmatrix} -T\mathbf{R}\mathbf{e}_3 + \mathbf{F}_{ext} \\ \mathbf{R}\boldsymbol{\omega}_\times\end{bmatrix} & \text{(15a)} \\ \Sigma_2 : \mathbf{J}\dot{\boldsymbol{\omega}} = -\boldsymbol{\omega}_\times\mathbf{J}\boldsymbol{\omega} + \boldsymbol{\Gamma} + \boldsymbol{\Gamma}_{ext} & \text{(15b)}\end{cases}$$

where $\mathbf{F}_{ext}$ and $\boldsymbol{\Gamma}_{ext}$ are the sum of all the acting forces and moments on the vehicle except the thrust force $T = c_T\sum_i \varpi_i^2$ and the torque $\boldsymbol{\Gamma} = \sum_i \lambda_i c_Q \varpi_i^2\mathbf{e}_3 - c_T\varpi_i^2\mathbf{d}_i^{\perp} \times \mathbf{e}_3$ generated by the rotors.

One can view $T \in \mathbb{R}^+$ and $\boldsymbol{\Gamma} \in \mathbb{R}^3$ as control inputs of System (15). For N mounted rotors, we can derive a linear mapping from the square of the propellers' angular velocity to the total thrust T and torque $\boldsymbol{\Gamma}$:

$$\begin{bmatrix} T \\ \boldsymbol{\Gamma} \end{bmatrix} = \begin{bmatrix} c_T & c_T & \dots & c_T \\ c_T\, d_{2,1} & c_T\, d_{2,2} & \dots & c_T\, d_{2,N} \\ -c_T\, d_{1,1} & -c_T\, d_{1,2} & \dots & -c_T\, d_{1,N} \\ \lambda_1\, c_Q & \lambda_2\, c_Q & \dots & \lambda_N\, c_Q \end{bmatrix} \begin{bmatrix} \varpi_1^2 \\ \varpi_2^2 \\ \vdots \\ \varpi_N^2 \end{bmatrix} \tag{16}$$

If $N = 4$ one can determine the desired angular rates of the rotors by inverting (16). When the UAV is actuated by more than four rotors, the set of equations (16) is overdetermined and the Moore-Penrose pseudo-inverse method can be used to determine the desired angular velocities of the propellers [23].

System (15) shows full actuation of the rotational dynamics and underactuation of the translational dynamics. For the rotational motion, exponential convergence of the angular velocity $\boldsymbol{\omega}$ to any bounded desired value $\boldsymbol{\omega}_d$ is easy to obtain, since the subsystem Σ_2 is fully actuated and the angular velocity vector $\boldsymbol{\omega}$ can be measured at high frequency from embedded gyrometers. A possible control solution is [27]

$$\boldsymbol{\Gamma} = -K_\omega(\boldsymbol{\omega} - \boldsymbol{\omega}_d) + \boldsymbol{\omega}_d \times \mathbf{J}\boldsymbol{\omega} + \mathbf{J}\dot{\boldsymbol{\omega}}_d, \tag{17}$$

with a sufficiently large diagonal positive gain matrix K_ω to dominate the disturbance torque $\boldsymbol{\Gamma}_{ext}$. From here on, all attention of control design can be given to the control of the subsystem Σ_1 using T and $\boldsymbol{\omega} \equiv \boldsymbol{\omega}_d$ as control inputs.

UAV Control Law

The control law applied to the UAV is designed such that the UAV's position x_s is stabilized at the desired value $\mathbf{x_d}$ defined by (6) regardless of the dynamics of the master joystick. This controller is inspired by the one proposed in [12]. Let us denote $\tilde{\mathbf{x}} = \mathbf{x} - \mathbf{x}_d$ and $\dot{\tilde{\mathbf{x}}} = \dot{\mathbf{x}} - \dot{\mathbf{x}}_d$, the position and velocity error variables, respectively. Define $\boldsymbol{\eta} := \mathbf{R}\mathbf{e}_3$, and

$$\gamma := h_p(|\tilde{\mathbf{x}}|)\tilde{\mathbf{x}} + h_v(|\dot{\tilde{\mathbf{x}}}|)\dot{\tilde{\mathbf{x}}} + \frac{\mathbf{F}_{ext}}{m} - \ddot{\mathbf{x}}_d, \tag{18}$$

with $h_p(\cdot)$ and $h_v(\cdot)$ some smooth bounded positive functions defined on $[0, +\infty)$ such that for some positive constants α_i, β_i (with index i being either p or v) (see e.g. [12])

$$|h_i(s^2)s| < \alpha_i,\ 0 < \frac{\partial}{\partial s}(h_i(s^2)s) < \beta_i,\ \forall s \in \mathbb{R}.$$

An example for the functions $h_p(\cdot)$ and $h_v(\cdot)$ is provided in Section 4.5.

Proposition 1. *Apply the control law*

$$\begin{cases} \mathbf{T} = m|\boldsymbol{\gamma}| \\ \boldsymbol{\omega}_{1,2} = \left(\mathbf{R}^\top \left(k_\eta \dfrac{\boldsymbol{\eta} \times \boldsymbol{\eta}_d}{(1+\boldsymbol{\eta}^\top \boldsymbol{\eta}_d)^2} - (\boldsymbol{\eta}_\times)^2(\boldsymbol{\eta}_d \times \dot{\boldsymbol{\eta}}_d)\right)\right)_{1,2} \end{cases} \tag{19}$$

to system (15a), *with* $\boldsymbol{\eta}_d := \frac{\gamma}{|\gamma|}$, k_η *a positive gain, and the notation* $\mathbf{u}_{1,2} = (u_1, u_2)^\top$, $\forall \mathbf{u} \in \mathbb{R}^3$. *Assume that*

- $\dot{\mathbf{x}}_d$, $\ddot{\mathbf{x}}_d$, $\mathbf{x}_d^{(3)}$ *are known and bounded;*
- γ *is always different from the null vector;*
- ω_3 *is bounded.*

Then, the equilibrium point $(\tilde{\mathbf{x}}, \dot{\tilde{\mathbf{x}}}, \boldsymbol{\eta}) = (0, 0, \boldsymbol{\eta}_d)$ *of the controlled system is almost-globally asymptotically stable provided that* $\boldsymbol{\eta}(0) \neq -\boldsymbol{\eta}_d(0)$.

Proof. Using (2) and (18) one verifies that the translational error dynamics satisfies

$$\begin{aligned} \ddot{\tilde{\mathbf{x}}} &= -h_p(|\tilde{\mathbf{x}}|)\tilde{\mathbf{x}} - h_v(|\dot{\tilde{\mathbf{x}}}|)\dot{\tilde{\mathbf{x}}} - \frac{1}{m_s} T\boldsymbol{\eta} + \gamma \\ &= -h_p(|\tilde{\mathbf{x}}|)\tilde{\mathbf{x}} - h_v(|\dot{\tilde{\mathbf{x}}}|)\dot{\tilde{\mathbf{x}}} + \boldsymbol{\varepsilon} \end{aligned} \tag{20}$$

with $\boldsymbol{\varepsilon} := |\gamma|(\boldsymbol{\eta} - \boldsymbol{\eta}_d)$. First, according to [12] one ensures the almost exponential convergence of $\boldsymbol{\eta}$ to $\boldsymbol{\eta}_d$. This is based on the analysis of the storage function $\mathcal{V}_\eta = 1 - \boldsymbol{\eta}^\top \boldsymbol{\eta}_d$ whose time-derivative along any solution to the controlled system satisfies (see [12])

$$\dot{\mathcal{V}}_\eta = -k_\eta \frac{|\boldsymbol{\eta} \times \boldsymbol{\eta}_d|^2}{(1 + \boldsymbol{\eta}^\top \boldsymbol{\eta}_d)^2}.$$

Due to the fact thatγ is bounded, it follows that ε is bounded and converges exponentially to zero. Consequently, application of Lemma 1 in [15] on Eq. (20) ensures the convergence of $\tilde{\mathbf{x}}$ and $\dot{\tilde{\mathbf{x}}}$ to zero.

3.3 Stability of the Teleoperation Loop

The stability of the teleoperated system is studied next. First, we show that the master system is input-to-state stable (see [13]) in the presence of bounded operator force $\mathbf{F}_h$ and environment force $\mathbf{F}_e$. Finally, for the free moving systems where the master and slave systems operate in free space, i.e. $\mathbf{F}_h = 0$ and $\mathbf{F}_e = \mathbf{F}_{err}$, we show that the joystick's position will be pushed back to the "position sphere" of radius $r_\star$ and the slave UAV will asymptotically stop. For simplicity, the effect of the membrane force $\mathbf{F}_{mem}$ is neglected in the stability analysis of the free moving systems case.

Proposition 2. *Consider the teleoperation system with the master system* (1) *controlled by the controller* (14) *and the slave system* (15a) *controlled by the controller* (19). *Assume that all assumptions in Proposition 1 are satisfied. Then, the results of Proposition 1 hold. Furthermore,*

- *In the case where the human and environment forces are bounded (i.e.* $\exists \alpha_h$ *and* α_e *such that* $|\mathbf{F}_h(t)| \leq \alpha_h$ *and* $|\mathbf{F}_e(t)| \leq \alpha_e$, $\forall t$*), the master system is input-to-stable stable (I.S.S.) with respect to* $\mathbf{F}_h$ *and* $\mathbf{F}_e$.

– *In the case of free moving systems (i.e.* $\mathbf{F}_h = 0$, $\mathbf{F}_e = \mathbf{F}_{err}$, $\mathbf{v}_{ob} = 0$*), one ensures the existence of a constant vector* $\mathbf{q}_\infty \in \mathbb{R}^3$ *such that* $|\mathbf{q}_\infty| \leq r_\star$ *and* $\lim_{t\to\infty} \mathbf{q}_m(t) = \mathbf{q}_\infty$*. Moreover, the UAV will asymptotically stop.*

Proof. Consider the following non-negative function

$$\begin{aligned}\mathcal{V}_m =& \frac{1}{2}(\dot{\mathbf{q}}_m + \boldsymbol{\Lambda}_1 \boldsymbol{\Psi}_m(\mathbf{q}_m))^\top \mathbf{M}_m(\dot{\mathbf{q}}_m + \boldsymbol{\Lambda}_1 \boldsymbol{\Psi}_m(\mathbf{q}_m)) \\ &+ 2\lambda_2 \mathbf{K}_m \int_0^{|\mathbf{q}_m(t)|} \max(0, s - r_\star) ds\end{aligned} \tag{21}$$

The time-derivative of $\mathcal{V}_m$ satisfies

$$\begin{aligned}\dot{\mathcal{V}}_m =& (\dot{\mathbf{q}}_m + \boldsymbol{\Lambda}_1 \boldsymbol{\Psi}_m(\mathbf{q}_m))^\top \mathbf{M}_m(\ddot{\mathbf{q}}_m + \boldsymbol{\Lambda}_1 \frac{\partial \boldsymbol{\Psi}_m(\mathbf{q}_m)}{\partial \mathbf{q}_m}\dot{\mathbf{q}}_m) \\ &+ \underbrace{\frac{1}{2}(\dot{\mathbf{q}}_m + \boldsymbol{\Lambda}_1 \boldsymbol{\Psi}_m)^\top \dot{\mathbf{M}}_m(\dot{\mathbf{q}}_m + \boldsymbol{\Lambda}_1 \boldsymbol{\Psi}_m)}_{=(\dot{\mathbf{q}}_m + \boldsymbol{\Lambda}_1 \boldsymbol{\Psi}_m(\mathbf{q}_m))^\top C_m(\dot{\mathbf{q}}_m + \boldsymbol{\Lambda}_1 \boldsymbol{\Psi}_m(\mathbf{q}_m))} \\ &+ \underbrace{k_X \mathbf{K}_m \max(0, |\mathbf{q}_m| - r_\star)\frac{\mathbf{q}_m^\top \dot{\mathbf{q}}_m}{|\mathbf{q}_m|}}_{=2\lambda_2 \boldsymbol{\Psi}_m(\mathbf{q}_m)^\top \dot{\mathbf{q}}_m} \\ =& (\dot{\mathbf{q}}_m + \boldsymbol{\Lambda}_1 \boldsymbol{\Psi}_m)^\top (\mathbf{F}_e + \mathbf{F}_h - \lambda_2(\boldsymbol{\Lambda}_1^{-1}\dot{\mathbf{q}}_m + \boldsymbol{\Psi}_m)) + 2\lambda_2 \boldsymbol{\Psi}_m^\top \dot{\mathbf{q}}_m \\ =& -\lambda_2 \dot{\mathbf{q}}_m^\top \boldsymbol{\Lambda}_1^{-1} \dot{\mathbf{q}}_m - \lambda_2 \boldsymbol{\Psi}_m(\mathbf{q}_m)^\top \boldsymbol{\Lambda}_1 \boldsymbol{\Psi}_m(\mathbf{q}_m) \\ &- (\dot{\mathbf{q}}_m + \boldsymbol{\Lambda}_1 \boldsymbol{\Psi}_m(\mathbf{q}_m))^\top (\mathbf{F}_e + \mathbf{F}_h)\end{aligned} \tag{22}$$

In view of (22) and the quadratic form of (21) there exists some positive constants α_1, α_2 such that

$$\dot{\mathcal{V}}_m \leq -\alpha_1 \mathcal{V}_m + \alpha_2(|\mathbf{F}_e| + |\mathbf{F}_h|)$$

From here, if the environment force $\mathbf{F}_e$ and the human force are bounded, then the function $\mathcal{V}_m$ remains bounded. This in turn implies the boundedness of the master state variables $\mathbf{q}_m$ and $\dot{\mathbf{q}}_m$. Besides, if $\mathbf{F}_e$ and $\mathbf{F}_h$ converge to zero (or are equal to zero), the application of the singular perturbation theorem (see e.g. [13]) ensures the convergence of $\mathcal{V}_m$ to zero which in turn implies the convergence of $\boldsymbol{\Psi}_m(\mathbf{q}_m)$ and $\dot{\mathbf{q}}_m$ to zero. This is the input-to-state stability property of the master loop with respect to $\mathbf{F}_h$ and $\mathbf{F}_e$.

Now, consider the case of free moving systems where the human does not act on the joystick (i.e. $\mathbf{F}_h = 0$) and the slave is not in contact with the environment (i.e. $\mathbf{v}_{ob} = 0$, $\mathbf{F}_e = \mathbf{F}_{err}$). As a result of Proposition 1, the UAV's position error remains bounded and converges to zero, i.e. $\lim_{t\to\infty} \tilde{\mathbf{x}}(t) = 0$. This is independent on the master joystick's dynamics. From here one deduces that the environment force $\mathbf{F}_e$ converges to zero since $\mathbf{F}_e = -\mathrm{sat}_{\Delta_1}(k_1 \tilde{\mathbf{x}})$. As a consequence, one deduces (as proved previously) that $\dot{\mathbf{q}}_m$ and $\boldsymbol{\Psi}_m(\mathbf{q}_m)$ converge to zero which means that the joystick's end effector converges to the "position sphere" and

Fig. 6. The Flybox hexacopter by Skybotix equipped with a Hokuyo UTM-30LN laser scanner. A mirror, mounted on top of the laser, is used to deflect some rays towards the ground for altitude estimation. The low-level autopilot and the high-level computer are inside the housing.

asymptotically stops. Then, by definition (3) the reference position x_r specified by the joystick also converges to a constant value. Since $\mathbf{x}_r$ tends to a constant value, its augmented value $\hat{\mathbf{x}}_r$ converges to it. Besides, since $\mathbf{v}_{ob} = 0$ one has $\dot{\hat{\mathbf{v}}}_{ob} = -\kappa\hat{\mathbf{v}}_{ob}$. This implies the exponential convergence of $\hat{\mathbf{v}}_{ob}$ to zero. From here, using the definition (6) of $\mathbf{x}_d$ one deduces that $\mathbf{x}_d$ tends to a constant value. Finally, the UAV controller ensures that it will asymptotically stop at $\mathbf{x}_d$.

4 Implementation

4.1 System Setup

The teleoperation setup consists of

- The ground station (GS) computer.
- The 3 DoF fully-actuated haptic joystick Novint Falcon.
- The hexacopter UAV platform Flybox by Skybotix.
- The 2D laser range scanner Hokuyo UTM-30LN.

The Flybox UAV, shown in Fig. 6, is equipped with a low-level (LL) autopilot and a high-level (HL) computer. The LL autopilot is built around a Cortex M3 32-bit microprocessor and is equipped with a custom-made IMU. It controls the UAVs attitude by tracking desired thrust vector $\boldsymbol{\eta}_d$ from the HL computer. The autopilot also provides attitude and inertial sensor information at 1 kHz to the HL computer. The HL computer is an off-the-shelf Atom 1.6 Ghz Single-Core computer running Ubuntu 12.04 and consuming less than 7 W.

The Hokuyo UTM-30LN laser scanner is rigidly fixed to the UAV and is connected to the HL computer via USB. The laser scanner provides 1080 points per scan up to 30 m in a 270 degree window at 40 Hz.

The ground station is connected to the HL computer using WiFi. The Robotic Operating System (ROS) [14] is used as a communication layer between GS and HL and to run the different components of the teleoperation scheme. All the system-critical tasks, such as obstacle avoidance and control algorithms, are run aboard the UAV, in order to ensure the proper operation of the UAV even in case of loss of communication link between GS and HL computer. The GS only interfaces the joystick in order to 1) provide the haptic feedback to the user and 2) send the position and velocity setpoints from the joystick to the HL computer.

4.2 Laser Preprocessing

The onboard 2D laser scanner is used as input to the obstacle avoidance scheme as well as the haptic rendering loop. Both algorithms require a spatial model of the environment. While there are algorithms available to build 3D models from 2D laser range data, they come with the drawback that they are either not suited for real-time use [19] or only offer a coarse spatial resolution [18]. As a trade-off between speed and spatial resolution, it was decided to generate a 2D environment model by assuming that the environment consists of vertical walls. Using this assumption, the laser range measurements can be projected down from the UAV frame onto the x-y plane in the inertial frame using the attitude information from the onboard IMU. It is assumed that the ground coincides with the x-y plane in the inertial frame. As a consequence, laser measurements intersecting the ground plane are rejected, as in the case when the UAV is tilted and flying close to ground.

4.3 Obstacle Avoidance

The repelling velocity of each obstacle feature in the x-y plane of the inertial frame is computed using Eq. 7. For the derivation of the function $\chi(d)$, we assume to have a circular UAV with radius r_u[2]. We set

$$\chi(d) = \begin{cases} v_{\max}\frac{d-d_\star}{d_{\text{crit}}-d_\star} & \text{if } d_{\text{crit}} < d < d_\star \\ v_{\max}\frac{d_{\text{crit}}-r_u}{\max(\epsilon,d-r_u)} & \text{if } d < d_{\text{crit}} \end{cases} \tag{23}$$

with $v_{\max}$ being the maximal allowed translational velocity of the trajectory generator and d_{crit} the distance at which we want the vehicle to stop. The term ϵ is present in (23) to prevent a division by zero. Let us quickly consider the 1D case where the UAV is perpendicularly approaching an infinitely long static vertical wall. Since the maximal allowed velocity is $v_{\max}$, the UAV will come to a stop at latest at d_{crit} because at this point $\chi(d_{\text{crit}}) = v_{\max}$ (see [4] for proof). However, since the UAV is operated in 3D, the average of all repulsive velocities

[2] When only considering obstacle avoidance during translational motion, this approach can be easily extended for non-circular UAVs by approximating them using an ellipsoid. However, in this case, obstacle avoidance during a yawing motion is not guaranteed with the current scheme.

are acting on the UAV. As a consequence, the UAV might get closer to the obstacle than d_{crit}. Therefore, when the $d < d_{\text{crit}}$, we shape $\chi(d)$ as a hyperbola approaching infinity as $d \to r_u$. This ensures that the position reference $\mathbf{x}_r$ will, under no circumstances, get closer to an obstacle than r_u.

Since the laser scanner only covers 270 degrees field of view around the UAV, special precautions have to be taken when the user commands a position change into this blind spot. In such a case, the UAV is rotated first until the demanded position setpoint lies in a field of view of 180 degrees. Only then is the UAV allowed to approach the setpoint. The reduction from 270 to 180 degrees is necessary for safety reasons since the UAV is not a point and could therefore hit an obstacle if no safety margin would be introduced.

4.4 Haptic Rendering

The haptic rendering loop runs at 1 kHz to provide the user with a believable haptic sensation. To interface the Novint Falcon haptic joystick, the open source library *HAPI* by SenseGraphics is used. The library provides an interface where we can set the sum of the force vectors $\mathbf{F}_e$ defined by (13) and $\mathbf{F}_m$ defined by (14) as input. Note that here the matrix $\mathbf{C}_m$ is set equal to zero for implementation simplicity. Finally, the total force vector is internally mapped into the corresponding motor torques of the joystick.

The library also provides an implementation of the god-object rendering algorithm. Using the projected laser measurements on the x-y plane in the inertial frame, it is straightforward to generate a polygonal 3D model with vertical surfaces.

4.5 Control

The LL autopilot developed by the company Skybotix tracks the defined thrust vector γ using Eq. 19. As a consequence, when implementing the position controller, the UAV can be considered a fully-actuated point-mass with 3 DoF force control inputs. This is a valid assumption, as long as the time scale separation between attitude and position controller is ensured. As a rule of thumb, the time constant of the attitude controller should be one magnitude larger than the time constant of the position controller.

Following Eq. 18, we define the thrust vector as

$$\gamma = \text{sat}_{\Delta_p}(k_p \tilde{\mathbf{x}}) + \text{sat}_{\Delta_v}(k_d \dot{\tilde{\mathbf{x}}}) + g\mathbf{e}_3 - \ddot{\mathbf{x}}_d \tag{24}$$

An integral term could be included in the expression. The interested reader is referred to [12]. The control gains are determined via a pole placement procedure performed on the linearized system of system (2) at hovering. Details on the gain-tuning process can be found in [20, Ch. 2].

The gain k_p of the trajectory generator relates to the responsiveness of the generator to position inputs. The gain k_d relates to the damping of the generator and is set to $k_d = 2\sqrt{k_p}$ to ensure critically-damped trajectory tracking without overshoot.

5 Experimental Results

The performance of the proposed teleoperation scheme is evaluated via several experiments in an indoor environment. For the stabilization of the UAV position, a Vicon motion tracking system is used. While the position of the UAV could be stabilized using the laser scanner, we use the Vicon system instead, since we want to evaluate the performance of the teleoperation scheme rather than the laser position estimator. For laser-based UAV stabilization we refer the reader to [22] and to the widely used open source implementation of the algorithms [24].

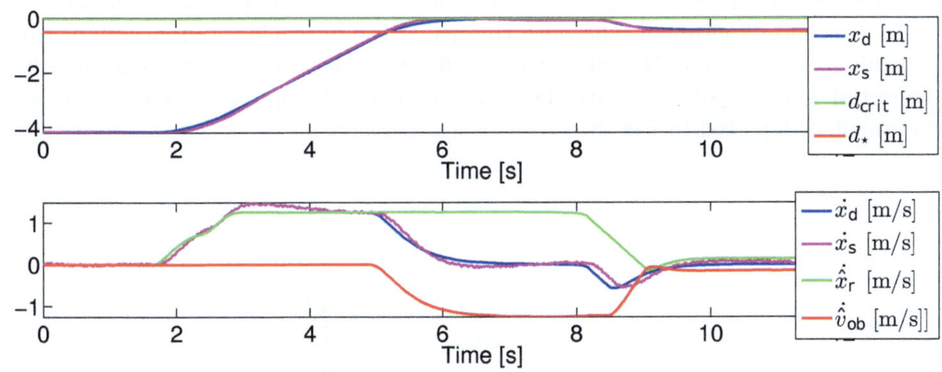

Fig. 7. 1D representation of UAV approaching a static vertical wall. The upper plot represents the approach of the UAV position setpoint $\mathbf{x}_d$ (blue) towards the wall at $\mathbf{x} = 1\,m$. The red and green line in the upper plot represent the distance $d_\star$ and d_{crit} from the wall. The middle plot shows the velocity input from the user $\hat{\dot{\mathbf{x}}}_r$ (green), the obstacle velocity $\hat{\dot{\mathbf{v}}}_{ob}$ (red) and the resulting reference velocity for the UAV $\dot{\mathbf{x}}_d$ (blue). The magenta line corresponds to the position $\mathbf{x}$ and velocity $\dot{\mathbf{x}}$ of the UAV.

The UAV velocity is estimated from the Vicon position data using a linear observer. The onboard laser scanner is used for both obstacle avoidance and haptic rendering. The numerical values used in the teleoperation scheme for the experiments are depicted in Table 5. A video recording of all experiments are on the homepage [25].

Table 1. Numerical values of the obstacle avoidance scheme used in the experiments

$d_\star$	d_{crit}	r_u	$v_{\max}$
$1.5\,m$	$1.0\,m$	$0.5\,m$	$2.0m/s$

In a first experiment, we evaluate the performance of the obstacle avoidance algorithm when approaching a vertical wall perpendicularly. The wall is at $\mathbf{x} = 1\,m$ in the Vicon frame. A 1D representation of the experiment is depicted in Fig. 7.

The experiment starts with the UAV approaching the wall from $4\,m$. At time $t = 5.0\,s$, the UAV setpoint gets closer than $d_\star$ to the obstacle. The obstacle avoidance starts injecting a repelling velocity $\dot{\hat{\mathbf{v}}}_{ob}$ in normal direction of the wall. Then, the translation of the UAV setpoint in direction of the wall comes to a stop at $t = 6.5\,s$ at distance d_{crit} without any oscillation. Now, the repelling velocity $\dot{\hat{\mathbf{v}}}_{ob}$ is counteracting the user defined velocity $\dot{\hat{\mathbf{x}}}_r$. At $t = 8.0\,s$, the user realeases the joystick. The UAV is pushed back to distance $d_\star$ because of the velocity term $\dot{\hat{\mathbf{v}}}_{ob}$. Concerning the performance of the position controller, the UAV is able to track the position reference without overshoot. The lowest plot shows the angular control error of thrust vector $\boldsymbol{\eta}_d$ and $\boldsymbol{\eta}$. Attitude control is asymptotically stable as required for the derivation of the position controller.

In a second experiment, the performance of the obstacle avoidance algorithm is evaluated when entering a narrow hallway. This is often problematic with potential-wall like obstacle avoidance approaches. A top-down view of the experiment is depicted in Fig. 8.

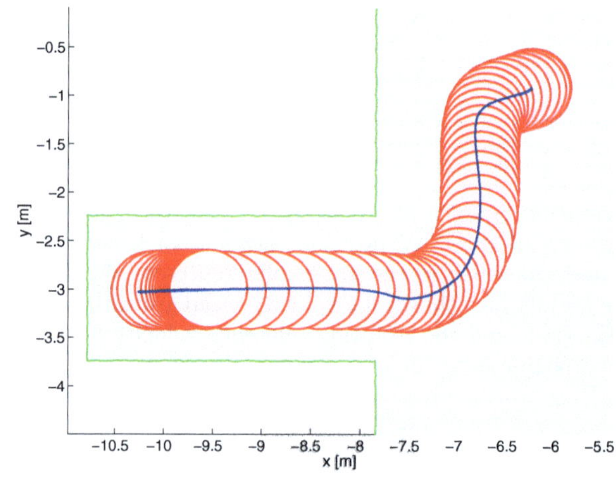

Fig. 8. Top-Down view of UAV entering a narrow hallway with a width of $1.5\,m$. The green line represents the laser scan taken at startup of the experiment. The red spheres represent snapshots of the position setpoint $\mathbf{x}_d$ in $0.5\,s$ intervals. The blue line is the trajectory of the UAV position $\mathbf{x}_s$.

As the UAV is approaching the wall, the obstacle avoidance starts injecting a repelling velocity $\dot{\hat{\mathbf{v}}}_{ob}$ in normal direction of the wall. As a consequence, the position setpoint is sliding along the wall until it enters the hallway opening. When the UAV is inside the hallway, it is able to fly at maximal velocity in direction along the hallway. This is due to the fact that the average repelling velocity has no component along the hallway when the UAV is completely in this one. When the end of the hallway is reached, the user releases the joystick. The UAV position setpoint stabilizes at the position sufficiently away from the obstacles.

We also evaluate the performance of the obstacle avoidance algorithm when exposed to moving human obstacles. For space reasons, we omit this experiment in the chapter and refer the reader to the video on [25].

6 Conclusions

In this chapter, a safe teleoperation scheme has been presented for a wide range of VTOL UAVs operated by untrained pilots. A force-feedback algorithm generates a force to a haptic joystick that enables the user to feel the texture of the environment. Although the joystick has a limited workspace, a novel mapping function enables the teleoperation of the UAV in an unlimited workspace in position control mode. The obstacle avoidance strategy presented autonomously modifies the position setpoint of the UAV independently of the pilot's commands. The stability analysis of the whole teleoperation loop is proven. Experiments showed the successful teleoperation of a UAV using an haptic joystick and a hexacopter UAV equipped with a 2D laser-range scanner.

Acknowledgment. This work was supported by the EU project EC FP7-ICT-231143 ECHORD. Additionally, the authors acknowledge the financial support by the SCAR ANR-ASTRID fund.

References

1. Mahony, R., Schill, F., Corke, P., Oh, Y.S.: A new framework for force feedback teleoperation of robotic vehicles based on optical flow. In: IEEE Int. Conf. on Robotics and Automation, pp. 1079–1085 (2009)
2. Schill, F., Mahony, R., Corke, P., Cole, L.: Visual Force Feedback Teleoperation of the Insectbot Using Optic Flow. In: Australian Conf. on Robotics and Automation (2009)
3. Lam, T.M., Boschloo, H.W., Mulder, M., van Passen, M.M.: Artificial Force Field for Haptic Feedback in UAV Tele-operation. IEEE Trans. on Systems, Man, and Cybernetics- Part A: Systems and Humans 39(6), 1316–1330 (2009)
4. Hua, M.-D., Rifai, H.: Obstacle Avoidance for Teleoperated Underactuated Aerial Vehicles using Telemetric Measurement. In: IEEE Conf. on Control and Decision, pp. 262–267 (2010)
5. Stramigioli, S., Mahony, R., Corke, P.: A Novel Approach to Haptic Teleoperation of Aerial Robot Vehicles. In: IEEE Int. Conf. on Robotics and Automation, pp. 5302–5308 (2010)
6. Rifai, H., Hua, M.-D., Hamel, T., Morin, P.: Haptic-based Bilateral Teleoperation of Underactuated Unmanned Aerial Vehicles. In: IFAC World Congress, Italy, pp. 13782–13788 (2011)
7. Mersha, A.Y., Stramigioli, S., Carloni, R.: Bilateral Teleoperation of Underactuated Unmanned Aerial Vehicles: The virtual Slave Concept. In: IEEE Int. Conf. on Robotics and Automation, Minnesota, pp. 4614–4620 (2012)
8. Lam, T.M., Mulder, M., Van Paassen, M.M.: Collision Avoidance in UAV Teleoperation with Time Delay. In: IEEE Int. Conf. on Systems, Man and Cybernetics, pp. 997–1002 (2007)

9. Mettler, B., Andersh, J., Papanikolopoulos, N.: A First Investigation into the Teleoperation of a miniature Rotorcraft. In: Khatib, O., Kumar, V., Pappas, G.J. (eds.) Experimental Robotics. STAR, vol. 54, pp. 191–199. Springer, Heidelberg (2009)
10. Suzuki, R., Matsumoto, T., Konno, A., Hoshino, Y., Go, K., Oosedo, A., Uchiyama, M.: Teleoperation of a Tail-sitter VTOL UAV. In: IEEE/RSJ Int. Conf. on Intelligent Robots and Systems, pp. 1618–1623 (2010)
11. Spong, M.W., Hutchinson, S., Vidyasagar, M.: Robot modeling and control. John Wiley & Sons, Inc. (2006)
12. Hua, M.-D., Hamel, T., Morin, P., Samson, C.: A Control Approach for Thrust-Propelled Underactuated Vehicles and its Application to VTOL Drones. IEEE Trans. on Automatic Control 59(8), 1837–1853 (2009)
13. Khalil, H.K.: Nonlinear Systems, 3rd edn. Prentice Hall (2002)
14. Quigley, M., Gerkey, B., Conley, K., Faust, J., Foote, T., Leibs, J., Berger, E., Wheeler, R., Ng, A.: ROS: An open-source robot operating system. In: ICRA Workshop on Open Source Software (2009)
15. Abdessameud, A., Tayebi, A.: Formation control of VTOL Unmanned Aerial Vehicles with communication delays. Automatica 47, 2383–2394 (2011)
16. Weiss, S., Achtelik, M., Kneip, L., Scaramuzza, D., Siegwart, R.: Intuitive 3D Maps for MAV Terrain Exploration and Obstacle Avoidance. Journal of Intelligent and Robotic Systems 61, 473–493 (2011)
17. Zilles, C.B.: A constraint-based god-object method for haptic display. In: IEEE Int. Conf. on Intelligent Robots ans Systems, pp. 146–151 (1995)
18. Dryanovski, I., Morris, W., Xiao, J.: Multi-Volume Occupancy Grids: an Efficient Probabilistic 3D Mapping Model for Micro Aerial Vehicles. In: IEEE Int. Conf. on Intelligent Robots ans Systems, pp. 1553–1559 (2010)
19. Holenstein, C., Zlot, R., Bosse, M.: Watertight surface reconstruction of caves from 3D laser data. In: IEEE Int. Conf. on Intelligent Robots ans Systems, pp. 3830–3837 (2011)
20. Hua, M.-D.: Contributions to the Automatic Control of Aerial Vehicles. Ph.D. thesis, University of Nice-Sophia Antipolis (2009)
21. Hua, M.-D., Hamel, T., Morin, P., Samson, C.: A control approach for thrust-propelled underactuated vehicles and its application to VTOL drones. IEEE Transactions on Automatic Control 54(8), 1837–1853 (2009)
22. Achtelik, M., Bachrach, A., He, R., Prentice, S., Roy, N.: Stereo vision and laser odometry for autonomous helicopters in GPS-denied indoor environments. In: Unmanned Systems Technology XI (2009)
23. Ducard, G., Hua, M.-D.: Discussion and practical aspects on control allocation for a multi-rotor helicopter. In: Conference on Unmanned Aerial Vehicle in Geomatics, pp. 1–6 (2011)
24. Dryanovski, I., Morris, W., Xiao, J.: An open-source pose estimation system for micro-air vehicles. In: IEEE Int. Conf. on Intelligent Robots ans Systems, pp. 4449–4454 (2011)
25. I3S Institute Homepage, `http://tinyurl.com/i3s-cnrs` (accessed 11:16 CET, May 30, 2013)
26. Hamel, T., Mahony, R., Lozano, R., Ostrowski, J.: Dynamic modelling and configuration stabilization for an X4-flyer. In: IFAC World Congress, pp. 200–212 (2002)
27. Rudin, K., Hua, M.-D., Ducard, G., Bouabdallah, S.: A Robust Attitude Controller and its Application to Quadrotor Helicopters. In: IFAC World Congress, pp. 10379–10384 (2011)

Multimodal Interfaces to Improve Therapeutic Outcomes in Robot-Assisted Rehabilitation

Loredana Zollo[1], Eugenia Papaleo[1], Luca Spedaliere[1], Eugenio Guglielmelli[1], Francisco Javier Badesa[2], Ricardo Morales[2], and Nicolas Garcia-Aracil[2]

[1] Laboratory of Biomedical Robotics and Biomicrosistems, Università Campus Bio-Medico di Roma
{l.zollo,e.papaleo,l.spedaliere,e.guglielmelli}@unicampus.it
[2] Virtual Reality and Robotics Laboratory, Universidad Miguel Hernández de Elche
{fbadesa,rmorales,nicolas.garcia}@umh.es

Abstract. This chapter intends to provide a description of the MAAT experiment in the framework of the ECHORD European project. The experiment is aimed at developing a novel robotic system for upper-limb rehabilitation, capable of maximizing patient motivation and involvement in the therapy and performing a continuous assessment of the progress of the patient recovery in a multimodal way. The key-issue of the MAAT approach is to include the patient in the control loop by means of multimodal patient data (biomechanical as well as physiological data) and an immersive virtual reality system. To this purpose, a bio-cooperative controller is developed that incorporates multimodal data and adaptively and dynamically change the complexity of the therapy and of the virtual environment in accordance with specific patient requirements and abilities. Two MAAT robotic platforms have been developed for the experimental validation of the proposed approach. They consist of the same multimodal interface and differ in the used robotic arm in charge of delivering the therapy. Preliminary experimental data on healthy subjects are reported in this chapter. The application to stroke patients is envisaged.

Keywords: Rehabilitation robotics, Human-robot interaction, Bio-cooperative control, Multi-modal interfaces.

1 Introduction

This chapter is focused on the description of the MAAT experiment, carried out in the framework of the European project ECHORD [1]. It is aimed at developing a robotic system for upper-limb rehabilitation able to maximize patient motivation and involvement in the therapy and performing a continuous assessment of the progress of the recovery from the functional and neurological viewpoint, with special attention to the issue of safety in human-robot interaction. The main novelty of the MAAT approach is to include the patient in the control loop and use multisensory data (i.e. biomechanical and physiological data) to adaptively and dynamically change the complexity of the therapy and real-time

F. Röhrbein et al. (eds.), *Gearing Up and Accelerating Cross-Fertilization between Academic and Industrial Robotics Research in Europe*, Springer Tracts in Advanced Robotics 94,
DOI: 10.1007/978-3-319-03838-4_16, © Springer International Publishing Switzerland 2014

display an immersive virtual reality environment according to specific patient requirements.

The rationale for the systematic application of robotics to rehabilitation directly originates from previous findings in medical science that clearly demonstrates how physical exercises, based on voluntary movements, are able to produce significant clinical results in motor recovery. Repetitive physical exercises actively executed by the patients promote the neurogenesis process [2] [3] and functional recovery after traumatic injury of the central nervous system [4].

The use of robotic devices as a possible rehabilitation strategy to achieve motor recovery can be justified because of its potential impact on better therapeutic treatment and motor learning. A number of studies have demonstrated the effectiveness of repetitive grasp and release exercises [5], constraint induced therapy for the paretic limb [6], increased intensity or duration of therapy including external manipulation [7] [8], bio-feedback [9], bilateral movement training [10] and robot-assisted therapy [10-16] in restoring motor function of the paretic upper limb, during acute and chronic stages of stroke recovery. Moreover, it is shown that the therapeutic approach needs to be well-structured and repetitive in order to promote cortical reorganization after stroke [18] [19].

The MAAT experiment intends to maximize the effectiveness of the robot-aided therapy by promoting the active role of the patients in the therapy through their inclusion in the control loop. To this purpose, the rehabilitation robotic system needs to be highly adaptive to patient specific needs, thus delivering a patient-tailored therapy. The ultimate twofold goal is: (i) To maximise patients motivation and involvement in the therapy by favouring their active role and promote functional recovery by performing assisted activities of daily living. In order to enhance the active role of the patient, by enabling even severely impaired patients to actively intervene in the decision and execution of a motor task, the robot has to assist the subject to carry out the part of the task that he/she is not able to autonomously perform, with a level of assistance that can be adapted to subject residual abilities. Moreover, in order to promote patient involvement, the system provides two kinds of feedback: visual feedback (e.g. the direct observation of patient movement and end-point target) and acoustic feedback (i.e. motivational and relaxing sounds, consistent with patient state). (ii) To continuously assess the progress of the recovery by means of a multimodal interface. Biomechanical and physiological data are used to extract a quantitative measure of patient global state and correspondingly change robot behaviour during the therapeutic treatment.

Two different versions of the multimodal robotic platform have been developed in the European countries involved in the experiment, i.e. Italy and Spain. Each of them consists of:

- one industrial robotic arm for robot-aided therapy administration,
- a set of wearable sensors for biomechanical performance (i.e. behavioral) and physiological analysis,
- a virtual reality system.

The two platforms incorporate the same components, except for the robot arm. Two different industrial robotic arms have been selected and analysed in this study in the experimental validation of the proposed MAAT approach.

The chapter is structured as follows. Section 1 explains the background and the pursued methodology; Section 2 illustrates the functional architecture of the MAAT system, explaining in detail the multimodal human-machine interface and the bio-cooperative controller. The experimental validation of the proposed system is presented in Section 3; conclusions and future work are illustrated in Section 4.

Background. Several examples of clinical studies based on the application of exoskeletal or end-effector robotic machines to stroke therapy can be found in the literature [20], e.g. the ARM Guide [21] [22], the CBM-MOTUS [23], the MEMOS [24], the MIME [15] [16] [25], the MIT-Manus [12] [26] and InMotion3 (i.e. MIT wrist robot) [27] for wrist rehabilitation.

Robot-aided rehabilitation therapy usually consists of a robot guiding the patient to accomplish a predetermined motor exercise, by typically waiting for the onset of voluntary patient motion. However, patient effort and/or physiological measures are not used to modulate robot behavior and are generally considered in the administration of the robotic therapy through a direct observation of the therapist. In addition, the therapist applies tactile, verbal and visual stimuli for motivating the patient and possibly achieving better therapeutic outcomes. With respect to the usual robot-aided rehabilitation, MAAT wants to focus on the specificity of the therapy and the adaptability of the robotic system to patient features in terms of motion capabilities as well as physiological properties, including fatigue and emotions. To do that, on one hand, the behaviour of robotic system needs to be changed during the therapy through the control system and, on the other hand, motivational and stimulating rewards can be provided by means of an immersive virtual reality and tactile and auditory stimulation systems.

A few examples of rehabilitation robotic systems incorporating patient performance or physiological signals in the robot control module can be found in the literature [27-35]. The first work dates back to several years ago [28]. It proposed a *performance-based* impedance control algorithm using the active power of patient and motion accuracy measures to define the level of robot assistance at the beginning of a new block of exercises; it was conceived for planar analytical exercises with the MIT-Manus robot. Performance measures were carried out on a set of reaching movements in different directions. In [29] a different approach was proposed to adjust robot joint stiffness, borrowed from studies on the biological mechanisms that regulate the elastic properties of the human arm. Later works by Novak and Koenig [32-33] proposed to use psychophysiological measurements of patient state and task performance analysis to adjust the difficulty of a task performed with the assistance of the Haptic Master robot; the results showed that patient psychophysiological measurements alone cannot be efficiently used to update the robot behavior; they can provide just supplementary information with respect to motor data. In the European project MIMICS, a system for robot-assisted gait rehabilitation was developed; it employed physiological measurements for assessing the level of patient frustration and/or boredom and accordingly adapting the virtual reality task difficulty or the challenge of the assigned exercise. A further work on the combination of patient physiological and motor measurements for changing the level of assistance is being published [37].

There is a clear evidence that the physiological activity associated with the affective state can be differentiated and systematically organized: the transition from one emotional state to another, e.g. from state of boredom to state of anxiety, is accompanied by dynamic shifts in indicators of Autonomic Nervous System (ANS) activity. Furthermore, the use of objective, accurate and repeatable metrics for assessing patient motor performance can be fruitfully exploited for real-time adapting robot behavior [38] [39].

Hence, the solution proposed in MAAT goes beyond the perceptual, reasoning and learning capabilities of the current robotic systems used in the robot-aided rehabilitation: the robotic controller developed within MAAT is able to adjust autonomy and complexity of the robotic therapy in accordance with the evaluation of patient global state and needs. The additional advantage of monitoring physiological signals is to guarantee that the user is trained inside a safe region where physiological values vary in an appropriate range.

2 Materials and Methods

The architecture of the MAAT system is modular, as shown in Figure 1, and organized as follows:

1. the therapist chooses and selects motor task features before starting the therapy;
2. a bio-cooperative controller, composed of the modules for high-level control and low-level robot control, receives inputs from the multimodal interface (in addition to therapist settings) and dynamically updates the virtual reality software and the controller of the robot arm based on the current measure of patient state;
3. a multimodal human-machine interface (i.e. HMI) real-time records patient physiological and behavioral (motion and force) data and stimulates the patient through encouraging audio and visual feedback provided through the virtual reality software;
4. the patient executes the task shown in the virtual reality environment assisted by the robotic device.

2.1 Multimodal HMI

The multimodal HMI is composed of:

- robot position sensors,
- robot force sensors,

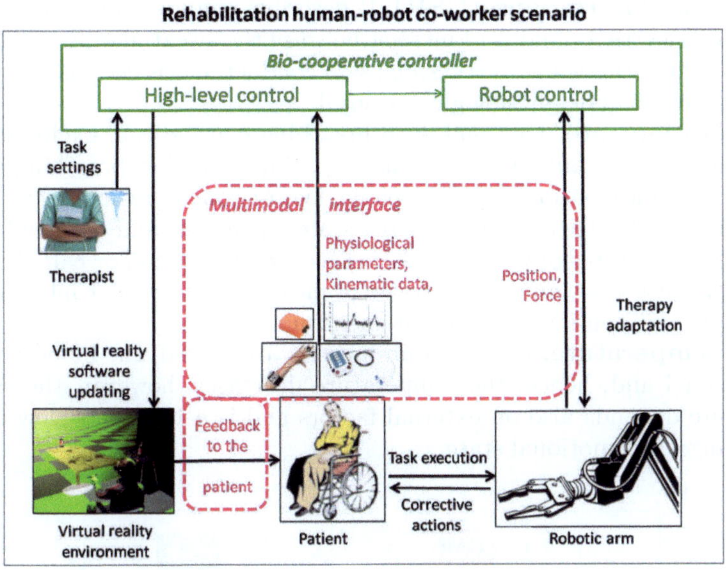

Fig. 1. MAAT system overview

- wearable magneto-inertial sensors,
- physiological sensors for measuring heart rate, skin temperature, galvanic skin response and respiration rate,
- a monitor with integrated speakers for the visual and auditory feedback provided by the virtual reality software.

HMI components can be grouped into three main modules for: (i) analysis of patient physiological state through the physiological sensors; (ii) analysis of patient motor behaviour through position, force and magneto-inertial sensors; (iii) auditory and visual feedback provided through the virtual reality software for increasing patient motivation.

Monitoring Patient Physiological State. It is well known that most physiological processes produce signals of several types: biochemical - in the form of hormones; electrical - in the form of current; physical - in the form of pressure and temperature. These changes are automatically controlled by the nervous system, which manages heart muscles, smooth muscle and various glands in our body. These bodily reactions, and corresponding bio-signals, can be measured and monitored. Moreover, the body reaction can be monitored using sensors and associated to emotions.

The module for monitoring patient physiological state, shown in Figure 2, carries out acquisition and processing of physiological signals to estimate user's physiological state. After a revision of the scientific literature, galvanic skin response, respiration rate, pulse and skin temperature were used in MAAT experiment to estimate user physiological state:

- **Galvanic skin response(GSR).** It describes changes in the skin ability to conduct electricity and is of interest because the sweat glands are controlled by the sympathetic nervous system; skin conductance is used as an indication of psychological or physiological arousal.
- **Respiration rate.** Fast and deep breathing can indicate excitement, such as anger or fear, but sometimes also joy. Rapid shallow breathing can indicate tense anticipation including panic, fear or concentration. Slow and deep breathing indicates a relaxed resting state while slow and shallow breathing can indicate states of withdrawal, passive like depression or calm happiness.
- **Pulse.** Pulse wave analysis is a way of measuring heartbeat intervals and is a simple and non-invasive measurement method.
- **Skin temperature.** Under strain muscles are tensed, the blood vessels are contracted and, hence, the temperature decreases; however, the skin temperature depends also on external factors and is a relatively slow indicator of changes in emotional state.

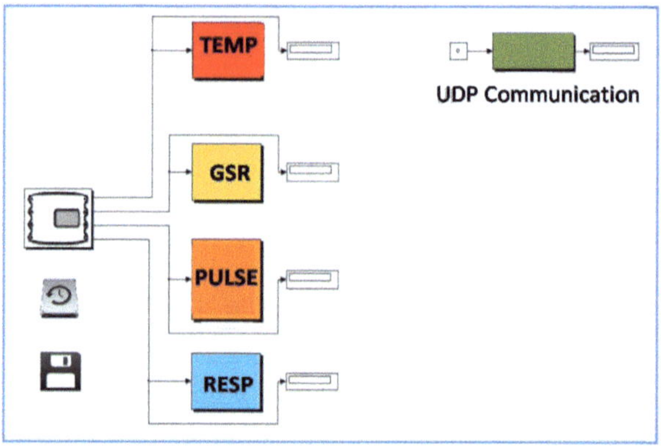

Fig. 2. Detail of MATLAB/Simulink scheme to acquire and monitor the physiological signals

Measuring Patient Biomechanical Performance. This module carries out a biomechanical evaluation of patient behaviour, using data acquired by robot position and force sensors and magneto-inertial sensors. It is composed of three parts (Figure 3): i) the first reconstructs the entire patient upper-limb kinematics, ii) the second part computes biomechanical indicators of patient performance, iii) the last one fuses the computed indicators for extracting patient global performance. These three parts are explained in detail in the following.

In order to evaluate patient biomechanics in a comprehensive way, it was chosen to reconstruct upper-limb kinematics in the joint space, in addition to kinematic information in the Cartesian space provided by the rehabilitation robot. In this way, patient motion performance can be measured through quantitative indicators in both joint space and Cartesian space and planning capability in both spaces could be investigated.

An inverse kinematics technique, based on the Augmented Jacobian matrix, was developed to reconstruct arm joint motion. It offers the main advantage of enabling an accurate reconstruction of joint motion in the space without resorting to obtrusive sensory systems. Indeed, it requires a minimum set of additional measures, with respect to hand position data provided by the robot, to solve redundancy of the 7-DoF upper-limb kinematic chain; they are: (a) hand orientation in the space, provided by a magneto-inertial sensor placed on the patient hand; (b) the elbow angle, needed to solve the kinematic redundancy and computed through upper-arm acceleration data provided by an accelerometer placed on patient upper-arm. Further details on the analytical procedure can be found in [40].

The quantitative evaluation of patient biomechanical behaviour is obtained through a set of indicators accounting for kinematic as well as dynamic aspects of patient motion. In order to update the bio-cooperative controller during the training, they need to be on-line computed [41].

The kinematic indicators are extracted from the sensory signals coming from the encoders, embedded in the robot, the magneto-inertial sensor and the accelerometer placed on the patient arm, after evaluating robot forward kinematics and patient joint angles. They are conceived to assess the following movement features: a) accuracy and direction, b) smoothness and c) inter-joint coordination.

In detail, the kinematic indicators are

- *Aiming angle* (α): it is the angle between the target direction and the direction of travel from the starting point up to peak speed point [39]; this index accounts for both movement accuracy and direction in the Cartesian space.
- *Mean-Arrest-Period-Ratio (MAPR)*: it is defined as the proportion of task duration where movement speed exceeds the 10% of the peak speed [42]; MAPR index is chosen for quantifying movement smoothness in the joint space, since stroke patients (who are the primary target population for the proposed system) are inclined to perform jerky movements, characterized by several stops.
- *Inter-joint coordination* ($q_{corr_{i,j}}$): it is defined as the correlation index between two upper-limb joint angles q_i and q_j [43]; this index accounts for the coordination between two arm joint angles, which is normally affected after stroke.

On the other hand, the dynamic indicators are extracted from the sensory signals provided by the robot torque sensors. They allow calculating the force exerted by the patient during the interaction with the robotic device. They are [39]:

- *Useful-Mean-Force (UMF)*: it is the amount of mean force along target direction.
- *Useful-Peak-Force (UPF)*: it is the amount of peak force along target direction.
- *Total-Work (TW)*: it is the total work expended during motion.
- *Useful-Work (UW)*: it is the amount of patient total work along target direction.

The computed biomechanical performance indicators are normalized with respect to their maximum, thus they result dimensionless; they are also adjusted so that they all increase with motor recovery. Then, they are used to estimate a unique value of behavioral state (to be sent to the bio-cooperative controller together with the physiological indicators), through a properly designed function. It is a weighted sum of the defined indicators in which weights were chosen through a "trial-and-error" approach: a weight configuration that allows discriminating among different quality levels of the executed movements was selected. Greater weights were given to performance indicators that are more indicative of a pathological condition versus a healthy condition.

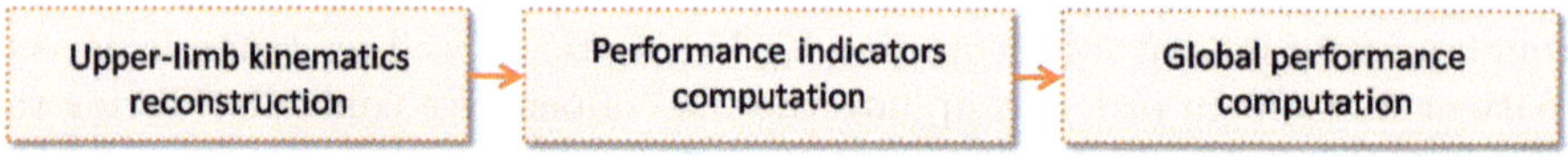

Fig. 3. Functional scheme for the analysis of patient performance

Auditory and Visual Feedback. In order to improve the interaction between user and system, an auditory feedback was integrated into the system: it provides encouraging words and sounds, which contribute to motivate the user during the task execution, and congratulatory or consolatory words on task completion. As regards the visual feedback, the upper-limb kinematic reconstruction (presented in Section 2.1), allows showing the user upper extremity movement in a 3D virtual environment(Figure 4).

Fig. 4. Detail of the 3D virtual environment. The correct position of the user arm is colored in red jointly with the current user position in order to show to the user the correctness of his movement.

2.2 Bio-cooperative Control System

The bio-cooperative control system incorporates three modules:

- patient-tailored interaction control,
- immersive virtual reality,
- modulation functions for the interaction control, on one hand, and for the immersive virtual reality system on the other.

Information extracted from the patient biomechanical and physiological monitoring are used to update robot interaction control as well as the immersive virtual reality system.

Patient-Tailored Interaction Control. The main goal of the controller is to assist patients (who are connected to the robot end-effector) when they are not able to accomplish the task, with a level of assistance that is tuned on the patient global state. The fundamental idea is to make the robot fully compliant when the patient is moving along the planned trajectory or, alternatively, stiffer and stiffer when the patient is moving away from it. In order to achieve that, a strategy based on impedance control in the Cartesian space was implemented. In particular, control parameters were changed based on patient state, thus enabling various "degrees of assistance" as required by a specific subject.

Robot control law can be written as follows [41]:

$$\boldsymbol{\tau}_{cmd} = J^T(\boldsymbol{q})\left[K\left(\boldsymbol{x}_p - \boldsymbol{x}\right) + \boldsymbol{FT}\right] + D\left(d\right) + \boldsymbol{f}_{dyn}\left(\boldsymbol{q}, \dot{\boldsymbol{q}}, \ddot{\boldsymbol{q}}\right)$$

where J^T is the transposed Jacobian matrix, K is the Cartesian Stiffness matrix, $\boldsymbol{x}_p$ is the vector of the desired Cartesian position, $\boldsymbol{x}$ is the vector of the actual Cartesian position, $D\left(d\right)$ is the damping term, $\boldsymbol{FT}$ is an eventual additional superposed Cartesian force, $\boldsymbol{q}$ is the vector of the actual joint positions and $\boldsymbol{f}_{dyn}$ is the dynamic model. Control gains and exercise execution time are varied based on the patient state, quantified by means of the physiological and biomechanical indicators.

For encouraging patient effort, a sort of error variability around the desired path can be permitted, introducing a "dead band", namely an area around the trajectory in which no assistance is provided [28], [44].

Initially 2D desired minimum-jerk trajectories were used, by analogy with [28]; then, in order to enable 3D motion as well as Activity of Daily Living (ADL) tasks, two different options were analyzed for determining the desired trajectories:

- A 3D minimum-jerk trajectory, well suited for patients with limited movement ability [28], and typically used in robot-aided rehabilitation therapy;
- Pre-recorded trajectories from unimpaired volunteers or during therapist-guided assistance [44], for functional tasks.

With the second option it is possible to perform also tasks related to ADLs: they are oriented to promote cortical reorganization and functional recovery, which could aid the resolution of motor learning problems (impaired sensorimotor associations, inability to generate internal models, etc.) [31].

Immersive Virtual Reality. The virtual environment provides activities of daily living (such as taking a glass, drinking and placing object on shelves) and game scenarios. Activities and games can be adjusted as a function of the evolution of the patient; in fact it is possible to modify three parameters:

- the size of the manipulated object that accounts for the accuracy of catching,
- the movement amplitude that is related to patient mobility,
- the time allotted to execute the task.

Modulation Functions. The block "Modify complexity of the therapy" of the bio-cooperative system (Figure 5) receives information about patient behaviour and physiological state from the monitoring systems described above and accordingly modulates control parameters. In this way, the assistance can be automatically changed on the basis of patient needs in the range indicated by the therapist. As mentioned, the modifiable control parameters are the values of proportional gains and the time duration of the tasks: for instance, if a patient completes the task quickly (with respect to the planned trajectory), the execution time of the successive task is opportunely reduced; while if the patient consistently deviates from the planned path, the robot stiffness is increased, in order to enhance the assistance. For both parameters, the modulation law is linear [41].

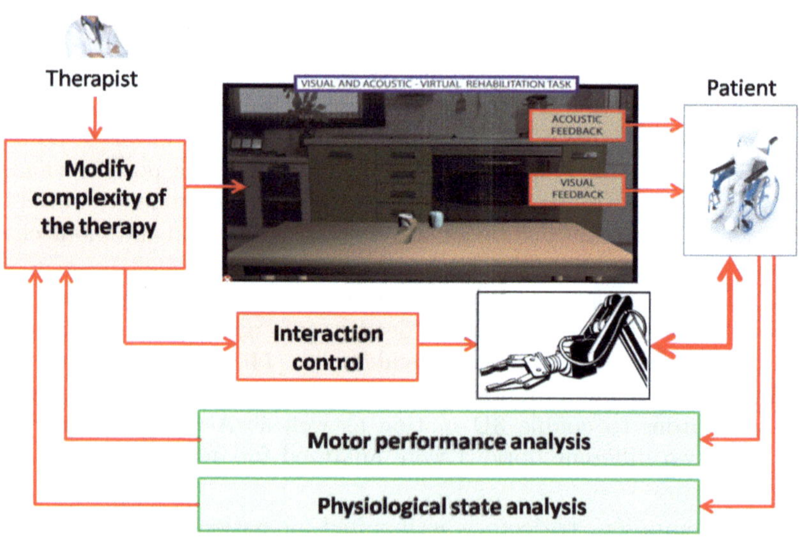

Fig. 5. Scheme of therapy adaptation

2.3 Experimental Setup

Two different versions of the multimodal robotic platforms were developed, the first in Italy and the latter one in Spain (see Figure 6). These two platforms share the multimodal interface for real-time monitoring and stimulation of the patient, and the bio-cooperative controller; they differ for the employed robotic arm used to deliver the therapy. They are the 7-DoF LightWeight Robot arm (LWR) from KUKA and a 7-DoF modular arm based on 7 PRL modules from SCHUNK.

As shown in Figure 6(a) each platform is composed of:

- a robot arm (the KUKA/LWR or the SCHUNK PLR modules) and low-level control for robotic therapy,
- the multimodal HMI,
- the virtual reality environment.

The multimodal HMI includes the following three subsystems:

1. The module for physiological monitoring, consisting of devices form g.tec Company:
   - a skin temperature sensor (g.TEMPsensor),
   - a respiration rate sensor (g.RESPsensor),
   - a galvanic skin response sensor (g.GSRsensor),
   - a pulse sensor (g.PULSEsensor),
   - EMG electrodes (AgCl and Ag/AgCl electrodes),
   - a biological signals' amplifier (g.USBamp),
   - a software for data storage and processing; it was developed in the experiment framework through the MATLAB/Simulink Software;

   Physiological signals were sampled at 256 Hz and different bandpass filters were used, depending on the characteristics of signals. Simulink was used for the online signal processing and a S-Function, based on a MATLAB API, was developed to connect the g.USBAmp with the Simulink interface.
2. The module for motor performance monitoring, consisting of:
   - robot position sensors,
   - robot torque sensors,
   - two magneto-inertial sensors (MTx sensors - Xsens); a further MTx sensor is added for providing hand orientation data in the case that the configuration used for attaching the patient hand to the robot end-effector does not allow directly obtaining these data from robot sensors,
   - a software for data acquisition and processing; it was developed in the experiment framework through the MATLAB/Simulink Software.

Both platforms use a purposely designed mounting flange for attaching user wrist to robot end-effector. Each flange integrates a holding magnet that allows a quick grasp between user and robot and ensures an instantaneous release through power supply interruption; this feature increases system safety and was specifically conceived for dangerous situations.

In order to further increase safety in the interaction, it was chosen to limit robot control commands so that, if the value in norm of the force exerted by the robot at the end-effector level exceeds a predetermined value, the control system immediately halts the movement.

Hardware and software components of the MAAT platforms were synchronized in order to allow simultaneous acquisition of all the monitoring systems

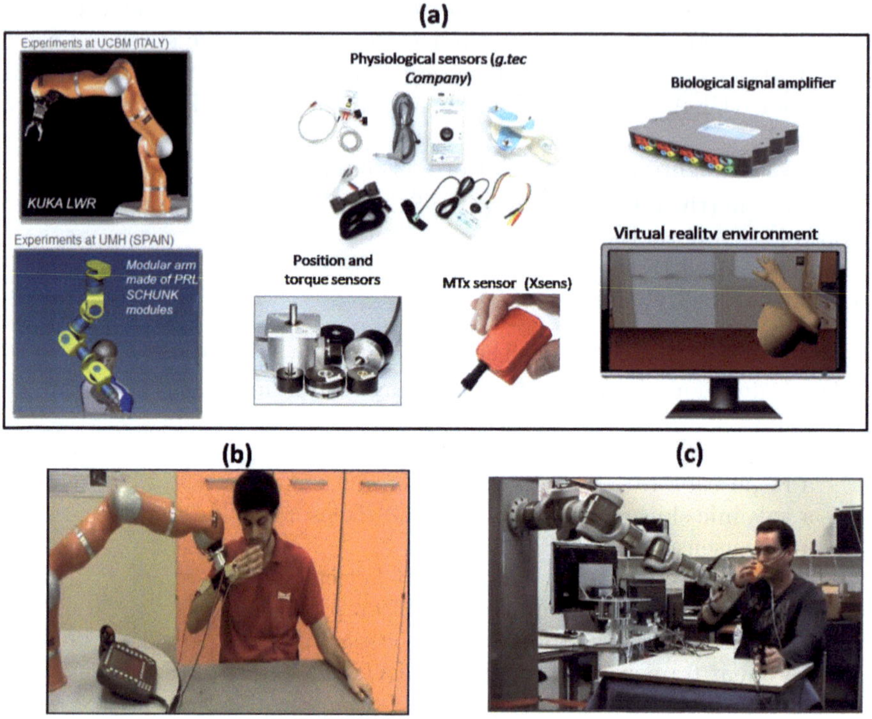

Fig. 6. Set-up of the two MAAT platforms

and the online processing of the collected data. Communication among all the modules was established via UDP protocol (Figure 7). C++ programming language was used to control the robot and MATLAB/Simulink was used to acquire data from sensors and compute physiological, kinematic and dynamic indicators [41], [45].

3 Experimental Validation and Results

3.1 Experimental Tests and Results

The experimental validation has the twofold objective of (i) validating the multimodal interface and assessing its benefit in terms of patient recovery and active involvement in the therapy; (ii) comparing the two robotic platforms and investigating their potentiality to be used as rehabilitation robotic machines for 3D therapy and ADL tasks. It is therefore expected to be able to identify improvements and/or modifications that the robots may need to address specific requirements coming from the application, and collaborate with the manufacturers to address specific changes. Activities of industrial exploitation of the MAAT results are expected in the very near future.

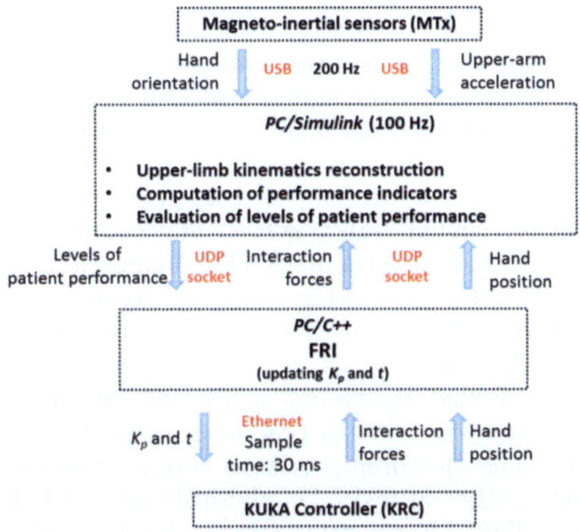

Fig. 7. Communication scheme of the MAAT platform at UCBM

Preliminary experimental tests on the two MAAT platforms were carried out in order to assess the operating mode of all the modules of the platforms, before applying the system to patient rehabilitation. For the experiments with human subjects, ethical approvals were obtained from the ethical committees of the involved universities. Volunteer subjects were asked to participate in two types of tasks:

- point-to-point movements, following 3D straight lines,
- ADL tasks such as *drinking* (i.e. to grasp a glass, bring it to the mouth and put it again on the table).

The participants were required to seat in front of a table and perform the task shown in a monitor while they were attached to robot end-effector.

All the experiments consisted of two main phases: the *evaluation* phase and the *robot-aided exercise* phase. During the first phase the robot was completely passive and the participants autonomously accomplished the tasks exploiting their residual capabilities; the evaluation phase was used to regulate values of stiffness and execution time of the interaction controller through the analysis of the patient state. In the latter phase, the robot assisted the patient during task execution in accordance with the selected assistance level.

Planned trajectories were minimum-jerk trajectories for the point-to-point movements and pre-recorded trajectories from healthy subjects for the drinking tasks.

The Module for Physiological State Analysis. Different experiments were carried out to evaluate user physiological responses to cognitive and/or physical tasks. The experiments were inspired by Novak and colleagues [46].

Experiment 1 consists of cognitive and/or physical tasks using the PHANTOM Omni (from SensAble Technologies). Fifteen volunteer subjects (11 males and 4 females) participated in the experiment. All were healthy, without any major cognitive or physical deficits. They were aged between 20 and 41 (mean age 28.0±6.6 years).

The hardware configuration was composed of signal monitoring and processing system, haptic device, virtual reality system and control system. The device used during the experiments was the PHANTOM Omni as a haptic input and a positioning output device to control the virtual rehabilitation task. PHANTOM is a well-known 6-degree-of-freedom (DOF) positioning and 3-degree-of-freedom force feedback haptic device.

The experiment consisted of different steps [46]: after experiment explanation, the subjects spent few minutes practicing tasks, then they stayed three minutes so that baseline measurements were obtained; spacing out by pauses, each subject executed different tasks, i.e. physical task = TASK 1, physical task with disturbance = TASK 2, cognitive task = TASK 3, physical + cognitive task = TASK 4, physical with disturbance + cognitive task = TASK 5. Each of tasks and pauses lasted three minutes; after each task the subject used a Self-Assessment Manikin (SAM) [47] for evaluating arousal and valence. In Figure 8, the setup of experiment is shown.

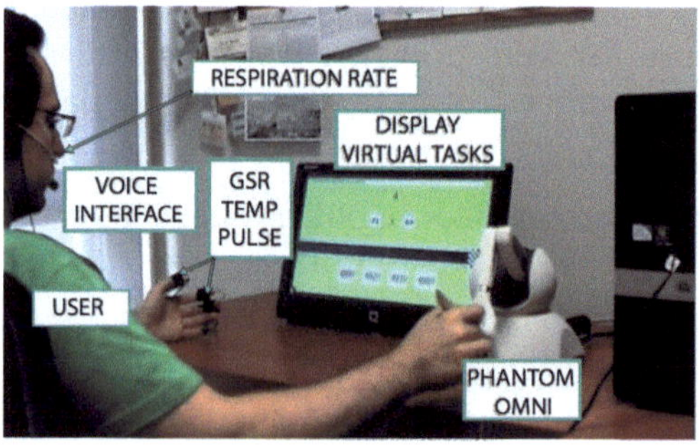

Fig. 8. Experiment 1: Virtual Task display, PHANTOM Omni device and equipment for physiological signals acquisition

The differences in physiological signals during the executed tasks are shown as box plots in Figure 9. For all the tasks, from tasks with purely physical load to tasks with cognitive load plus physical load, pulse and skin conductance level showed a significant difference between task with different physical load even when tasks have cognitive load as well (Task 4 and 5). Thus, a combination of pulse and skin conductance level appears to be a robust method of physical and cognitive workload estimation in virtual rehabilitation interacting with haptic robots.

The key result of Experiment 1 is that it is possible to use physiological responses to real-time estimate different subject's physiological state during a

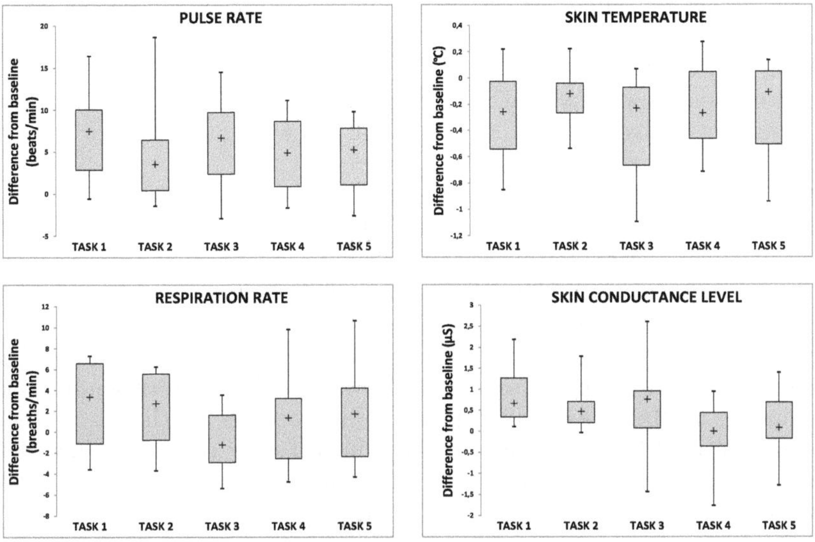

Fig. 9. Statistical values of changes in physiological signals as a response to different tasks. Task 1: Physical, Task 2: Physical with disturbance, Task 3: Cognitive, Task 4: Physical + cognitive, Task 5: Physical with disturbance + cognitive.

virtual task. Moreover, a combination of pulse and skin conductance level seems to be a robust method of physical and cognitive workload estimation in virtual rehabilitation tasks.

Experiment 2 involves cognitive and/or physical tasks using the PUPArm. A similar experiment was carried out with a force-controlled planar robot designed by the Nbio research research group at Universidad Miguel Hernandez; results are reported in [37]. This experiment led to the same conclusions of Experiment 1: it is possible to use physiological responses to estimate different subject's physiological state during a virtual task assisted by a haptic robot; moreover, a combination of pulse and skin conductance level seems to be a robust method of physical and cognitive workload estimation in virtual rehabilitation tasks assisted by a haptic robot.

The Module for Motor Performance Analysis. The tests for the evaluation of the module for the analysis of motor performance consisted of: i) 2D movements performed in 8 different directions, spaced 45 degrees apart, from North (N) to North-West (NW) (as in the traditional 'clock game' [48], where the tasks consist of point-to-point movements along 8 radial directions, starting from the circle center), and ii) 3D tasks carried out in 4 different directions, spaced 90 degrees apart. Consistently with the robot workspace and with the traditional 'clock game', each point-to-point movement had a length of 0.14 m.

Subjects were required to exhibit i) a *healthy behavior* (aimed at evaluating system response when the user performed a correct motion) and ii) a *simulated post-stroke behavior* (aimed at testing the module for motor performance analysis in case of impaired movements). In the second case, subjects had to try emulating motion difficulty of stroke patients, i.e. intentionally failing to perform the motion, especially that involving elbow extension (which is very often reduced in post-stroke patients [49]), going in the wrong direction or repeatedly stopping during the task execution.

Table 1 shows the values of some of the computed performance indicators for one of the tested subjects; they are relative to each task direction of the 2D and 3D tasks. Data concerning the *training* phase of the experiment for both *healthy behavior* and *simulated post-stroke behavior* are reported. As can be noticed, performance indicators are higher in the case of healthy motion than in the case of simulated pathological behavior. As regards the simulated post-stroke behavior, indicators change with movement direction depending on the predominance of elbow flexion or extension: in directions South (S), South-West (SW) and West (W) of the planar tasks, which mainly involved elbow flexion, the value of the indicators are greater than in the other directions (involving elbow extension).

Table 1. Values of performance indicators in *training* phase

2D tasks

task	α	*MAPR*	$q_{corr_{1,4}}$	$q_{corr_{2,4}}$	*UMF*	*UPF*
Healthy behavior						
N	0.96	0.81	0.85	1.00	0.70	0.78
NE	0.98	0.79	0.86	1.00	0.76	0.75
E	0.99	0.90	0.97	0.98	0.85	0.87
SE	0.98	0.73	0.99	0.99	0.66	0.66
S	0.98	0.89	0.56	1.00	0.69	0.72
SW	0.99	0.92	0.92	1.00	0.78	0.79
W	0.99	0.94	0.80	0.89	0.98	0.99
NW	0.96	0.91	0.99	0.99	0.87	0.86
Simulated post-stroke behavior						
N	0.70	0.20	0.37	0.14	0.49	0.29
NE	0.67	0.58	0.61	0.98	0.49	0.008
E	0.16	0.61	0.76	0.88	0.51	0.58
SE	0.61	0.59	0.76	0.99	0.48	0.42
S	0.90	0.96	0.30	0.98	0.50	0.78
SW	0.87	0.79	0.23	0.98	0.62	0.19
W	0.85	0.86	0.005	0.92	0.61	0.95
NW	0.45	0.57	0.006	0.98	0.49	0.32

3D tasks

task	α	*MAPR*	$q_{corr_{1,4}}$	$q_{corr_{2,4}}$	*UMF*	*UPF*
Healthy behavior						
NE	0.88	0.75	0.91	1.00	0.85	0.96
SE	0.84	0.75	0.59	1.00	0.96	0.98
SW	0.78	0.71	0.90	0.81	0.72	0.79
NW	0.73	0.82	0.92	1.00	0.84	0.76
Simulated post-stroke behavior						
NE	0.29	0.77	0.01	0.52	0.55	0.80
SE	0.52	0.16	0.93	0.91	0.43	0.30
SW	0.78	0.81	0.25	0.09	0.62	0.76
NW	0.32	0.31	0.91	1.00	0.65	0.78

Table 2. Updated values of K (N/m), t (s) and performance indicators (*robot-aided exercise* phase)

2D tasks								
task	$K[N/m]$	$t[s]$	α	$MAPR$	$q_{corr_{1,4}}$	$q_{corr_{2,4}}$	UMF	UPF
Simulated post-stroke behavior								
N	300	10	1.00	0.92	0.99	1.00	0.32	0.28
NE	300	7.5	0.92	0.82	0.94	1.00	0.16	0.13
E	1000	7.5	0.96	0.91	0.99	0.96	0.24	0.22
SE	300	7.5	0.99	1.00	0.91	0.99	0.22	0.21
S	300	5	0.97	0.89	0.97	1.00	0.56	0.64
SW	300	7.5	0.97	0.78	0.90	1.00	0.68	0.82
W	300	7.5	0.99	0.86	0.79	0.75	0.78	0.81
NW	1000	7.5	1.00	0.73	0.99	1.00	0.50	0.63
3D tasks								
task	$K[N/m]$	$t[s]$	α	$MAPR$	$q_{corr_{1,4}}$	$q_{corr_{2,4}}$	UMF	UPF
Simulated post-stroke behavior								
NE	1000	7.5	0.97	0.80	0.99	0.91	0.05	0.03
SE	300	10	0.85	0.86	0.87	1.00	0.31	0.30
SW	300	7.5	0.93	0.87	0.004	0.95	0.07	0.05
NW	300	7.5	0.90	0.71	0.94	0.98	0.18	0.18

The Adaptation of Robot Behavior. Results about the adaptation of robot behavior are reported in the Table 2 for the same subject of Table 1. These data are relative to the *robot-aided exercise* phase, i.e. after updating control parameters, in the case of *simulated post-stroke behavior* and represent the updated values of K and t and the new values of the performance indicators.

As can be noticed, kinematic indicators significantly increased, while the dynamic indicators (which quantify the amount of force directed towards the target) did not considerably grow, except for the case of movements requiring elbow flexion (S, SW, W). In the case, subjects tended to be moved by the robot.

Figure 10 shows the Cartesian positions of the hand during the 2D point-to-point tasks executed by one of the tested subjects; both behaviors are reported (i.e. healthy and simulated post-stroke). In the left figure, related to the *training* phase, the blue line represents the healthy motion while the red line the simulated post-stroke motion. In the right figure, related to the *robot-aided exercise* phase (i.e. after updating control parameters), the trajectories executed simulating a post-stroke behavior are represented through a green line. The dotted black line in both figures is the desired motion.

As expected, after changing robot parameters, the trajectories performed simulating a post-stroke behaviour get closer the desired paths thanks to robot assistance; the assistance level was computed on the basis of subject performance measured in the previous set of tasks (red lines).

In order to validate the proposed adaptive interaction control system, also a set of ADL tasks (drinking tasks) were carried out; three types of tests were executed using different levels of robot assistance.

In the first test, subjects, interacting with the robot, were almost totally passive and correspondingly robot stiffness was high (1000 N/m). On the left

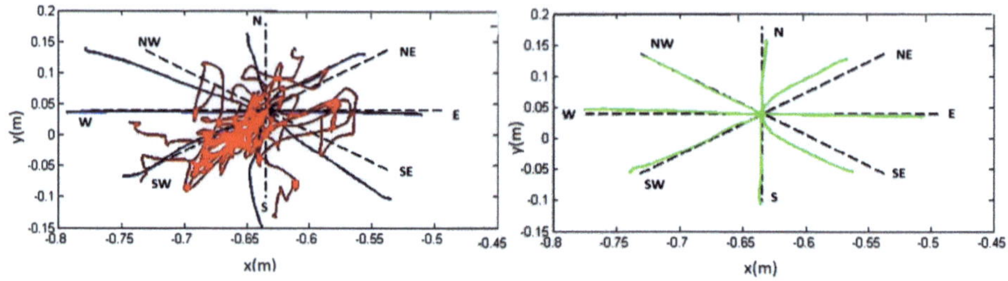

Fig. 10. Trajectory of the hand of one of the tested subjects during 2D tasks in *training* phase (left) and *robot-aided exercise* phase (right)

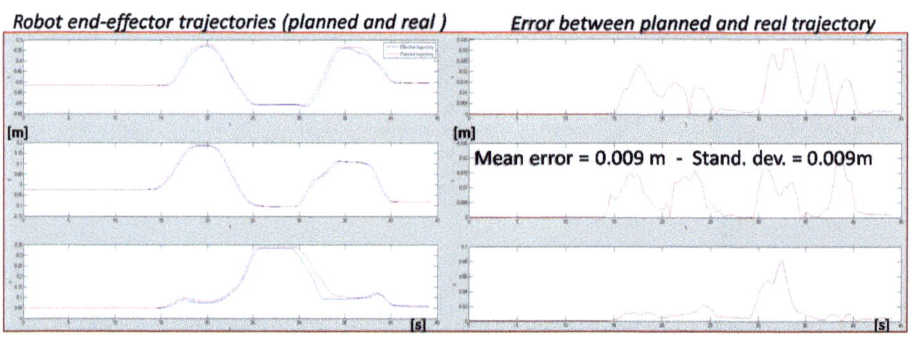

Fig. 11. Left: Time course of real robot end-effector trajectory (blue line) and planned trajectory (red line) for $K = 1000$ N/m. Right: Error between planned and real trajectory.

of Figure 11 the planned trajectory and the real trajectory of the robot end-effector, connected to the wrist of one of the tested subjects, are reported. The corresponding Cartesian position error is shown in the right part of the same figure. In such a case (i.e. a robot with high stiffness guiding a participant along the desired trajectory) the dynamic indicators can discriminate between different patients and different performance: for instance, Figure 12 shows, for two different subjects A and B, the values of the force exerted by each subject while moving towards the target (i.e. the glass on the table). The force values were negative because the manipulator guided the subjects (and not viceversa); however, it can be observed that performance of subject A were better than performance of subject B.

In the second test, subjects were assisted by the robot with a medium value of Cartesian stiffness (300 N/m). As expected, the Cartesian position error (shown in Figure 13 for one of the tested subjects) was greater with respect to the previous test and, in fact, the adaptation of the control parameters led to an increase of stiffness value (and therefore to an increase of the provided assistance).

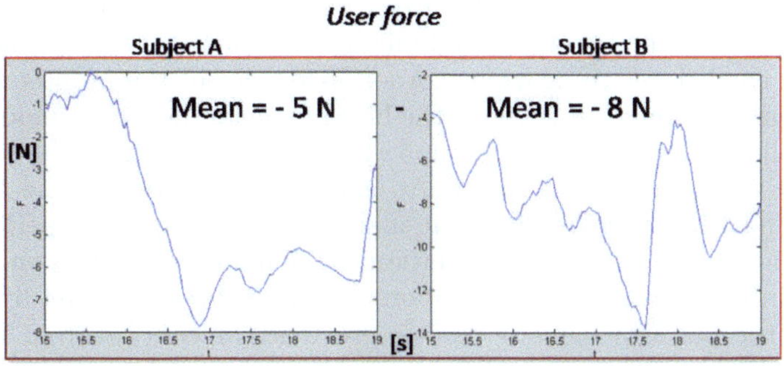

Fig. 12. Force measured during the drinking task for two different subjects (A and B), while they took the glass

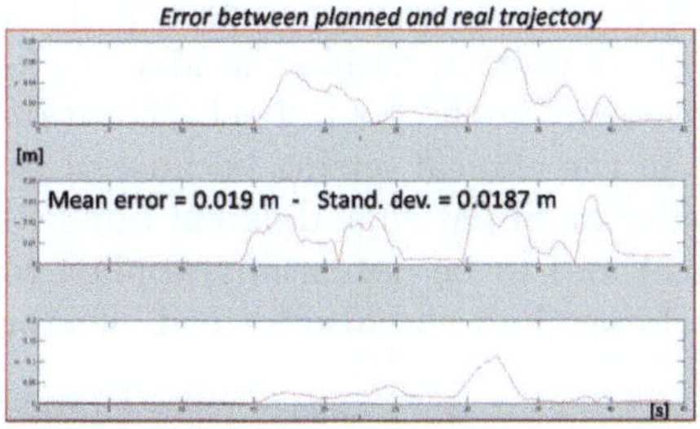

Fig. 13. Error between planned trajectory and real trajectory ($K = 300$ N/m)

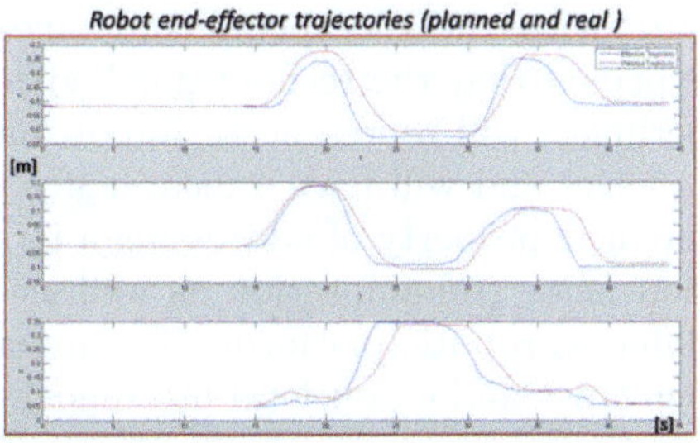

Fig. 14. Time course of real robot end-effector trajectory (blue line) and planned trajectory (red line) for $K = 100$ N/m

Finally, the drinking task was executed in a natural way by the enrolled healthy subjects, thus requiring very low assistance from the robot ($K = 100$ N/m). Figure 14, relative to one of the subjects, shows that he was able to complete the task more quickly than the planned trajectory, without the need of assistance. In this case, the consequent adaptation of the control parameters led to a decrease of the execution time.

4 Conclusions and Future Work

The MAAT experiment provided two robotic platforms for upper-limb robot-aided rehabilitation, with two main important features: (i) multimodal HMI for monitoring patients' state (thus characterizing their behaviour in terms of physiological and biomechanical features) and assessing the recovery process; (ii) a biocooperative controller that, based on patient monitoring, is able to adapt robot behaviour to subject specific needs (thus maximising patient involvement and active role in the therapy). Preliminary experimental tests on healthy subjects were carried out to provide a proof-of-concept of the proposed approach; they were mainly aimed at showing the benefits of adding a multimodal interface to a rehabilitation robotic machine and including the patient in the control loop.

Current efforts are aimed at: (i) An extensive validation of the MAAT system with healthy subjects aimed at assessing and comparing performance of the two platforms. (ii) Clinical validation of the multimodal HMI on post-stroke patients; in this case, the 7-DoF robot arms have been replaced by 2-DoF rehabilitation robot arms already used in previous clinical trials (i.e. MIT-Manus and PUPARM).

However, for completing the experimental validation of the two complete MAAT platforms and carrying out a comparative analysis between them and with respect to other machines in the literature, an extensive clinical experimentation on post-stroke subjects is planned as an immediate follow-up for evaluating the therapeutic outcomes of the proposed training approach. It is expected that the technological advancements developed within the MAAT experiment will represent an important achievement in the field of rehabilitation robotics and will have a meaningful impact in the clinical setting because of its intrinsic property of addressing a large population of patients, different application fields (rehabilitation as well as assistive) and to be potentially applied to different robotic platforms (for upper limb, lower limb and also for end-effector machines and exoskeletal machines).

Acknowledgements. This work was supported by the EU Project EC FP7-ICT-231143 ECHORD.

References

1. European Clearing House for Open Robotics Development, EU-funded project within the Seventh Framework Program, GA 231143 (2009-2013), `www.echord.info`
2. Kempermann, G., Van Praag, H., Gage, F.H.: Activity-dependent regulation of neuronal plasticity and self repair. Prog. Brain Res. 127, 35–48 (2000)
3. Pellegrino, G., Pellegrino, G., Tombini, M., Assenza, G., Bravi, M., Sterzi, S., Giacobbe, V., Zollo, L., Guglielmelli, E., Cavallo, G., Vernieri, F., Tecchio, F.: Interhemispheric coupling changes associate with motor improvements after robotic stroke rehabilitation. Restorative Neurology and Neuroscience 30, 497–510 (2012)
4. Jones, T.A., Chu, C.J., Grande, L.A., Gregory, A.D.: Motor Skills Training Enhances Lesion-Induced Structural Plasticity in the Motor Cortex of Adult Rats. The Journal of Neuroscience 19, 10153–10163 (1999)
5. Butesch, C., Hummelsheim, H., Denzler, P., Mauritz, K.H.: Repetitive training of isolated movements improves the outcome of motor rehabilitation of the centrally paretic hand. J. Neurol. Sci. 130, 59–68 (1995)
6. van der Lee, J.H., Wagenaar, R.C., Lankhorst, G.J., Vogelaar, T.W., Deville, W.L., Bouter, L.M.: Forced use of the upper extremity in chronic stroke patients: Results from a single-blind randomized clinical trial. Stroke 30, 2369–2375 (1999)

7. Dam, M., Tonin, P., Casson, S., Ermani, M., Pizzolato, G., Iaia, V., Battistin, L.: The effects of long-term rehabilitation therapy on poststroke hemiplegic patients. Stroke 24, 1186–1191 (1993)
8. Morales, R., Badesa, F.J., Garcia-Aracil, N., Sabater, J.M., Perez-Vidal, C.: Pneumatic robotic systems for upper limb rehabilitation. Medical and Biological Engineering and Computing 49, 1145–1156 (2011)
9. Wolf, S.L., Binder-Macloud, S.A.: Electromyographic biofeedback applications to the hemiplegic patient: Changes in upper extremity neuromuscular and functibnal status. Journal of the American Physical Therapy Association 63, 1393–1403 (1983)
10. Mudie, M.H., Matyas, T.A.: Can simultaneous bilateral movement involve the undamaged hemisphere in reconstruction of neural networks damaged by stroke? Disabil. Rehabil. 22, 23–37 (2000)
11. Aisen, M.L., Krebs, H.I., Hogan, N., McDowell, F., Volpe, B.T.: The effect of robot-assisted therapy and rehabilitative training on motor recovery following stroke. Arch. Neurol. 54, 443–446 (1997)
12. Krebs, H.I., Hogan, N., Aisen, M.L., Volpe, B.T.: Robot-aided Neurorihabilitation. IEEE Trans. Rehabil. Eng. 6, 75–87 (1998)
13. Krebs, H.I., Volpe, B.T., Ferraro, M., Fasoli, S., Palazzolo, J., Rohrer, B., Edelstein, L., Hogan, N.: Robot-aided neuro-rehabilitation: from evidence-based to science-based rehabilitation. Top Stroke Rehabil. 8, 54–70 (2002)
14. Volpe, B.T., Krebs, H.I., Hogan, N., Edelstein, L., Diels, C.M., Aisen, M.L.: Robot training enhanced motor outcome in patients with stroke maintained over 3 years. Neurology 53, 1874–1876 (1999)
15. Burgar, C.G., Lum, P.S., Shor, P.C., Van der Loos, M.: Development of robots for rehabilitation therapy: the Palo Alto VA/Stanford experience. J. Reh. Res. Dev. 37, 663–673 (2000)
16. Lum, P.S., Burgar, C.G., Shor, P.C., Majmundar, M., Van der Loos, M.: Robot-Assisted Movement Training Compared With Conventional Therapy Techniques for the Rehabilitation of Upper-Limb Motor Function After Stroke. Arch. Phys. Med. Rehabil. 83, 952–959 (2002)
17. Fasoli, S.D., Krebs, H.I., Stein, J., Frontera, W.R., Hogan, N.: Effects of Robotic Therapy on Motor Impairment and Recovery in Chronic Stroke. Archives of Physical Medicine and Rehabilitation 84, 477–482 (2003)
18. Nudo, R.J., Friel, K.: Cortical plasticity after stroke: implications for rehabilitation. Rev. Neurol. 155, 713–717 (1999)
19. Staines, W.R., McIlroy, W.E., Graham, S.J., Black, S.E.: Bilateral movement enhances ipsilesional cortical activity in acute stroke: A pilot functional MRI study. Neurology 56, 401–404 (2001)
20. Prange, G.B., Jannink, M.J.A., Groothuis-Oudshoorn, C.G.M., Hermens, H.J., IJzerman, M.J.: Systematic review of the effect of robot-aided therapy on recovery of the hemiparetic arm after stroke. Journal of Rehabilitation Research and Development 43, 171–184 (2006)
21. Reinkensmeyer, D.J., Kahn, L.E., Averbuch, M., McKenna-Cole, A., Schmit, B.D., Rymer, W.Z.: Understanding and treating arm movement impairment after chronic brain injury:progress with the ARM guide. J. Rehabil. Res. Dev. 37, 653–662 (2000)
22. Kahn, L.E., Zygman, M.L., Rymer, W.Z., Reinkensmeyer, D.: Robot-assisted reaching exercise promotes arm movement recovery in chronic hemiparetic stroke: a randomized controlled pilot study. Journal of NeuroEngineering and Rehabilitation 3, 12–23 (2006)

23. Zollo, L., Accoto, D., Torchiani, F., Formica, D., Guglielmelli, E.: Design of a Planar Robotic Machine for Neuro-rehabilitation. In: International Conference on Robotics and Automation, ICRA 2008, Pasadena, CA (2008)
24. Micera, S., Carrozza, M.C., Guglielmelli, E., Cappiello, G., Zaccone, F., Freschi, C., Colombo, R., Mazzone, A., Delconte, C., Pisano, F., Minuco, G., Dario, P.: A Simple Robotic System for Neurorehabilitation 19, 271–284 (2005)
25. Lum, P.S., Burgar, C.G., Shor, P.C.: Evidence for Improved Muscle Activation Patterns After Retraining of Reaching Movements with the MIME Robotic System in Subjects with Post-Stroke Hemiparesis. IEEE Transactions on Neural Systems and Rehabilitation Engineering, 186–194 (2004)
26. Volpe, B.T., Krebs, H.I., Hogan, N., Edelstein, O.L., Diels, C., Aisen, M.: A novel approach to stroke rehabilitation: robotaided sensorimotor stimulation. Neurology 54, 1938–1944 (2000)
27. Formica, D., Charles, S.K., Zollo, L., Guglielmelli, E., Hogan, N., Krebs, H.I.: The passive stiffness of the wrist and forearm. Journal of Neurophysiology 108, 1158–1166 (2012)
28. Krebs, H.I., Palazzolo, J.J., Dipietro, L., Ferraro, M., Krol, J., Rannekleiv, K., Volpe, B.T., Hogan, N.: Rehabilitation robotics: performance-based progressive robot-assisted therapy. Autonomous Robots 15, 7–20 (2003)
29. Formica, D., Zollo, L., Guglielmelli, E.: Torque-dependent compliance control in the joint space for robot-mediated motor therapy. Journal of Dynamic Systems, Measurement, and Control 128, 152–158 (2006)
30. Kan, P., Huq, R., Hoey, J., Goetschalckx, R., Mihailidis, A.: The development of an adaptive upper-limb stroke rehabilitation robotic system. Journal of Neuroengineering and Rehabilitation 8, 33–50 (2011)
31. Balasubramanian, S., Zhang, H., Buchanan, S., Austin, H., Herman, R., He, J.: Cooperative and active assistance based interactive therapy. In: 2010 IEEE/ICME International Conference on Complex Medical Engineering (CME), pp. 311–315 (2010)
32. Novak, D., Mihelj, M., Ziherl, J., Olensek, A., Munih, M.: Psychophysiological measurements in a biocooperative feedback loop for upper extremity rehabilitation. IEEE Transactions on Neural Systems and Rehabilitation Engineering 19, 400–410 (2011)
33. Koenig, A., Novak, D., Omlin, X., Pulfer, M., Perreault, E., Zimmerli, L., Mihelj, M., Riener, R.: Real-time closed-loop control of cognitive load in neurological patients during robot-assisted gait training. IEEE Transactions on Neural Systems and Rehabilitation Engineering 19, 453–464 (2011)
34. Duschau-Wicke, A., Caprez, A., Riener, R.: Patient cooperative control increases active participation of individuals with sci during robot-aided gait training. Journal of NeuroEngineering and Rehabilitation 7 (2010)
35. Riener, R., Munih, M.: Guest editorial special section on rehabilitation via biocooperative control. IEEE Transactions on Neural Systems and Rehabilitation Engineering 18, 337–338 (2010)
36. Rodriguez Guerrero, C., Fraile Marinero, J., Perez Turiel, J., Rivera Farina, P.: Bio cooperative robotic platform for motor function recovery of the upper limb after stroke. EMBC Annual International Conference of the IEEE 31, 4472–4475 (2010)
37. Badesa, F.J., Morales, R., Garcia-Aracil, N., Sabater, J.M., Perez-Vidal, C., Fernandez, E.: Multimodal Interfaces to Improve Therapeutic Outcomes in Robot-Assisted Rehabilitation. IEEE Transactions on Systems, Man, and Cybernetics, Part C: Applications and Reviews 42, 1152–1158 (2012)

38. Zollo, L., Rossini, L., Bravi, M., Magrone, G., Sterzi, S., Guglielmelli, E.: Quantitative evaluation of upper-limb motor control in robot-aided rehabilitation. Medical and Biological Engineering and Computing 9, 1131–1144 (2011)
39. Zollo, L., Gallotta, E., Guglielmelli, E., Sterzi, S.: Robotic Technologies and Rehabilitation: New Tools for Upper-limb Therapy and Assessment in Chronic Stroke. European Journal of Physical and Rehabilitation Medicine 47, 223–236 (2011)
40. Papaleo, E., Zollo, L., Sterzi, S., Guglielmelli, E.: An inverse kinematics algorithm for upper-limb joint reconstruction during robot-aided motor therapy. In: BIOROB-IEEE/RAS-EMBS International Conference on Biomedical Robotics and Biomechatronics, pp. 1983–1988 (2012)
41. Papaleo, E., Zollo, L., Spedaliere, L., Guglielmelli, G.: Patient-Tailored Adaptive Robotic System for Upper-Limb Rehabilitation. In: IEEE International Conference on Robotics and Automation (ICRA 2013), Karlsruhe, Germany, May 6-10, pp. 3860–3865 (2013)
42. Rohrer, B., Fasoli, S., Krebs, H.I., Hughes, R., Volpe, B.T., Frontera, W.R., Stein, J., Hogan, N.: Movement smoothness changes during stroke recovery. J. Neurosci. 22, 8297–8304 (2002)
43. Dipietro, L., Krebs, H.I., Fasoli, S.E., Volpe, B.T., Stein, J., Bever, C., Hogan, N.: Changing motor synergies in chronic stroke. J. Neurophysiol. 98, 757–768 (2007)
44. Marchal-Crespo, L., Reinkensmeyer, D.J.: Review of control strategies for robotic movement training after neurologic injury. Journal of Neuro Engineering and Rehabilitation 6, 20–35 (2009)
45. Morales, R., Badesa, F.J., Rodriguez, J., Garcia-Aracil, N., Prez, C., Mara-Azorn, J.: A Platform for Researching on Multimodal Robot-Assited Rehabilitation Therapies. In: BIOROB 2012 - IEEE/RAS-EMBS International Conference on Biomedical Robotics and Biomechatronics, Rome, Italy (2012)
46. Novak, D., Ziherl, J., Olensek, A., Milavec, M., Podobnik, J., Mihelj, M., Munih, M.: Psychophysiological responses to robotic rehabilitation tasks in stroke. IEEE Transactions on Neural Systems and Rehabilitation Engineering 18, 351–361 (2010)
47. Bradley, M.M., Lang, P.: Measuring emotion: the self-assessment manikin and the semantic differential. Journal of Behavior Therapy and Experimental Psychiatry 25, 49–59 (1994)
48. Krebs, H.I., Hogan, N., Volpe, B.T., Aisen, M.L., Edelstein, L., Diel, C.: Overview of clinical trials with MIT-MANUS: a robot-aided neuro-rehabilitation facility. Technology and Health Care 7, 419–423 (1999)
49. Levin, M.F., Selles, R.W., Verheul, M.H., Meijer, O.G.: Deficits in the coordination of agonist and antagonist muscles in stroke patients: implications for normal motor control. Brain Res. 853, 352–369 (2000)

Author Index

MIX
Papier aus verantwortungsvollen Quellen
Paper from responsible sources
FSC® C105338

If you have any concerns about our products,
you can contact us on
ProductSafety@springernature.com

In case Publisher is established outside the EU,
the EU authorized representative is:
Springer Nature Customer Service Center GmbH
Europaplatz 3, 69115 Heidelberg, Germany

Printed by Libri Plureos GmbH
in Hamburg, Germany